# SCHAUM'S OUTLINE OF

# THEORY AND PROBLEMS

## of Plane and Spherical

# TRIGONOMETRY

•

**BY**

## FRANK AYRES, JR., Ph. D.

*Professor and Head, Department of Mathematics*

*Dickinson College*

•

# SCHAUM'S OUTLINE SERIES

## McGRAW-HILL BOOK COMPANY

*New York, St. Louis, San Francisco, Toronto, Sydney*

ISBN 07-002651-3

15  SH SH  7 5 4 3 2 1 0 6

# Preface

This book is designed as an aid to those who are studying Trigonometry for the first time by providing a collection of completely solved, representative problems. At the same time, the arrangement of the material makes it a convenient manual for those who wish to review the fundamental principles and applications.

The book, while complete in itself, is not written in formal textbook style. Each chapter contains a summary of the necessary definitions and theorems, followed by a set of graded solved problems. The proofs of theorems and the derivations of all formulas are included among the solved problems. These, in turn, are followed by a set of supplementary problems with answers.

The numerical aspects of Plane Trigonometry have been treated thoroughly. Equal attention has been given to non-logarithmic and logarithmic solutions of both right and oblique triangles. The applications are numerous and in wide variety. The figures have been carefully drawn and labeled for greater usefulness, and answers have been rounded off consistent with the given data.

Simple trigonometric identities and equations require a knowledge of elementary algebra. The problems here have been carefully selected, the solutions have been spelled out in great detail, and all arranged to illustrate clearly the algebraic processes involved as well as the use of the basic trigonometric relations.

The chapters dealing with Spherical Trigonometry are preceded by a chapter on Solid Geometry. The theory and formulas for the solution of right and oblique spherical triangles are covered rather completely and include the use of haversine and right triangle methods in solving oblique triangles. Applications consist of problems involving distance and direction on the earth's surface and certain problems relative to the celestial sphere.

FRANK AYRES, JR.

Carlisle, Pa.
September, 1954

# Contents

# CHAPTER 1

# Angles and Arc Length

TRIGONOMETRY, as the word implies, is concerned with the measurement of the parts of a triangle. Plane trigonometry, considered in the next several chapters, is restricted to triangles lying in planes. Spherical trigonometry deals with certain triangles which lie on spheres.

The science of trigonometry is based on certain ratios, called trigonometric functions, to be defined in the next chapter. The early applications of the trigonometric functions were to surveying, navigation, and engineering. These functions also play an important role in the study of all sorts of vibratory phenomena — sound, light, electricity, etc. As a consequence, a considerable portion of the subject matter is concerned properly with a study of the properties of and relations among the trigonometric functions.

THE PLANE ANGLE *XOP* is formed by the two intersecting half lines *OX* and *OP*. The point *O* is called the *vertex* and the half lines are called the sides of the angle.

More often, a plane angle is to be thought of as generated by revolving (in a plane) a half line from the initial position *OX* to a terminal position *OP*. Then *O* is again the vertex, *OX* is called the *initial side*, and *OP* is called the *terminal* side of the angle.

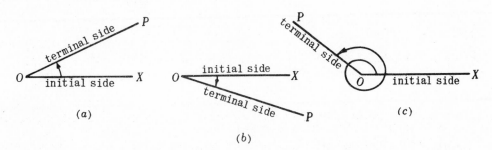

(a)                    (b)                    (c)

An angle, so generated, is called *positive* if the direction of rotation (indicated by a curved arrow) is counterclockwise and *negative* if the direction of rotation is clockwise. The angle is positive in Figures (a) and (c), and negative in Figure (b).

MEASURES OF ANGLES.

**A.** A *degree* (°) is defined as the measure of the central angle subtended by an arc of a circle equal to 1/360 of the circumference of the circle.

A *minute* (') is 1/60 of a degree; a *second* (") is 1/60 of a minute.

EXAMPLE 1.  a) $\frac{1}{4}(36°24') = 9°6'$          b) $\frac{1}{2}(127°24') = \frac{1}{2}(126°84') = 63°42'$

c) $\frac{1}{2}(81°15') = \frac{1}{2}(80°75') = 40°37.5'$ or $40°37'30''$

d) $\frac{1}{4}(74°29'20'') = \frac{1}{4}(72°149'20'') = \frac{1}{4}(72°148'80'') = 18°37'20''$

1

*B.* A *radian* (rad) is defined as the measure of the central angle subtended by an arc of a circle equal to the radius of the circle.

The circumference of a circle = $2\pi$(radius) and subtends an angle of $360°$. Then $2\pi$ radians $= 360°$, from which we obtain

$$1 \text{ radian} = \frac{180°}{\pi} = 57.296° = 57°17'45''\qquad\text{and}$$

$$1 \text{ degree} = \frac{\pi}{180}\text{ radian} = 0.017453\text{ rad, approx.,}$$

where $\pi = 3.14159$.

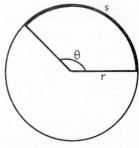

**EXAMPLE 2.** *a)* $\frac{7}{12}\pi$ rad $= \frac{7\pi}{12}\cdot\frac{180°}{\pi} = 105°$  *b)* $50° = 50\cdot\frac{\pi}{180}$ rad $= \frac{5\pi}{18}$ rad.

(See Problems 1-3.)

*C.* A *mil*, used in military science, is defined as the measure of the central angle subtended by an arc of a circle equal to 1/6400 of the circumference of the circle. The name is derived from the fact that, approximately,

$$1 \text{ mil} = \frac{1}{1000}\text{ radian.}$$

Since 6400 mils $= 360°$, $\quad 1$ mil $= \dfrac{360°}{6400} = \dfrac{9°}{160}\quad$ and $\quad 1° = \dfrac{160}{9}$ mils.

(See Problems 14-16.)

## ARC LENGTH.

*A.* On a circle of radius $r$, a central angle of $\theta$ radians intercepts an arc of length

$$s = r\theta,$$

that is,

arc length = radius × central angle in radians.

(Note. $s$ and $r$ may be measured in any convenient unit of length but they must be expressed in the same unit.)

**EXAMPLE 3.** *a)* On a circle of radius 30 in., the length of arc intercepted by a central angle of 1/3 radian is

$$s = r\theta = 30\left(\frac{1}{3}\right) = 10 \text{ in.}$$

*b)* On the same circle a central angle of $50°$ intercepts an arc of length

$$s = r\theta = 30\left(\frac{5\pi}{18}\right) = \frac{25\pi}{3} \text{ in.}$$

*c)* On the same circle an arc of length $1\frac{1}{2}$ ft subtends a central angle

$$\theta = \frac{s}{r} = \frac{18}{30} = \frac{3}{5} \text{ rad,}\quad\text{when } s \text{ and } r \text{ are expressed in inches,}$$

or   $\theta = \dfrac{s}{r} = \dfrac{3/2}{5/2} = \dfrac{3}{5}$ rad, when $s$ and $r$ are expressed in feet.

(See Problems 4-13.)

*B.* If the central angle is relatively small, the length of the intercepted arc may be taken as a close approximation of the length of its chord.

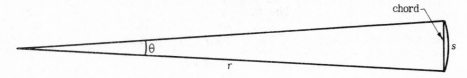

Now since $\theta$ rad $= 1000\theta$ mils and $s = r\theta = \dfrac{r}{1000}(1000\theta)$, it follows that

$$\text{length of chord} = \frac{r}{1000}(\text{central angle in mils}), \quad \text{approx.}$$

For military purposes this is written as $W = Rm$, where $m$ is the central angle expressed in mils, $R$ is the radius (range) expressed in thousands of yards, and $W$ is the chord (width) expressed in yards.

(See Problems 17-19.)

## SOLVED PROBLEMS

1. Express each of the following angles in radian measure:
   *a)* $30^{\circ}$,　*b)* $135^{\circ}$,　*c)* $25^{\circ}30'$,　*d)* $42^{\circ}24'35''$.

   Since $1^{\circ} = \dfrac{\pi}{180}$ radian $= 0.017453$ rad,

   *a)*　　$30^{\circ} = 30 \times \dfrac{\pi}{180}$ rad $= \dfrac{\pi}{6}$ rad or $0.5236$ rad,

   *b)*　　$135^{\circ} = 135 \times \dfrac{\pi}{180}$ rad $= \dfrac{3\pi}{4}$ rad or $2.3562$ rad,

   *c)*　　$25^{\circ}30' = 25.5^{\circ} = 25.5 \times \dfrac{\pi}{180}$ rad $= 0.4451$ rad,

   *d)* $42^{\circ}24'35'' = 42^{\circ} + \left(\dfrac{24 \times 60 + 35}{3600}\right)^{\circ} = 42.41^{\circ} = 42.41 \times \dfrac{\pi}{180}$ rad $= 0.7402$ rad.

2. Express each of the following angles in degree measure:
   *a)* $\pi/3$ rad,　*b)* $5\pi/9$ rad,　*c)* $2/5$ rad,　*d)* $4/3$ rad.

   Since $1$ rad $= \dfrac{180^{\circ}}{\pi} = 57^{\circ}17'45''$,

   *a)* $\dfrac{\pi}{3}$ rad $= \dfrac{\pi}{3} \times \dfrac{180^{\circ}}{\pi} = 60^{\circ}$,　　　　　　*b)* $\dfrac{5\pi}{9}$ rad $= \dfrac{5\pi}{9} \times \dfrac{180^{\circ}}{\pi} = 100^{\circ}$,

   *c)* $\dfrac{2}{5}$ rad $= \dfrac{2}{5} \times \dfrac{180^{\circ}}{\pi} = \dfrac{72^{\circ}}{\pi}$ or $\dfrac{2}{5}(57^{\circ}17'45'') = 22^{\circ}55'6''$,

   *d)* $\dfrac{4}{3}$ rad $= \dfrac{4}{3} \times \dfrac{180^{\circ}}{\pi} = \dfrac{240^{\circ}}{\pi}$ or $\dfrac{4}{3}(57^{\circ}17'45'') = 76^{\circ}23'40''$.

3. A wheel is turning at the rate 48 rpm (revolutions per minute or rev/min). Express this angular speed in *a)* rev/sec, *b)* rad/min, *c)* rad/sec.

a)  48 rev/min  =  $\dfrac{48}{60}$ rev/sec  =  $\dfrac{4}{5}$ rev/sec

b)  Since  1 rev = $2\pi$ rad,  48 rev/min  =  $48(2\pi)$ rad/min  =  301.6 rad/min.

c)  48 rev/min  =  $\dfrac{4}{5}$ rev/sec  =  $\dfrac{4}{5}(2\pi)$ rad/sec  =  5.03 rad/sec

or  48 rev/min  =  $96\pi$ rad/min  =  $\dfrac{96\pi}{60}$ rad/sec  =  5.03 rad/sec.

4. The minute hand of a clock is 12 in. long. How far does the tip of the hand move during 20 min?

During 20 min the hand moves through an angle $\theta = 120^0 = 2\pi/3$ rad and the tip of the hand moves over a distance $s = r\theta = 12(2\pi/3) = 8\pi$ in. = 25.1 in.

5. A central angle of a circle of radius 30 in. intercepts an arc of 6 in.  Express the central angle $\theta$ in radians and in degrees.

$$\theta = \frac{s}{r} = \frac{6}{30} = \frac{1}{5} \text{ rad} = 11^0 27' 33''$$

6. A railroad curve is to be laid out on a circle. What radius should be used if the track is to change direction by $25^0$ in a distance of 120 ft ?

We are required to find the radius of a circle on which a central angle $\theta = 25^0 = 5\pi/36$ rad intercepts an arc of 120 ft.  Then

$$r = \frac{s}{\theta} = \frac{120}{5\pi/36} = \frac{864}{\pi} \text{ ft} = 275 \text{ ft}.$$

7. A train is moving at the rate 8 miles per hour (mi/hr) along a piece of circular track of radius 2500 ft.  Through what angle does it turn in one minute ?

Since  8 mi/hr = $\dfrac{8(5280)}{60}$ ft/min = 704 ft/min, the train passes over an arc of length $s =$ 704 ft in 1 min.  Then  $\theta = \dfrac{s}{r} = \dfrac{704}{2500}$ = 0.2816 rad or $16^0 8'$.

8. Assuming the earth to be a sphere of radius 3960 miles, find the distance of a point in latitude $36^0$ N from the equator.

Since  $36^0 = \dfrac{\pi}{5}$ radian,  $s = r\theta = 3960(\dfrac{\pi}{5})$ = 2488 miles.

9. Two cities 270 miles apart lie on the same meridian.  Find their difference in latitude.

$$\theta = \frac{s}{r} = \frac{270}{3960} = \frac{3}{44} \text{ rad} \text{ or } 3^0 54.4'.$$

10. A wheel 4 ft in diameter is rotating at 80 rpm. Find the distance (in ft) traveled by a point on the rim in one second, that is, the linear speed of the point (in ft/sec).

$$80 \text{ rpm} = 80(\frac{2\pi}{60}) \text{ rad/sec} = \frac{8\pi}{3} \text{ rad/sec}.$$

Then in 1 sec the wheel turns through an angle $\theta = 8\pi/3$ rad and a point on the wheel will travel a distance $s = r\theta = 2(8\pi/3)$ ft = 16.8 ft.  The linear velocity is 16.8 ft/sec.

11. Find the diameter of a pulley which is driven at 360 rpm by a belt moving at 40 ft/sec.

$$360 \text{ rev/min} = 360\left(\frac{2\pi}{60}\right) \text{ rad/sec} = 12\pi \text{ rad/sec.}$$

Then in 1 sec the pulley turns through an angle $\theta = 12\pi$ rad and a point on the rim travels a distance $s = 40$ ft.

$$d = 2r = 2\left(\frac{s}{\theta}\right) = 2\left(\frac{40}{12\pi}\right) \text{ ft} = \frac{20}{3\pi} \text{ ft} = 2.12 \text{ ft.}$$

12. A point on the rim of a turbine wheel of diameter 10 ft moves with a linear speed 45 ft/sec. Find the rate at which the wheel turns (angular speed) in rad/sec and in rev/sec.

In 1 sec a point on the rim travels a distance $s = 45$ ft. Then in 1 sec the wheel turns through an angle $\theta = s/r = 45/5 = 9$ radians and its angular speed is 9 rad/sec.

Since 1 rev = $2\pi$ rad or 1 rad = $\frac{1}{2\pi}$ rev, 9 rad/sec = $9\left(\frac{1}{2\pi}\right)$ rev/sec = 1.43 rev/sec.

13. Determine the speed of the earth (in mi/sec) in its course around the sun. Assume the earth's orbit to be a circle of radius 93,000,000 miles and 1 year = 365 days.

In 365 days the earth travels a distance of $2\pi r = 2(3.14)(93,000,000)$ miles.

In 1 second it will travel a distance $s = \frac{2(3.14)(93,000,000)}{365(24)(60)(60)}$ miles = 18.5 miles. Its speed is 18.5 mi/sec.

14. Express each of the following angles in mils: *a*) 18°, *b*) 16°20′, *c*) 0.22 rad, *d*) 1.6 rad.

Since 1° = $\frac{160}{9}$ mils and 1 rad = 1000 mils,

*a*) 18° = $18\left(\frac{160}{9}\right)$ mils = 320 mils,       *b*) 16°20′ = $\frac{49}{3}\left(\frac{160}{9}\right)$ mils = 290 mils,

*c*) 0.22 rad = 0.22(1000) mils = 220 mils,     *d*) 1.6 rad = 1.6(1000) mils = 1600 mils.

15. Express each of the following angles in degrees and in radians: *a*) 40 mils, *b*) 100 mils.

Since 1 mil = $\frac{9°}{160}$ = 0.001 rad,

*a*) 40 mils = $40\left(\frac{9°}{160}\right)$ = 2°15′   and 40 mils = 40(0.001) rad = 0.04 rad,

*b*) 100 mils = $100\left(\frac{9°}{160}\right)$ = 5°37.5′ and 100 mils = 100(0.001) rad = 0.1 rad.

16. Show that 1 mil = 0.001 radian, approximately.

$$1 \text{ mil} = \frac{2\pi}{6400} \text{ rad} = \frac{3.14159}{3200} \text{ rad} = 0.00098175 \text{ rad or, approximately, 0.001 rad.}$$

17. At 5000 yd range a battery subtends an angle of 15 mils. Find the width of the battery.

$$R = \frac{5000}{1000} = 5, \quad m = 15, \quad \text{and} \quad W = Rm = 5(15) = 75 \text{ yd.}$$

**18.** A ship 360 ft long is found to subtend an angle of 40 mils at an observation post on shore. Find the distance from shore to ship.

   $W = 360$ ft $= 120$ yd,   $m = 40$,   and   $R = W/m = 120/40 = 3$.
   The required distance is 3000 yd.

**19.** A shell is observed to burst 200 yd to the left of the target. What angular correction should be made in aiming the gun, if the range is  $a$) 5000 yd  and  $b$) 7500 yd?

   $a$) The correction is  $m = W/R = 200/5 = 40$ mils, to the right.

   $b$) The correction is  $m = W/R = 200/7.5 = 27$ mils, to the right.

## SUPPLEMENTARY PROBLEMS

**20.** Express each of the following in radian measure:
   $a$) $25^\circ$,   $b$) $160^\circ$,   $c$) $75^\circ 30'$,   $d$) $112^\circ 40'$,   $e$) $12^\circ 12' 20''$.

   *Ans.*   $a$) $5\pi/36$ rad or 0.4363 rad      $c$) $151\pi/360$ rad or 1.3177 rad      $e$) 0.2130 rad
   $b$) $8\pi/9$ rad or 2.7925 rad      $d$) $169\pi/270$ rad or 1.9664 rad

**21.** Express each of the following in degree measure:
   $a$) $\pi/4$ rad,   $b$) $7\pi/10$ rad,   $c$) $5\pi/6$ rad,   $d$) $1/4$ rad,   $e$) $7/5$ rad.

   *Ans.*   $a$) $45^\circ$,   $b$) $126^\circ$,   $c$) $150^\circ$,   $d$) $14^\circ 19' 26''$,   $e$) $80^\circ 12' 51''$

**22.** On a circle of radius 24 inches, find the length of arc subtended by a central angle $a$) of $2/3$ rad,   $b$) of $3\pi/5$ rad,   $c$) of $75^\circ$,   $d$) of $130^\circ$.

   *Ans.*   $a$) 16 in.,   $b$) $14.4\pi$ or 45.2 in.,   $c$) $10\pi$ or 31.4 in.,   $d$) $52\pi/3$ or 54.5 in.

**23.** A circle has a radius of 30 in. How many radians are there in an angle at the center subtended by an arc $a$) of 30 in.,   $b$) of 20 in.,   $c$) of 50 in.?

   *Ans.*   $a$) 1 rad,   $b$) $2/3$ rad,   $c$) $5/3$ rad

**24.** Find the radius of the circle for which an arc 15 inches long subtends an angle  $a$) of 1 rad, $b$) of $2/3$ rad,   $c$) of 3 rad,   $d$) of $20^\circ$,   $e$) of $50^\circ$.

   *Ans.*   $a$) 15 in.,   $b$) 22.5 in.,   $c$) 5 in.,   $d$) 43.0 in.,   $e$) 17.2 in.

**25.** The end of a 40 in. pendulum describes an arc of 5 in. Through what angle does the pendulum swing?   *Ans.* $1/8$ rad or $7^\circ 9' 43''$

**26.** A train is traveling at the rate 12 mi/hr on a curve of radius 3000 ft. Through what angle has it turned in one minute?   *Ans.* 0.352 rad or $20^\circ 10'$

**27.** A reversed curve on a railroad track consists of two circular arcs. The central angle of one is $20^\circ$ with radius 2500 ft and the central angle of the other is $25^\circ$ with radius 3000 ft. Find the total length of the two arcs.   *Ans.* $6250\pi/9$ ft or 2182 ft

**28.** A flywheel of radius 10 in. is turning at the rate 900 rpm. How fast does a point on the rim travel in ft/sec?   *Ans.* 78.5 ft/sec

**29.** An automobile tire is 30 in. in diameter. How fast (rpm) does the wheel turn on the axle when the automobile maintains a speed of 45 mph?   *Ans.* 504 rpm

30. In grinding certain tools the linear velocity of the grinding surface should not exceed 6000 ft/sec. Find the maximum number of revolutions per second  a) of a 12 in. (diameter) emery wheel,  b) of an 8 in. wheel.
    *Ans.*  a) $6000/\pi$ rev/sec or 1910 rev/sec,  b) 2865 rev/sec

31. If an automobile wheel 32 in. in diameter rotates at 800 rpm, what is the speed of the car in mph?  *Ans.* 76.2 mph

32. Express each of the following angles in mils:  a) $45^\circ$,  b) $10^\circ 15'$,  c) 0.4 rad,  d) 0.06 rad.
    *Ans.*  a) 800 mils,  b) 182 mils,  c) 400 mils,  d) 60 mils

33. Express each of the following in degree and in radian measure:  a) 25 mils,  b) 60 mils,  c) 110 mils.   *Ans.* a) $1^\circ 24'$ and 0.025 rad,  b) $3^\circ 22'$ and 0.06 rad,  c) $6^\circ 11'$ and 0.11 rad

34. The side of a hangar 1750 yd distant subtends an angle of 40 mils.  How long is it?
    *Ans.* 70 yd

35. A balloon 120 ft long is directly overhead. If it subtends an angle of 50 mils, how high is it?  *Ans.* 800 yd

36. From a boat at sea, the angle of elevation of the top of a cliff is measured as 12 mils.  If the cliff is known to be 90 ft high, how far is the boat from the cliff?  *Ans.* 2500 yd

37. A hill, known to be 180 ft high, subtends an angle of 30 mils from a point on a level plain. From the same point, the angle of elevation of a machine gun nest on the side of the hill is found to be 12 mils. How far is the nest above the base of the hill?  *Ans.* 72 ft

# CHAPTER 2

# Trigonometric Functions of a General Angle

**NUMBER SCALE.** A *directed line* is a line on which one direction is taken as positive and the other as negative. The positive direction is indicated by an arrowhead.

A *number scale* is established on a directed line by choosing a point $O$ (see Fig. 2-A) called the *origin* and a unit of measure $OA = 1$. On this scale $B$ is 4 units to the right of $O$ (that is, in the positive direction from $O$) and $C$ is 2 units to the left of $O$ (that is, in the negative direction from $O$).

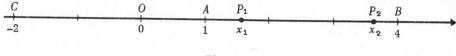

Fig. 2-A

The directed distance $OB = +4$ and the directed distance $OC = -2$. It is important to note that, since the line is directed, $OB \neq BO$ and $OC \neq CO$. The directed distance $BO = -4$, being measured contrary to the indicated positive direction, and the directed distance $CO = +2$. Then $CB = CO + OB = 2 + 4 = 6$ and $BC = BO + OC = -4 + (-2) = -6$.

**A RECTANGULAR COORDINATE SYSTEM** in a plane consists of two number scales (called *axes*), one horizontal and the other vertical, whose point of intersection (*origin*) is the origin on each scale. It is customary to choose the positive direction on each axis as indicated in the figure, that is, positive to the right on the horizontal axis or *x*-axis and positive upward on the vertical or *y*-axis. For convenience, we shall assume the same unit of measure on each axis.

By means of such a system the position of any point $P$ in the plane is given by its (directed) distances, called *coordinates*, from the axes. The *x*-coordinate or *abscissa* of a point $P$ (see Fig. 2-B) is the directed distance $BP = OA$ and the *y*-coordinate or *ordinate* is the directed distance $AP = OB$. A point $P$ with abscissa $x$ and ordinate $y$ will be denoted by $P(x,y)$.

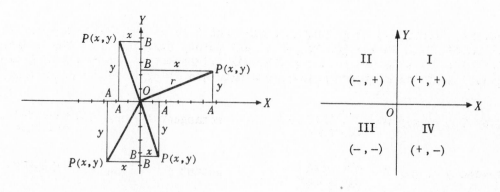

Fig. 2-B    Fig. 2-C

The axes divide the plane into four parts, called *quadrants*, which are numbered I, II, III, IV. The numbered quadrants, together with the signs of the coordinates of a point in each, are shown in Fig. 2-C.

The undirected distance $r$ of any point $P(x, y)$ from the origin, called the *distance of P* or the radius vector of $P$, is given by

$$r = \sqrt{x^2 + y^2}.$$

Thus, with each point in the plane, we associate three numbers: $x, y, r$.

See Problems 1-3.

**ANGLES IN STANDARD POSITION.** With respect to a rectangular coordinate system, an angle is said to be *in standard position* when its vertex is at the origin and its initial side coincides with the positive $x$-axis.

An angle is said to be a *first quadrant angle* or to be *in the first quadrant* if, when in standard position, its terminal side falls in that quadrant. Similar definitions hold for the other quadrants. For example, the angles 30°, 59°, and –330° are first quadrant angles; 119° is a second quadrant angle; –119° is a third quadrant angle; –10° and 710° are fourth quadrant angles.

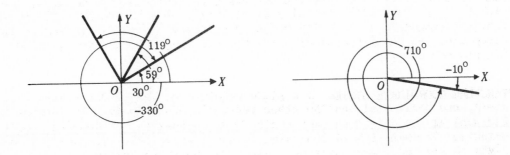

Two angles which, when placed in standard position, have coincident terminal sides are called *coterminal angles*. For example, 30° and –330°, –10° and 710° are pairs of coterminal angles. There are an unlimited number of angles coterminal with a given angle. (See Problem 4.)

The angles 0°, 90°, 180°, 270°, and all angles coterminal with them are called *quadrantal angles*.

**TRIGONOMETRIC FUNCTIONS OF A GENERAL ANGLE.** Let $\theta$ be an angle (not quadrantal) in standard position and let $P(x, y)$ be any point, distinct from the origin, on the terminal side of the angle. The six trigonometric functions of $\theta$ are defined, in terms of the abscissa, ordinate and distance of $P$, as follows:

sine $\theta$  = sin $\theta$ = $\dfrac{\text{ordinate}}{\text{distance}} = \dfrac{y}{r}$          cotangent $\theta$ = cot $\theta$ = $\dfrac{\text{abscissa}}{\text{ordinate}} = \dfrac{x}{y}$

cosine $\theta$ = cos $\theta$ = $\dfrac{\text{abscissa}}{\text{distance}} = \dfrac{x}{r}$          secant $\theta$   = sec $\theta$ = $\dfrac{\text{distance}}{\text{abscissa}} = \dfrac{r}{x}$

tangent $\theta$ = tan $\theta$ = $\dfrac{\text{ordinate}}{\text{abscissa}} = \dfrac{y}{x}$          cosecant $\theta$  = csc $\theta$ = $\dfrac{\text{distance}}{\text{ordinate}} = \dfrac{r}{y}$

As an immediate consequence of these definitions, we have the so-called *Reciprocal Relations*:

$$\sin \theta = 1/\csc \theta \qquad \tan \theta = 1/\cot \theta \qquad \sec \theta = 1/\cos \theta$$
$$\cos \theta = 1/\sec \theta \qquad \cot \theta = 1/\tan \theta \qquad \csc \theta = 1/\sin \theta$$

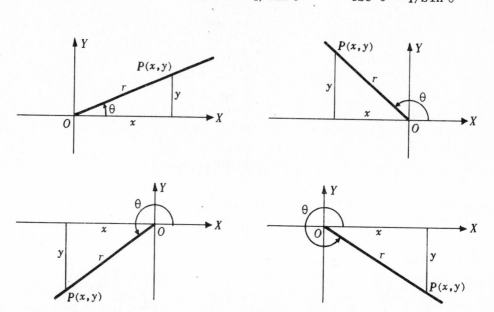

It is evident from the figures that the values of the trigonometric functions of θ change as θ changes. In Problem 5 it is shown that the values of the functions of a given angle θ are independent of the choice of the point P on its terminal side.

**ALGEBRAIC SIGNS OF THE FUNCTIONS.** Since r is always positive, the signs of the functions in the various quadrants depend upon the signs of x and y. To determine these signs one may visualize the angle in standard position or use some device as shown in the accompanying figure in which only the functions having positive signs are listed. (See Problem 6.)

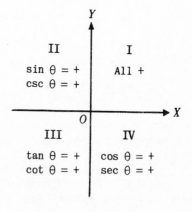

When an angle is given, its trigonometric functions are uniquely determined. When, however, the value of one function of an angle is given, the angle is not uniquely determined. For example, if sin θ = ½, then θ = 30°, 150°, 390°, 510°, ····. In general, two possible positions of the terminal side are found — for example, the terminal sides of 30° and 150° in the above illustration. The exceptions to this rule occur when the angle is quadrantal.     (See Problems 7-15.)

**TRIGONOMETRIC FUNCTIONS OF QUADRANTAL ANGLES.** For a quadrantal angle, the terminal side coincides with one of the axes. A point P, distinct from the origin, on the terminal side has either $x = 0, y \neq 0$ or $x \neq 0, y = 0$. In either case, two of the six functions will not be defined. For example, the terminal side of the angle 0° coincides with the positive x-axis and the ordinate of P is 0. Since the ordinate occurs in the denominator of the ratio defining the cotangent and cosecant, these functions are not defined. Certain authors indicate this by writing $\cot 0° = \infty$ and others write $\cot 0° = \pm\infty$. The following results are obtained in Problem 16.

| angle $\theta$ | $\sin\theta$ | $\cos\theta$ | $\tan\theta$ | $\cot\theta$ | $\sec\theta$ | $\csc\theta$ |
|---|---|---|---|---|---|---|
| $0°$ | 0 | 1 | 0 | $\pm\infty$ | 1 | $\pm\infty$ |
| $90°$ | 1 | 0 | $\pm\infty$ | 0 | $\pm\infty$ | 1 |
| $180°$ | 0 | $-1$ | 0 | $\pm\infty$ | $-1$ | $\pm\infty$ |
| $270°$ | $-1$ | 0 | $\pm\infty$ | 0 | $\pm\infty$ | $-1$ |

## SOLVED PROBLEMS

1. Using a rectangular coordinate system, locate the following points and find the value of $r$ for each: $A(1,2)$, $B(-3,4)$, $C(-3, -3\sqrt{3})$, $D(4,-5)$.

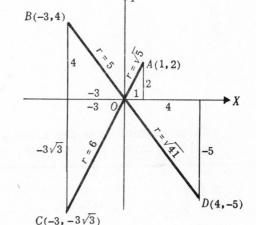

For $A$: $r = \sqrt{x^2 + y^2} = \sqrt{1+4} = \sqrt{5}$

For $B$: $r = \sqrt{9 + 16} = 5$

For $C$: $r = \sqrt{9 + 27} = 6$

For $D$: $r = \sqrt{16 + 25} = \sqrt{41}$

2. Determine the missing coordinate of $P$ in each of the following:

a) $x = 2$, $r = 3$, $P$ in the first quadrant;
b) $x = -3$, $r = 5$, $P$ in the second quadrant;
c) $y = -1$, $r = 3$, $P$ in the third quadrant;
d) $x = 2$, $r = \sqrt{5}$, $P$ in the fourth quadrant;
e) $x = 3$, $r = 3$;
f) $y = -2$, $r = 2$;   g) $x = 0$, $r = 2$, $y$ positive;   h) $y = 0$, $r = 1$, $x$ negative.

a) Using the relation $x^2 + y^2 = r^2$, we have $4 + y^2 = 9$; then $y^2 = 5$ and $y = \pm\sqrt{5}$. Since $P$ is in the first quadrant, the missing coordinate is $y = \sqrt{5}$.

b) Here $9 + y^2 = 25$, $y^2 = 16$, and $y = \pm 4$. Since $P$ is in the second quadrant, the missing coordinate is $y = 4$.

c) We have $x^2 + 1 = 9$, $x^2 = 8$, and $x = \pm 2\sqrt{2}$. Since $P$ is in the third quadrant, the missing coordinate is $x = -2\sqrt{2}$.

d) $y^2 = 5 - 4$ and $y = \pm 1$. Since $P$ is in the fourth quadrant, the missing coordinate is $y = -1$.

e) Here $y^2 = r^2 - x^2 = 9 - 9 = 0$ and the missing coordinate is $y = 0$.

f) $x^2 = r^2 - y^2 = 0$ and $x = 0$.   g) $y^2 = r^2 - x^2 = 4$ and $y = 2$ is the missing coordinate.

h) $x^2 = r^2 - y^2 = 1$ and $x = -1$ is the missing coordinate.

3. In what quadrants may $P(x,y)$ be located if

a) $x$ is positive and $y \neq 0$ ?        c) $y/r$ is positive ?        e) $y/x$ is positive ?

b) $y$ is negative and $x \neq 0$ ?        d) $r/x$ is negative ?

    *a*) In the first quadrant when **y** is positive and in the fourth quadrant when **y** is negative.

    *b*) In the fourth quadrant when *x* is positive and in the third quadrant when *x* is negative.

    *c*) In the first and second quadrants.    *d*) In the second and third quadrants.

    *e*) In the first quadrant when both *x* and *y* are positive and in the third quadrant when **both *x*** and *y* are negative.

**4.** *a*) Construct the following angles in standard position and determine those which are coterm**inal**:
$$125°, \quad 210°, \quad -150°, \quad 385°, \quad 930°, \quad -370°, \quad -955°, \quad -870°.$$
    *b*) Give five other angles coterminal with $125°$.

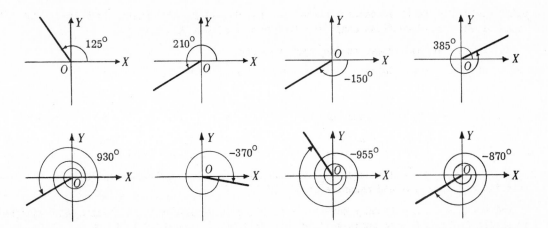

    *a*) The angles $125°$ and $-955° = 125° - 3\cdot360°$ are coterminal.  The angles $210°$, $-150° = 210° - 360°$, $930° = 210° + 2\cdot360°$, and $-870° = 210° - 3\cdot360°$ are coterminal.

    *b*) $485° = 125° + 360°$, $\quad 1205° = 125° + 3\cdot360°$, $\quad 1925° = 125° + 5\cdot360°$, $\quad -235° = 125° - 360°$, $-1315° = 125° - 4\cdot360°$ are coterminal with $125°$.

**5.** Show that the values of the trigonometric functions of an angle $\theta$ do not depend upon the choice of the point $P$ selected on the terminal side of the angle.

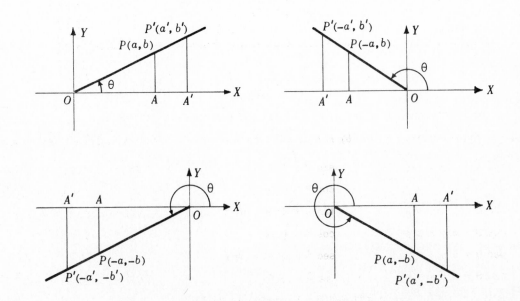

On the terminal side of each of the angles of the figures above, let $P$ and $P'$ have coordinates as indicated, and denote the distances $OP$ and $OP'$ by $r$ and $r'$ respectively. Drop the perpendiculars $AP$ and $A'P'$ to the $x$-axis. In each of the figures, the triangles $OAP$ and $OA'P'$, having sides $a,b,r$ and $a',b',r'$ respectively, are similar; thus,

1) $\quad b/r = b'/r', \quad a/r = a'/r', \quad b/a = b'/a', \quad a/b = a'/b', \quad r/a = r'/a', \quad r/b = r'/b'.$

Since the ratios are the trigonometric ratios for the first quadrant angle, the values of the functions of any first quadrant angle are independent of the choice of $P$.

From 1) it follows that

$b/r = b'/r', \quad -a/r = -a'/r', \quad b/-a = b'/-a', \quad -a/b = -a'/b', \quad r/-a = r'/-a', \quad r/b = r'/b'.$

Since these are the trigonometric ratios for the second quadrant angle, the values of the functions of any second quadrant angle are independent of the choice of $P$.

It is left for the reader to consider the cases

$-b/r = -b'/r', \quad -a/r = -a'/r', \quad \text{etc.}, \qquad \text{and} \qquad -b/r = -b'/r', \quad a/r = a'/r', \text{ etc.}$

**6.** Determine the signs of the functions sine, cosine, and tangent in each of the quadrants.

$\sin \theta = y/r$. Since $y$ is positive in quadrants I,II and negative in quadrants III,IV, while $r$ is always positive, $\sin \theta$ is positive in quadrants I,II and negative in quadrants III,IV.

$\cos \theta = x/r$. Since $x$ is positive in quadrants I,IV and negative in II,III, $\cos \theta$ is positive in quadrants I,IV and negative in II,III.

$\tan \theta = y/x$. Since $x$ and $y$ have the same signs in quadrants I, III and opposite signs in quadrants II,IV, $\tan \theta$ is positive in quadrants I,III and negative in II,IV.

**7.** Determine the values of the trigonometric functions of angle $\theta$ (smallest positive angle in standard position) if $P$ is a point on the terminal side of $\theta$ and the coordinates of $P$ are:
a) $P(3,4)$, b) $P(-3,4)$, c) $P(-1,-3)$.

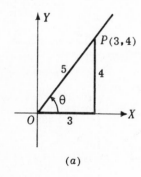 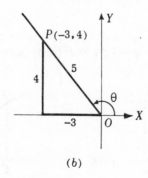 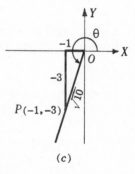

| (a) | (b) | (c) |

a) $r = \sqrt{3^2 + 4^2} = 5$

$\sin \theta = y/r = 4/5$

$\cos \theta = x/r = 3/5$

$\tan \theta = y/x = 4/3$

$\cot \theta = x/y = 3/4$

$\sec \theta = r/x = 5/3$

$\csc \theta = r/y = 5/4$

b) $r = \sqrt{(-3)^2 + 4^2} = 5$

$\sin \theta = 4/5$

$\cos \theta = -3/5$

$\tan \theta = 4/-3 = -4/3$

$\cot \theta = -3/4$

$\sec \theta = 5/-3 = -5/3$

$\csc \theta = 5/4$

c) $r = \sqrt{(-1)^2 + (-3)^2} = \sqrt{10}$

$\sin \theta = -3/\sqrt{10} = -3\sqrt{10}/10$

$\cos \theta = -1/\sqrt{10} = -\sqrt{10}/10$

$\tan \theta = -3/-1 = 3$

$\cot \theta = -1/-3 = 1/3$

$\sec \theta = \sqrt{10}/-1 = -\sqrt{10}$

$\csc \theta = \sqrt{10}/-3 = -\sqrt{10}/3$

Note the reciprocal relationships. For example, in b)
$\sin \theta = 1/\csc \theta = 4/5, \quad \cos \theta = 1/\sec \theta = -3/5, \quad \tan \theta = 1/\cot \theta = -4/3, \text{ etc.}$

**8.** In what quadrant will θ terminate, if
    *a*) sin θ and cos θ are both negative?    *c*) sin θ is positive and secant θ is negative?
    *b*) sin θ and tan θ are both positive?    *d*) sec θ is negative and tan θ is negative?

    *a*) Since sin θ = $y/r$ and cos θ = $x/r$, both $x$ and $y$ are negative. (Recall that $r$ is always positive.) Thus, θ is a third quadrant angle.

    *b*) Since sin θ is positive, $y$ is positive; since tan θ = $y/x$ is positive, $x$ is also positive. Thus, θ is a first quadrant angle.

    *c*) Since sin θ is positive, $y$ is positive; since sec θ is negative, $x$ is negative. Thus, θ is a second quadrant angle.

    *d*) Since sec θ is negative, $x$ is negative; since tan θ is negative, $y$ is then positive. Thus, θ is a second quadrant angle.

**9.** In what quadrants may θ terminate, if
    *a*) sin θ is positive?    *b*) cos θ is negative?    *c*) tan θ is negative?    *d*) sec θ is positive?

    *a*) Since sin θ is positive, $y$ is positive.
    Then $x$ may be positive or negative and θ is a first or second quadrant angle.

    *b*) Since cos θ is negative, $x$ is negative.
    Then $y$ may be positive or negative and θ is a second or third quadrant angle.

    *c*) Since tan θ is negative, either $y$ is positive and $x$ is negative or $y$ is negative and $x$ is positive. Thus, θ may be a second or fourth quadrant angle.

    *d*) Since sec θ is positive, $x$ is positive. Thus, θ may be a first or fourth quadrant angle.

**10.** Find the values of cos θ and tan θ, given sin θ = 8/17 and θ in quadrant I.

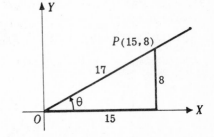

    Let $P$ be a point on the terminal line of θ. Since sin θ = $y/r$ = 8/17, we take $y = 8$ and $r = 17$. Since θ is in quadrant I, $x$ is positive; thus

$$x = \sqrt{r^2 - y^2} = \sqrt{(17)^2 - (8)^2} = 15.$$

    To draw the figure, locate the point $P(15,8)$, join it to the origin, and indicate the angle θ. Then

      cos θ = $x/r$ = 15/17  and  tan θ = $y/x$ = 8/15.

    The choice of $y = 8$, $r = 17$ is one of convenience. Note that 8/17 = 16/34 and we might have taken $y = 16$, $r = 34$. Then $x = 30$, cos θ = 30/34 = 15/17 and tan θ = 16/30 = 8/15. (See Problem 5.)

**11.** Find the values of sin θ and tan θ, given cos θ = 5/6.

    Since cos θ is positive, θ is in quadrant I or IV.

    Since cos θ = $x/r$ = 5/6, we take $x = 5$, $r = 6$; $y = \pm\sqrt{(6)^2 - (5)^2} = \pm\sqrt{11}$.

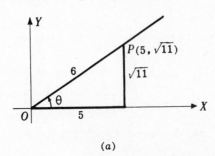

(a)

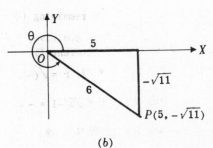

(b)

*a*) For θ in quadrant I (Figure *a*) we have $x = 5$, $y = \sqrt{11}$, $r = 6$;  then

$$\sin\theta = y/r = \sqrt{11}/6 \quad\text{and}\quad \tan\theta = y/x = \sqrt{11}/5.$$

*b*) For θ in quadrant IV (Figure *b*) we have $x = 5$, $y = -\sqrt{11}$, $r = 6$;  then

$$\sin\theta = y/r = -\sqrt{11}/6 \quad\text{and}\quad \tan\theta = y/x = -\sqrt{11}/5.$$

**12.** Find the values of sin θ and cos θ, given tan θ = –3/4.

Since $\tan\theta = y/x$ is negative, θ is in quadrant II (take $x = -4$, $y = 3$) or in quadrant IV (take $x = 4$, $y = -3$).   In either case  $r = \sqrt{16 + 9} = 5$.

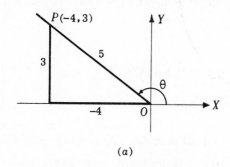          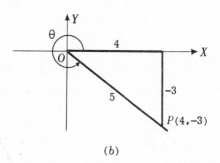

(*a*)                                (*b*)

*a*) For θ in quadrant II (Figure *a*),   $\sin\theta = y/r = 3/5$   and   $\cos\theta = x/r = -4/5$.

*b*) For θ in quadrant IV (Figure *b*),   $\sin\theta = y/r = -3/5$  and  $\cos\theta = x/r = 4/5$.

**13.** Find sin θ, given cos θ = –4/5 and that tan θ is positive.

Since $\cos\theta = x/r$ is negative, $x$ is negative.   Since also $\tan\theta = y/x$ is positive, $y$ must be negative.   Then θ is in quadrant III.   (See Figure *c* below.)

Take $x = -4$, $r = 5$;  then $y = -\sqrt{5^2 - (-4)^2} = -3$.   Thus, $\sin\theta = y/r = -3/5$.

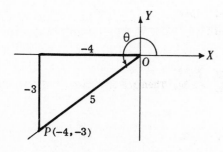          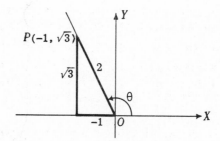

Fig.(*c*) Prob. 13                    Fig.(*d*) Prob. 14

**14.** Find the values of the remaining functions of θ, given sin θ = $\sqrt{3}/2$ and cos θ = –1/2.

Since $\sin\theta = y/r$ is positive, $y$ is positive.   Since $\cos\theta = x/r$ is negative, $x$ is negative.   Thus, θ is in quadrant II.   (See Figure *d* above.)

Taking $x = -1$, $y = \sqrt{3}$, $r = \sqrt{(-1)^2 + (\sqrt{3})^2} = 2$,  we have

$$\tan\theta = y/x = \sqrt{3}/-1 = -\sqrt{3} \qquad\qquad \cot\theta = 1/\tan\theta = -1/\sqrt{3} = -\sqrt{3}/3$$

$$\sec\theta = 1/\cos\theta = -2 \qquad\qquad\quad \csc\theta = 1/\sin\theta = 2/\sqrt{3} = 2\sqrt{3}/3.$$

**15.** Determine the values of $\cos \theta$ and $\tan \theta$ if $\sin \theta = m/n$, a negative fraction.

Since $\sin \theta$ is negative, $\theta$ is in quadrant III or IV.

*a*) In quadrant III :   Take $y = m$, $r = n$, $x = -\sqrt{n^2 - m^2}$ ; then

$$\cos \theta = x/r = -\sqrt{n^2 - m^2}/n \qquad \text{and} \qquad \tan \theta = y/x = -m/\sqrt{n^2 - m^2} \, .$$

*b*) In quadrant IV :   Take $y = m$, $r = n$, $x = +\sqrt{n^2 - m^2}$ ; then

$$\cos \theta = x/r = \sqrt{n^2 - m^2}/n \qquad \text{and} \qquad \tan \theta = y/x = m/\sqrt{n^2 - m^2} \, .$$

**16.** Determine the values of the trigonometric functions of *a*) $0^\circ$,  *b*) $90^\circ$,  *c*) $180^\circ$,  *d*) $270^\circ$.

Let $P$ be any point (not 0) on the terminal side of $\theta$. When $\theta = 0^\circ$, $x = r, y = 0$; when $\theta = 90^\circ$, $x = 0, y = r$; when $\theta = 180^\circ$, $x = -r, y = 0$; when $\theta = 270^\circ$, $x = 0, y = -r$.

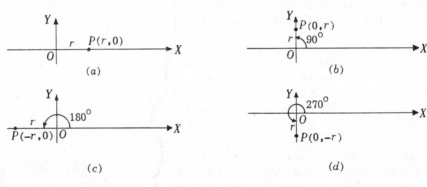

*a*) $\theta = 0^\circ$; $x = r$, $y = 0$

$\sin 0^\circ = y/r = 0/r = 0$
$\cos 0^\circ = x/r = r/r = 1$
$\tan 0^\circ = y/x = 0/r = 0$
$\cot 0^\circ = x/y = \pm\infty$
$\sec 0^\circ = r/x = r/r = 1$
$\csc 0^\circ = r/y = \pm\infty$

*b*) $\theta = 90^\circ$; $x = 0$, $y = r$

$\sin 90^\circ = y/r = r/r = 1$
$\cos 90^\circ = x/r = 0/r = 0$
$\tan 90^\circ = y/x = \pm\infty$
$\cot 90^\circ = x/y = 0/r = 0$
$\sec 90^\circ = r/x = \pm\infty$
$\csc 90^\circ = r/y = r/r = 1$

*c*) $\theta = 180^\circ$; $x = -r$, $y = 0$

$\sin 180^\circ = y/r = 0/r = 0$
$\cos 180^\circ = x/r = -r/r = -1$
$\tan 180^\circ = y/x = 0/-r = 0$
$\cot 180^\circ = x/y = \pm\infty$
$\sec 180^\circ = r/x = r/-r = -1$
$\csc 180^\circ = r/y = \pm\infty$

*d*) $\theta = 270^\circ$; $x = 0$, $y = -r$

$\sin 270^\circ = y/r = -r/r = -1$
$\cos 270^\circ = x/r = 0/r = 0$
$\tan 270^\circ = y/x = \pm\infty$
$\cot 270^\circ = x/y = 0/-r = 0$
$\sec 270^\circ = r/x = \pm\infty$
$\csc 270^\circ = r/y = r/-r = -1$

It has been noted that $\cot 0^\circ$ and $\csc 0^\circ$ are not defined since division by zero is **never** permitted. In Figure (*e*) below, take $\theta$ a small positive angle in standard position and on its terminal side take $P(x,y)$ at a distance $r$ from 0. Now $x$ is slightly less than $r$ and $y$ is positive and very small; then $\cot \theta = x/y$ and $\csc \theta = r/y$ are positive and very large. Next let $\theta$ decrease toward $0^\circ$ (that is, $OP$ turns toward $OX$) with $P$ remaining at a distance $r$ from 0. Now $x$ increases but is always smaller than $r$ while $y$ decreases but remains greater than 0. Then $\cot \theta$ and $\csc \theta$ become larger and larger. (To see this, take $r = 1$ and compute $\csc \theta$ when

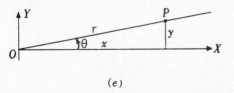

(*e*)

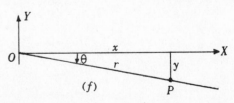

(*f*)

$y$ = 0.1, 0.01, 0.001, $\cdots\cdots$.) This state of affairs is indicated by writing cot $0^0$ = + $\infty$ and csc $0^0$ = + $\infty$. Note that while the sign = is used, we do not really mean that "cot $0^0$ equals"; we mean that as a small positive angle becomes smaller and smaller, the cotangent of the angle becomes a larger and larger positive number.

Next suppose, as in Figure $(f)$ above, that $\theta$ is numerically small and negative and take $P(x,y)$ on its terminal side at a distance $r$ from O. Then $x$ is positive and slightly smaller than $r$ while $y$ is negative and numerically small; cot $\theta$ and csc $\theta$ are negative and numerically large. As $\theta$ increases toward $0^0$, cot $\theta$ and csc $\theta$ remain negative but become larger and larger numerically. This is indicated by writing cot $0^0$ = $-\infty$ and csc $0^0$ = $-\infty$.

**17.** Evaluate: $a$) sin $0^0$ + 2 cos $0^0$ + 3 sin $90^0$ + 4 cos $90^0$ + 5 sec $0^0$ + 6 csc $90^0$

$\qquad\qquad$ $b$) sin $180^0$ + 2 cos $180^0$ + 3 sin $270^0$ + 4 cos $270^0$ $-$ 5 sec $180^0$ $-$ 6 csc $270^0$

$a$) 0 + 2(1) + 3(1) + 4(0) + 5(1) + 6(1) = 16
$b$) 0 + 2($-$1) + 3($-$1) + 4(0) $-$ 5($-$1) $-$ 6($-$1) = 6

**18.** Using a protractor, construct an angle of $20^0$ in standard position. With O as center describe an arc of radius 10 units meeting the terminal side in $P$. From $P$ drop a perpendicular to the $x$-axis, meeting it in $A$. By actual measurement, $OA$ = 9.4 and $AP$ = 3.4, and $P$ has coordinates (9.4, 3.4). Then

sin $20^0$ = 3.4/10 = 0.34,$\qquad$ cot $20^0$ = 9.4/3.4 = 2.8,
cos $20^0$ = 9.4/10 = 0.94,$\qquad$ sec $20^0$ = 10/9.4 = 1.1,
tan $20^0$ = 3.4/9.4 = 0.36,$\qquad$ csc $20^0$ = 10/3.4 = 2.9.

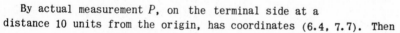

$(g)$

**19.** Obtain the trigonometric functions of $50^0$ as in Problem 18. Refer to Figure $(g)$.

By actual measurement $P$, on the terminal side at a distance 10 units from the origin, has coordinates (6.4, 7.7). Then

$\qquad\qquad$ sin $50^0$ = 7.7/10 = 0.77,$\qquad$ cot $50^0$ = 6.4/7.7 = 0.83,
$\qquad\qquad$ cos $50^0$ = 6.4/10 = 0.64,$\qquad$ sec $50^0$ = 10/6.4 = 1.6,
$\qquad\qquad$ tan $50^0$ = 7.7/6.4 = 1.2,$\qquad$ csc $50^0$ = 10/7.7 = 1.3.

## SUPPLEMENTARY PROBLEMS

20. State the quadrant in which each angle terminates and the signs of the sine, cosine, and tangent of each angle.

   a) $125^{\circ}$,  b) $75^{\circ}$,  c) $320^{\circ}$,  d) $212^{\circ}$,  e) $460^{\circ}$,  f) $750^{\circ}$,  g) $-250^{\circ}$,  h) $-1000^{\circ}$.

   Ans.  a) II; +,−,−  b) I; +,+,+  c) IV; −,+,−  d) III; −,−,+  e) II  f) I  g) II  h) I

21. In what quadrant will θ terminate if
   a) sin θ and cos θ are both positive?       e) tan θ is positive and sec θ is negative?
   b) cos θ and tan θ are both positive?       f) tan θ is negative and sec θ is positive?
   c) sin θ and sec θ are both negative?       g) sin θ is positive and cos θ is negative?
   d) cos θ and cot θ are both negative?       h) sec θ is positive and csc θ is negative?

   Ans.  a) I,  b) I,  c) III,  d) II,  e) III,  f) IV,  g) II,  h) IV

22. Denote by θ the smallest positive angle whose terminal side passes through the given point and find the trigonometric functions of θ:

   a) $P(-5,12)$,  b) $P(7,-24)$,  c) $P(2,3)$,  d) $P(-3,-5)$.

   Ans.  a) 12/13, −5/13, −12/5, −5/12, −13/5, 13/12
   b) −24/25, 7/25, −24/7, −7/24, 25/7, −25/24
   c) $3/\sqrt{13}$, $2/\sqrt{13}$, 3/2, 2/3, $\sqrt{13}/2$, $\sqrt{13}/3$
   d) $-5/\sqrt{34}$, $-3/\sqrt{34}$, 5/3, 3/5, $-\sqrt{34}/3$, $-\sqrt{34}/5$

23. Find the values of the trigonometric functions of θ, given:

   a) sin θ = 7/25       d) cot θ = 24/7       g) tan θ = 3/5       j) csc θ = $-2/\sqrt{3}$
   b) cos θ = −4/5       e) sin θ = −2/3       h) cot θ = $\sqrt{6}/2$
   c) tan θ = −5/12      f) cos θ = 5/6        i) sec θ = $-\sqrt{5}$

   Ans.  a) I: 7/25, 24/25, 7/24, 24/7, 25/24, 25/7
          II: 7/25, −24/25, −7/24, −24/7, −25/24, 25/7

   b) II: 3/5, −4/5, −3/4, −4/3, −5/4, 5/3;   III: −3/5, −4/5, 3/4, 4/3, −5/4, −5/3

   c) II: 5/13, −12/13, −5/12, −12/5, −13/12, 13/5
      IV: −5/13, 12/13, −5/12, −12/5, 13/12, −13/5

   d) I: 7/25, 24/25, 7/24, 24/7, 25/24, 25/7
      III: −7/25, −24/25, 7/24, 24/7, −25/24, −25/7

   e) III: −2/3, $-\sqrt{5}/3$, $2/\sqrt{5}$, $\sqrt{5}/2$, $-3/\sqrt{5}$, −3/2
      IV: −2/3, $\sqrt{5}/3$, $-2/\sqrt{5}$, $-\sqrt{5}/2$, $3/\sqrt{5}$, −3/2

   f) I: $\sqrt{11}/6$, 5/6, $\sqrt{11}/5$, $5/\sqrt{11}$, 6/5, $6/\sqrt{11}$
      IV: $-\sqrt{11}/6$, 5/6, $-\sqrt{11}/5$, $-5/\sqrt{11}$, 6/5, $-6/\sqrt{11}$

   g) I: $3/\sqrt{34}$, $5/\sqrt{34}$, 3/5, 5/3, $\sqrt{34}/5$, $\sqrt{34}/3$
      III: $-3/\sqrt{34}$, $-5/\sqrt{34}$, 3/5, 5/3, $-\sqrt{34}/5$, $-\sqrt{34}/3$

   h) I: $2/\sqrt{10}$, $\sqrt{3}/\sqrt{5}$, $2/\sqrt{6}$, $\sqrt{6}/2$, $\sqrt{5}/\sqrt{3}$, $\sqrt{10}/2$
      III: $-2/\sqrt{10}$, $-\sqrt{3}/\sqrt{5}$, $2/\sqrt{6}$, $\sqrt{6}/2$, $-\sqrt{5}/\sqrt{3}$, $-\sqrt{10}/2$

   i) II: $2/\sqrt{5}$, $-1/\sqrt{5}$, −2, −1/2, $-\sqrt{5}$, $\sqrt{5}/2$;   III: $-2/\sqrt{5}$, $-1/\sqrt{5}$, 2, 1/2, $-\sqrt{5}$, $-\sqrt{5}/2$

   j) III: $-\sqrt{3}/2$, −1/2, $\sqrt{3}$, $1/\sqrt{3}$, −2, $-2/\sqrt{3}$;   IV: $-\sqrt{3}/2$, 1/2, $-\sqrt{3}$, $-1/\sqrt{3}$, 2, $-2/\sqrt{3}$

24. Evaluate each of the following:

   a) tan $180^{\circ}$ − 2 cos $180^{\circ}$ + 3 csc $270^{\circ}$ + sin $90^{\circ}$ = 0.

   b) sin $0^{\circ}$ + 3 cot $90^{\circ}$ + 5 sec $180^{\circ}$ − 4 cos $270^{\circ}$ = −5.

# CHAPTER 3

# Trigonometric Functions of an Acute Angle

TRIGONOMETRIC FUNCTIONS OF AN ACUTE ANGLE. In dealing with any right triangle, it will be convenient (see Fig. 3-A) to denote the vertices as $A, B, C$ = vertex of the right angle, to denote the angles of the triangle as $A, B, C = 90°$, and the sides opposite the angles as $a, b, c$ respectively. With respect to angle $A$, $a$ will be called the *opposite side* and $b$ will be called the *adjacent side*; with respect to angle $B$, $a$ will be called the *adjacent side* and $b$ the *opposite side*. Side $c$ will always be called the *hypotenuse*.

If now the right triangle is placed in a coordinate system (Fig. 3-B) so that angle $A$ is in standard position, the point $B$ on the terminal side of angle $A$ has coordinates $(b, a)$ and distance $c = \sqrt{a^2 + b^2}$. Then the trigonometric functions of angle $A$ may be defined in terms of the sides of the right triangle, as follows:

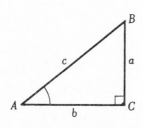

Fig. 3-A

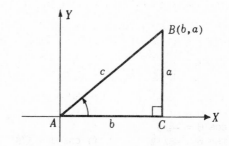

Fig. 3-B

$$\sin A = \frac{a}{c} = \frac{\text{opposite side}}{\text{hypotenuse}} \qquad \cot A = \frac{b}{a} = \frac{\text{adjacent side}}{\text{opposite side}}$$

$$\cos A = \frac{b}{c} = \frac{\text{adjacent side}}{\text{hypotenuse}} \qquad \sec A = \frac{c}{b} = \frac{\text{hypotenuse}}{\text{adjacent side}}$$

$$\tan A = \frac{a}{b} = \frac{\text{opposite side}}{\text{adjacent side}} \qquad \csc A = \frac{c}{a} = \frac{\text{hypotenuse}}{\text{opposite side}}$$

TRIGONOMETRIC FUNCTIONS OF COMPLEMENTARY ANGLES. The acute angles $A$ and $B$ of the right triangle $ABC$ are complementary, that is, $A + B = 90°$. From Fig. 3-A, we have

$$\sin B = b/c = \cos A \qquad \cot B = a/b = \tan A$$
$$\cos B = a/c = \sin A \qquad \sec B = c/a = \csc A$$
$$\tan B = b/a = \cot A \qquad \csc B = c/b = \sec A$$

These relations associate the functions in pairs — sine and cosine, tangent and cotangent, secant and cosecant — each function of a pair being called the *cofunction* of the other. Thus, any function of an acute angle is equal to the corresponding cofunction of the complementary angle.

TRIGONOMETRIC FUNCTIONS OF 30°, 45°, 60°.  The following results are obtained in Problems 8-9.

| Angle θ | sin θ | cos θ | tan θ | cot θ | sec θ | csc θ |
|---------|-------|-------|-------|-------|-------|-------|
| 30° | $\frac{1}{2}$ | $\frac{1}{2}\sqrt{3}$ | $\frac{1}{3}\sqrt{3}$ | $\sqrt{3}$ | $\frac{2}{3}\sqrt{3}$ | 2 |
| 45° | $\frac{1}{2}\sqrt{2}$ | $\frac{1}{2}\sqrt{2}$ | 1 | 1 | $\sqrt{2}$ | $\sqrt{2}$ |
| 60° | $\frac{1}{2}\sqrt{3}$ | $\frac{1}{2}$ | $\sqrt{3}$ | $\frac{1}{3}\sqrt{3}$ | 2 | $\frac{2}{3}\sqrt{3}$ |

PROBLEMS 10-16 illustrate a number of simple applications of the trigonometric functions. For this purpose the following table will be used.

| Angle θ | sin θ | cos θ | tan θ | cot θ | sec θ | csc θ |
|---------|-------|-------|-------|-------|-------|-------|
| 15° | 0.26 | 0.97 | 0.27 | 3.7 | 1.0 | 3.9 |
| 20° | 0.34 | 0.94 | 0.36 | 2.7 | 1.1 | 2.9 |
| 30° | 0.50 | 0.87 | 0.58 | 1.7 | 1.2 | 2.0 |
| 40° | 0.64 | 0.77 | 0.84 | 1.2 | 1.3 | 1.6 |
| 45° | 0.71 | 0.71 | 1.0 | 1.0 | 1.4 | 1.4 |
| 50° | 0.77 | 0.64 | 1.2 | 0.84 | 1.6 | 1.3 |
| 60° | 0.87 | 0.50 | 1.7 | 0.58 | 2.0 | 1.2 |
| 70° | 0.94 | 0.34 | 2.7 | 0.36 | 2.9 | 1.1 |
| 75° | 0.97 | 0.26 | 3.7 | 0.27 | 3.9 | 1.0 |

## SOLVED PROBLEMS

1. Find the values of the trigonometric functions of the acute angles of the right triangle $ABC$, given $b = 24$ and $c = 25$.

Since $a^2 = c^2 - b^2 = (25)^2 - (24)^2 = 49$, $a = 7$.  Then

$$\sin A = \frac{\text{opposite side}}{\text{hypotenuse}} = \frac{7}{25} \qquad \cot A = \frac{\text{adjacent side}}{\text{opposite side}} = \frac{24}{7}$$

$$\cos A = \frac{\text{adjacent side}}{\text{hypotenuse}} = \frac{24}{25} \qquad \sec A = \frac{\text{hypotenuse}}{\text{adjacent side}} = \frac{25}{24}$$

$$\tan A = \frac{\text{opposite side}}{\text{adjacent side}} = \frac{7}{24} \qquad \csc A = \frac{\text{hypotenuse}}{\text{opposite side}} = \frac{25}{7}$$

and

$$\sin B = 24/25 \qquad\qquad \cot B = 7/24$$

$$\cos B = 7/25 \qquad\qquad \sec B = 25/7$$

$$\tan B = 24/7 \qquad\qquad \csc B = 25/24$$

**2.** Find the values of the trigonometric functions of the acute angles of the right triangle $ABC$, given $a = 2$, $c = 2\sqrt{5}$.

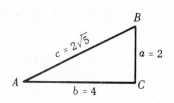

Since $b^2 = c^2 - a^2 = (2\sqrt{5})^2 - (2)^2 = 20 - 4 = 16$, $b = 4$. Then

$\sin A = 2/2\sqrt{5} = \sqrt{5}/5 = \cos B$ $\qquad$ $\cot A = 4/2 = 2 = \tan B$

$\cos A = 4/2\sqrt{5} = 2\sqrt{5}/5 = \sin B$ $\qquad$ $\sec A = 2\sqrt{5}/4 = \sqrt{5}/2 = \csc B$

$\tan A = 2/4 = 1/2 = \cot B$ $\qquad$ $\csc A = 2\sqrt{5}/2 = \sqrt{5} = \sec B$

**3.** Find the values of the trigonometric functions of the acute angle $A$, given $\sin A = 3/7$.

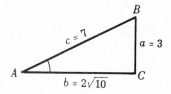

Construct the right triangle $ABC$ having $a = 3$, $c = 7$ and $b = \sqrt{7^2 - 3^2} = 2\sqrt{10}$ units. Then

$\sin A = 3/7$ $\qquad\qquad$ $\cot A = 2\sqrt{10}/3$

$\cos A = 2\sqrt{10}/7$ $\qquad\qquad$ $\sec A = 7/2\sqrt{10} = 7\sqrt{10}/20$

$\tan A = 3/2\sqrt{10} = 3\sqrt{10}/20$ $\qquad$ $\csc A = 7/3$

**4.** Find the values of the trigonometric functions of the acute angle $B$, given $\tan B = 1.5$.

Refer to Fig. $(a)$ below. Construct the right triangle $ABC$ having $b = 15$ and $a = 10$ units. (Note that $1.5 = 3/2$ and a right triangle with $b = 3$, $a = 2$ will serve equally well.)

Then $\quad c = \sqrt{a^2 + b^2} = \sqrt{10^2 + 15^2} = 5\sqrt{13}$ and

$\sin B = 15/5\sqrt{13} = 3\sqrt{13}/13$ $\qquad\qquad$ $\cot B = 2/3$

$\cos B = 10/5\sqrt{13} = 2\sqrt{13}/13$ $\qquad\qquad$ $\sec B = 5\sqrt{13}/10 = \sqrt{13}/2$

$\tan B = 15/10 = 3/2$ $\qquad\qquad\qquad$ $\csc B = 5\sqrt{13}/15 = \sqrt{13}/3.$

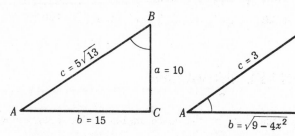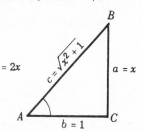

$\qquad$ Fig. $(a)$ Prob. 4 $\qquad\qquad$ Fig. $(b)$ Prob. 5 $\qquad\qquad$ Fig. $(c)$ Prob. 6

**5.** If $A$ is acute and $\sin A = 2x/3$, determine the values of the remaining functions.

Construct the right triangle $ABC$ having $a = 2x < 3$ and $c = 3$, as in Fig. $(b)$ above.

Then $\quad b = \sqrt{c^2 - a^2} = \sqrt{9 - 4x^2}$ and

$$\sin A = \frac{2x}{3}, \quad \cos A = \frac{\sqrt{9-4x^2}}{3}, \quad \tan A = \frac{2x}{\sqrt{9-4x^2}}, \quad \cot A = \frac{\sqrt{9-4x^2}}{2x}, \quad \sec A = \frac{3}{\sqrt{9-4x^2}}, \quad \csc A = \frac{3}{2x}.$$

**6.** If $A$ is acute and $\tan A = x = x/1$, determine the values of the remaining functions.

Construct the right triangle $ABC$ having $a = x$ and $b = 1$, as in Fig. $(c)$ above. Then $c = \sqrt{x^2 + 1}$ and

$$\sin A = \frac{x}{\sqrt{x^2+1}}, \quad \cos A = \frac{1}{\sqrt{x^2+1}}, \quad \tan A = x, \quad \cot A = \frac{1}{x}, \quad \sec A = \sqrt{x^2+1}, \quad \csc A = \frac{\sqrt{x^2+1}}{x}.$$

**7.** If $A$ is an acute angle:  
    *a)* Why is $\sin A < 1$?  
    *b)* When is $\sin A = \cos A$?  
    *c)* Why is $\sin A < \csc A$?  
    *d)* Why is $\sin A < \tan A$?  
    *e)* When is $\sin A < \cos A$?  
    *f)* When is $\tan A > 1$?

In any right triangle $ABC$:

*a)* Side $a <$ side $c$;  therefore  $\sin A = a/c < 1$.  
*b)* $\sin A = \cos A$ when $a/c = b/c$; then $a = b$, $A = B$, and $A = 45^\circ$.  
*c)* $\sin A < 1$ (above) and $\csc A = 1/\sin A > 1$.  
*d)* $\sin A = a/c$, $\tan A = a/b$, and $b < c$;  therefore $a/c < a/b$ or $\sin A < \tan A$.  
*e)* $\sin A < \cos A$ when $a < b$; then $A < B$ or $A < 90^\circ - A$, and $A < 45^\circ$.  
*f)* $\tan A = a/b > 1$ when $a > b$; then $A > B$ and $A > 45^\circ$.

**8.** Find the values of the trigonometric functions of $45^\circ$.

In any isosceles right triangle $ABC$, $A = B = 45^\circ$ and $a = b$.  
Let $a = b = 1$; then  $c = \sqrt{1+1} = \sqrt{2}$  and

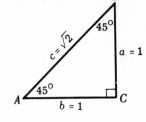

$$\sin 45^\circ = 1/\sqrt{2} = \tfrac{1}{2}\sqrt{2} \qquad \cot 45^\circ = 1$$
$$\cos 45^\circ = 1/\sqrt{2} = \tfrac{1}{2}\sqrt{2} \qquad \sec 45^\circ = \sqrt{2}$$
$$\tan 45^\circ = 1/1 = 1 \qquad \csc 45^\circ = \sqrt{2}.$$

**9.** Find the values of the trigonometric functions of $30^\circ$ and $60^\circ$.

In any equilateral triangle $ABD$, each angle is $60^\circ$. The bisector of any angle, as $B$, is the perpendicular bisector of the opposite side. Let the sides of the equilateral triangle be of length 2 units. Then in the right triangle $ABC$, $AB = 2$, $AC = 1$, and $BC = \sqrt{2^2 - 1^2} = \sqrt{3}$.

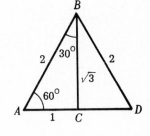

$$\sin 30^\circ = 1/2 = \cos 60^\circ \qquad \cot 30^\circ = \sqrt{3} = \tan 60^\circ$$
$$\cos 30^\circ = \sqrt{3}/2 = \sin 60^\circ \qquad \sec 30^\circ = 2/\sqrt{3} = 2\sqrt{3}/3 = \csc 60^\circ$$
$$\tan 30^\circ = 1/\sqrt{3} = \sqrt{3}/3 = \cot 60^\circ \qquad \csc 30^\circ = 2 = \sec 60^\circ$$

**10.** When the sun is $20^\circ$ above the horizon, how long is the shadow cast by a building 150 ft high?

In Fig. $(d)$ below, $A = 20^\circ$ and $CB = 150$.  Then  $\cot A = AC/CB$  and

$$AC = CB \cot A = 150 \cot 20^\circ = 150(2.7) = 405 \text{ ft.}$$

Fig. $(d)$ Prob. 10      Fig. $(e)$ Prob. 11      Fig. $(f)$ Prob. 12

**11.** A tree 100 ft tall casts a shadow 120 ft long. Find the angle of elevation of the sun.

In Fig. $(e)$ above, $CB = 100$ and $AC = 120$.  Then  $\tan A = CB/AC = 100/120 = 0.83$ and $A = 40^\circ$.

**12.** A ladder leans against the side of a building with its foot 12 ft from the building. How far from the ground is the top of the ladder and how long is the ladder if it makes an angle of $70^\circ$ with the ground?

From Fig. $(f)$ above, $\tan A = CB/AC$; then $CB = AC \tan A = 12 \tan 70^\circ = 12(2.7) = 32.4$. The top of the ladder is 32 ft above the ground.

$\sec A = AB/AC$; then $AB = AC \sec A = 12 \sec 70^\circ = 12(2.9) = 34.8$. The ladder is 35 ft long.

13. From the top of a lighthouse, 120 ft above the sea, the angle of depression of a boat is $15°$. How far is the boat from the lighthouse?

In Fig.(g) below, the right triangle $ABC$ has $A = 15°$ and $CB = 120$; then

$$\cot A = AC/CB \quad \text{and} \quad AC = CB \cot A = 120 \cot 15° = 120(3.7) = 444 \text{ ft.}$$

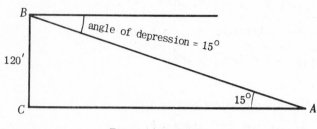

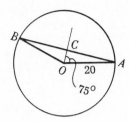

Fig. (g)  Prob. 13                    Fig. (h)  Prob. 14

14. Find the length of the chord of a circle of radius 20 in. subtended by a central angle of $150°$.

In Fig.(h) above, $OC$ bisects $\angle AOB$. Then $BC = AC$ and $OAC$ is a right triangle. In $\triangle OAC$,

$$\sin \angle COA = AC/OA \quad \text{and} \quad AC = OA \sin \angle COA = 20 \sin 75° = 20(0.97) = 19.4.$$

Then $BA = 38.8$ and the length of the chord is 39 in.

15. Find the height of a tree if the angle of elevation of its top changes from $20°$ to $40°$ as the observer advances 75 ft toward its base.  See Fig.(i) below.

In the right triangle $ABC$,  $\cot A = AC/CB$;  then  $AC = CB \cot A$  or  $DC + 75 = CB \cot 20°$.

In the right triangle $DBC$,  $\cot D = DC/CB$;  then  $DC = CB \cot 40°$.

Then    $DC = CB \cot 20° - 75 = CB \cot 40°$,    $CB(\cot 20° - \cot 40°) = 75$,

$$CB(2.7 - 1.2) = 75, \quad \text{and} \quad CB = 75/1.5 = 50 \text{ ft.}$$

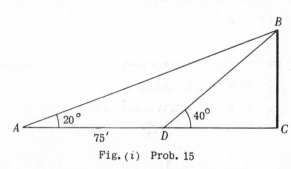

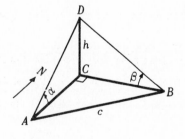

Fig. (i)  Prob. 15                    Fig. (j)  Prob. 16

16. A tower standing on level ground is due north of point $A$ and due west of point $B$, a distance $c$ ft from $A$. If the angles of elevation of the top of the tower as measured from $A$ and $B$ are $\alpha$ and $\beta$ respectively, find the height $h$ of the tower.

In the right triangle $ACD$ of Fig.(j) above, $\cot \alpha = AC/h$; and in the right triangle $BCD$, $\cot \beta = BC/h$. Then $AC = h \cot \alpha$ and $BC = h \cot \beta$.

Since $ABC$ is a right triangle,  $(AC)^2 + (BC)^2 = c^2 = h^2(\cot \alpha)^2 + h^2(\cot \beta)^2$  and

$$h = \frac{c}{\sqrt{(\cot \alpha)^2 + (\cot \beta)^2}}.$$

17. If holes are to be spaced regularly on a circle, show that the distance $d$ between the centers of two successive holes is given by  $d = 2r \sin \dfrac{180°}{n}$,  where $r$ = radius of the circle and $n$ =

number of holes.  Find $d$ when $r = 20$ in. and $n = 4$.

Let $A$ and $B$ be the centers of two consecutive holes on the circle of radius $r$ and center $O$. Let the bisector of the angle $O$ of the triangle $AOB$ meet $AB$ at $C$.  In right triangle $AOC$,

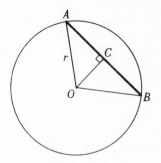

$$\sin \angle AOC = AC/r = \tfrac{1}{2}d/r = d/2r.$$

Then $\quad d = 2r \sin \angle AOC = 2r \sin \tfrac{1}{2} \angle AOB$

$$= 2r \sin \tfrac{1}{2}(360^{\circ}/n) = 2r \sin \frac{180^{\circ}}{n}.$$

When $r = 20$ and $n = 4$, $d = 2 \cdot 20 \sin 45^{\circ} = 2 \cdot 20 \cdot \dfrac{\sqrt{2}}{2} = 20\sqrt{2}$ in.

## SUPPLEMENTARY PROBLEMS

18. Find the values of the trigonometric functions of the acute angles of the right triangle $ABC$, given: $a$) $a = 3$, $b = 1$;  $b$) $a = 2$, $c = 5$;  $c$) $b = \sqrt{7}$, $c = 4$.

    *Ans.*   $a$) $A$: $3/\sqrt{10}$, $1/\sqrt{10}$, $3$, $1/3$, $\sqrt{10}$, $\sqrt{10}/3$;   $B$: $1/\sqrt{10}$, $3/\sqrt{10}$, $1/3$, $3$, $\sqrt{10}/3$, $\sqrt{10}$
         $b$) $A$: $2/5$, $\sqrt{21}/5$, $2/\sqrt{21}$, $\sqrt{21}/2$, $5/\sqrt{21}$, $5/2$;   $B$: $\sqrt{21}/5$, $2/5$, $\sqrt{21}/2$, $2/\sqrt{21}$, $5/2$, $5/\sqrt{21}$
         $c$) $A$: $3/4$, $\sqrt{7}/4$, $3/\sqrt{7}$, $\sqrt{7}/3$, $4/\sqrt{7}$, $4/3$;   $B$: $\sqrt{7}/4$, $3/4$, $\sqrt{7}/3$, $3/\sqrt{7}$, $4/3$, $4/\sqrt{7}$

19. Which is the greater and why:   $a$) $\sin 55^{\circ}$ or $\cos 55^{\circ}$?     $c$) $\tan 15^{\circ}$ or $\cot 15^{\circ}$?
                                 $b$) $\sin 40^{\circ}$ or $\cos 40^{\circ}$?     $d$) $\sec 55^{\circ}$ or $\csc 55^{\circ}$?
    Hint: Consider a right triangle having as acute angle the given angle.
    *Ans.*   $a$) $\sin 55^{\circ}$,   $b$) $\cos 40^{\circ}$,   $c$) $\cot 15^{\circ}$,   $d$) $\sec 55^{\circ}$

20. Find the value of each of the following.

    $a$) $\sin 30^{\circ} + \tan 45^{\circ}$

    $b$) $\cot 45^{\circ} + \cos 60^{\circ}$

    $c$) $\sin 30^{\circ} \cos 60^{\circ} + \cos 30^{\circ} \sin 60^{\circ}$

    $d$) $\cos 30^{\circ} \cos 60^{\circ} - \sin 30^{\circ} \sin 60^{\circ}$

    $e$) $\dfrac{\tan 60^{\circ} - \tan 30^{\circ}}{1 + \tan 60^{\circ} \tan 30^{\circ}}$

    $f$) $\dfrac{\csc 30^{\circ} + \csc 60^{\circ} + \csc 90^{\circ}}{\sec 0^{\circ} + \sec 30^{\circ} + \sec 60^{\circ}}$

    *Ans.*   $a$) $3/2$,   $b$) $3/2$,   $c$) $1$,   $d$) $0$,   $e$) $1/\sqrt{3}$,   $f$) $1$.

21. A man drives 500 ft along a road which is inclined $20^{\circ}$ to the horizontal. How high above his starting point is he?   *Ans.* 170 ft

22. A tree broken over by the wind forms a right triangle with the ground.  If the broken part makes an angle of $50^{\circ}$ with the ground and if the top of the tree is now 20 ft from its base, how tall was the tree?   *Ans.* 56 ft

23. Two straight roads intersect to form an angle of $75^{\circ}$.  Find the shortest distance from one road to a gas station on the other road 1000 ft from the junction.   *Ans.* 970 ft

24. Two buildings with flat roofs are 60 ft apart.  From the roof of the shorter building, 40 ft in height, the angle of elevation to the edge of the roof of the taller building is $40^{\circ}$. How high is the taller building?   *Ans.* 90 ft

25. A ladder, with its foot in the street, makes an angle of $30^{\circ}$ with the street when its top rests on a building on one side of the street and makes an angle of $40^{\circ}$ with the street when its top rests on a building on the other side of the street. If the ladder is 50 ft long, how wide is the street?   *Ans.* 82 ft

26. Find the perimeter of an isosceles triangle whose base is 40 in. and whose base angle is $70^{\circ}$.   *Ans.* 156 in.

# CHAPTER 4

# Tables of Trigonometric Functions

## SOLUTION OF RIGHT TRIANGLES

**APPROXIMATE VALUES OF THE FUNCTIONS** of acute angles are given in Tables of Natural Trigonometric Functions. These tables, as published in texts, differ in several respects. Some give the values of the six functions; others are restricted to the functions sine, cosine, tangent, and cotangent; some give the values to four digits while others give values to four decimal places. We shall use the latter table here. (In case the values of secant and cosecant are not included in your table, reference to these functions is to be deleted.)

## FOUR PLACE TABLE OF NATURAL TRIGONOMETRIC FUNCTIONS

**WHEN THE ANGLE IS LESS THAN 45°**, the angle is found in the left hand column of the table and the function is read at the top of the page. When the angle is greater than 45°, the angle is found in the right hand column and the function is read at the bottom of the page.

**TO FIND THE VALUE OF A TRIGONOMETRIC FUNCTION** of a given acute angle. If the angle contains a number of degrees only or a number of degrees and a multiple of 10′, the value of the function is read directly from the table.

    EXAMPLE 1. Find sin 24°40′.

        Opposite 24°40′ (<45°) in the left hand column read the entry 0.4173 in the column labeled Sin at the top of the page.

    EXAMPLE 2. Find cos 72°.

        Opposite 72° (>45°) in the right hand column read the entry 0.3090 in the column labeled Cos at the bottom of the page.

    EXAMPLE 3. *a*) tan 55°20′ = 1.4460. Read *up* the page since 55°20′ > 45°.
               *b*) cot 41°50′ = 1.1171. Read down the page since 41°50′ < 45°.

If the number of minutes in the given angle is not a multiple of 10, as in 24°43′, interpolate between the values of the functions of the two nearest angles (24°40′ and 24°50′) using the method of proportional parts.

    EXAMPLE 4. Find sin 24°43′.

        We find
$$\begin{aligned} \sin 24°40′ &= 0.4173 \\ \sin 24°50′ &= \underline{0.4200} \\ \text{Difference for } 10′ &= \overline{0.0027} = \text{tabular difference} \end{aligned}$$

Correction = difference for 3′ = 0.3(0.0027) = 0.00081 or 0.0008 when rounded off to four decimal places.

As the angle increases, the sine of the angle increases; thus,
$$\sin 24°43′ = 0.4173 + 0.0008 = 0.4181.$$

If a five place table is available, the value 0.41813 can be read directly from the table and then rounded off to 0.4181.

EXAMPLE 5. Find cos 64°26'.

We find

$$\begin{array}{rl}
\cos 64°20' &= 0.4331 \\
\cos 64°30' &= \underline{0.4305} \\
\text{Tabular difference} &= 0.0026
\end{array}$$

Correction = 0.6(0.0026) = 0.00156 or 0.0016 to four decimal places.

As the angle increases, the cosine of the angle decreases. Thus

$$\cos 64°26' = 0.4331 - 0.0016 = 0.4315.$$

To save time, we should proceed as follows in Example 4.

a) Locate sin 24°40' = 0.4173. For the moment, disregard the decimal point and use only the sequence 4173.

b) Find (mentally) the tabular difference 27, that is, the difference between the sequence 4173 corresponding to 24°40' and the sequence 4200 corresponding to 24°50'.

c) Find 0.3(27) = 8.1 and round off to the nearest integer. This is the correction.

d) Add (since sine) the correction to 4173, obtaining 4181. Then
$$\sin 24°43' = 0.4181.$$

When, as in the above example, we interpolate from the smaller angle to the larger: 1) The correction is added in finding sine, tangent, and secant. 2) The correction is subtracted in finding cosine, cotangent, and cosecant.

See also Problem 1.

**TO FIND THE ANGLE WHOSE FUNCTION IS GIVEN.** The process is a reversal of that given above.

EXAMPLE 6. Reading directly from the table, we find

$$0.2924 = \sin 17°, \qquad 2.7725 = \tan 70°10'.$$

EXAMPLE 7. Find $A$, given sin $A$ = 0.4234.

The given value is not an entry in the table. We find, however,

$$\begin{array}{ll}
0.4226 = \sin 25°\ 0' & \qquad 0.4226 = \sin 25°0' \\
\underline{0.4253 = \sin 25°10'} & \qquad \underline{0.4234 = \sin A} \\
\text{Tabular diff.} = 0.0027 & \qquad 0.0008 = \text{partial difference}
\end{array}$$

Correction $= \dfrac{0.0008}{0.0027}(10') = \dfrac{8}{27}(10') \doteq 3'$, to the nearest minute.

Adding (since sine) the correction, we have   $25°0' + 3' = 25°3' = A$.

EXAMPLE 8. Find $A$, given cot $A$ = 0.6345.

We find

$$\begin{array}{ll}
0.6330 = \cot 57°40' & \qquad 0.6330 = \cot 57°40' \\
\underline{0.6371 = \cot 57°30'} & \qquad \underline{0.6345 = \cot A} \\
\text{Tabular diff.} = 0.0041 & \qquad 0.0015 = \text{partial difference}
\end{array}$$

Correction $= \dfrac{0.0015}{0.0041}(10') = \dfrac{15}{41}(10') = 4'$, to the nearest minute.

Subtracting (since cot) the correction, we have

$$57°40' - 4' = 57°36' = A.$$

To save time, we should proceed as follows in Example 7:

a) Locate the next smaller entry, $0.4226 = \sin 25°0'$. For the moment use only the sequence 4226.

b) Find the tabular difference, 27.

c) Find the partial difference, 8, between 4226 and the given sequence 4234.

d) Find $\dfrac{8}{27}(10') = 3'$ and add to $25°0'$.  See Problem 3.

ERRORS IN COMPUTED RESULTS arise from:

a) Errors in the given data. These errors are always present in data resulting from measurements.

b) The use of tables of trigonometric functions. The entries in such tables are usually approximations of never ending decimals.

A measurement recorded as 35 feet means that the result is correct to the nearest foot, that is, the true length is between 34.5 and 35.5 feet. Similarly, a recorded length of 35.0 ft means that the true length is between 34.95 and 35.05 ft; a recorded length of 35.8 ft means that the true length is between 35.75 and 35.85 ft; a recorded length of 35.80 ft means that the true length is between 35.795 and 35.805 ft; and so on.

SIGNIFICANT DIGITS. In the number 35 there are two significant digits, 3 and 5. They are also the significant digits in 3.5, 0.35, 0.035, 0.0035 but not in 35.0, 3.50, 0.350, 0.0350. In the numbers 35.0, 3.50, 0.350, 0.0350 there are three significant digits, 3, 5, and 0. This is another way of saying that 35 and 35.0 are not the same measurement.

It is impossible to determine the significant figures in a measurement recorded as 350, 3500, 35000, ..... For example, 350 may mean that the true result is between 345 and 355 or between 349.5 and 350.5.

ACCURACY IN COMPUTED RESULTS. A computed result should not show more decimal places than that shown in the least accurate of the measured data. Of importance here are the following relations giving comparable degrees of accuracy in lengths and angles:

a) Distances expressed to 2 significant digits and angles expressed to the nearest degree.

b) Distances expressed to 3 significant digits and angles expressed to the nearest 10'.

c) Distances expressed to 4 significant digits and angles expressed to the nearest 1'.

d) Distances expressed to 5 significant digits and angles expressed to the nearest 0.1'.

## SOLVED PROBLEMS

1.  a) $\sin 56°34' = 0.8345$ ;   $8339 + 0.4(16) = 8339 + 6$

b) $\cos 19°45' = 0.9412$ ;   $9417 - 0.5(10) = 9417 - 5$

c) $\tan 77°12' = 4.4016$ ;   $43897 + 0.2(597) = 43897 + 119$

d) $\cot 40°36' = 1.1667$ ;   $11708 - 0.6(68) = 11708 - 41$

e) $\sec 23°47' = 1.0928$ ;   $10918 + 0.7(14) = 10918 + 10$

f) $\csc 60°4' = 1.1539$ ;   $11547 - 0.4(19) = 11547 - 8$

2. If the correction is 6.5, 13.5, 10.5, etc., we shall round off so that the *final* result is even.

$a)$ sin $28°37'$ = 0.4790 ;   4772 + 0.7(25) = 4772 + 17.5

$b)$ cot $65°53'$ = 0.4476 ;   4487 − 0.3(35) = 4487 − 10.5

$c)$ cos $35°25'$ = 0.8150 ;   8158 − 0.5(17) = 8158 − 8.5

$d)$ sec $39°35'$ = 1.2976 ;   12960 + 0.5(31) = 12960 + 15.5

3. $a)$ sin $A$ = 0.6826,  $A$ = $43°3'$ ;    $43°0'$ + $\frac{6}{21}$(10') = $43°0'$ + 3'

$b)$ cos $A$ = 0.5957,  $A$ = $53°26'$ ;    $53°30'$ − $\frac{9}{24}$(10') = $53°30'$ − 4'

$c)$ tan $A$ = 0.9470,  $A$ = $43°26'$ ;    $43°20'$ + $\frac{35}{55}$(10') = $43°20'$ + 6'

$d)$ cot $A$ = 1.7580,  $A$ = $29°38'$ ;    $29°40'$ − $\frac{24}{119}$(10') = $29°40'$ − 2'

$e)$ sec $A$ = 2.3198,  $A$ = $64°28'$ ;    $64°20'$ + $\frac{110}{140}$(10') = $64°20'$ + 8'

$f)$ csc $A$ = 1.5651,  $A$ = $39°43'$ ;    $39°50'$ − $\frac{40}{55}$(10') = $39°50'$ − 7'

4. Solve the right triangle in which $A$ = $35°10'$ and $c$ = 72.5.

Solution:  $B$ = $90°$ − $35°10'$ = $54°50'$.

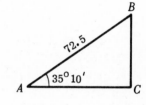

$a/c$ = sin $A$,    $a$ = $c$ sin $A$ = 72.5(0.5760) = 41.8

$b/c$ = cos $A$,    $b$ = $c$ cos $A$ = 72.5(0.8175) = 59.3

Check:  $a/b$ = tan $A$,    $a$ = $b$ tan $A$ = 59.3(0.7046) = 41.8

5. Solve the right triangle in which $a$ = 24.36, $A$ = $58°53'$.

Solution:  $B$ = $90°$ − $58°53'$ = $31°7'$.

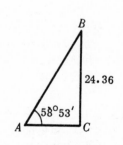

$b/a$ = cot $A$,    $b$ = $a$ cot $A$ = 24.36(0.6036) = 14.70.

$c/a$ = csc $A$,    $c$ = $a$ csc $A$ = 24.36(1.1681) = 28.45, or

$a/c$ = sin $A$,    $c$ = $a$/sin $A$ = 24.36/0.8562  = 28.45.

Check:  $b/c$ = cos $A$,    $b$ = $c$ cos $A$ = 28.45(0.5168) = 14.70.

6. Solve the right triangle $ABC$ in which $a$ = 43.9, $b$ = 24.3.

Solution:  tan $A$ = $\frac{43.9}{24.3}$ = 1.8066 ;   $A$ = $61°2'$,  $B$ = $90°$ − $A$ = $28°58'$.

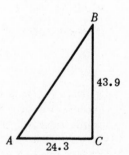

$c/a$ = csc $A$,    $c$ = $a$ csc $A$ = 43.9(1.1430) = 50.2, or

$a/c$ = sin $A$,    $c$ = $a$/sin $A$ = 43.9/0.8749  = 50.2.

Check:  $c/b$ = sec $A$,    $c$ = $b$ sec $A$ = 24.3(2.0649) = 50.2, or

$b/c$ = cos $A$,    $c$ = $b$/cos $A$ = 24.3/0.4843  = 50.2.

7. Solve the right triangle $ABC$ in which $b$ = 15.25, $c$ = 32.68.

Solution:  sin $B$ = $\frac{15.25}{32.68}$ = 0.4666 ;   $B$ = $27°49'$,  $A$ = $90°$ − $B$ = $62°11'$.

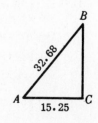

$a/b$ = cot $B$,    $a$ = $b$ cot $B$ = 15.25(1.8953) = 28.90

Check:  $a/c$ = cos $B$,    $a$ = $c$ cos $B$ = 32.68(0.8844) = 28.90

**8.** The base of an isosceles triangle is 20.4 and the base angles are 48°40′. Find the equal sides and the altitude of the triangle.

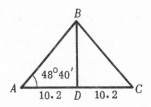

In the figure, $BD$ is perpendicular to $AC$ and bisects it.

In the right triangle $ABD$,

$$AB/AD = \sec A, \qquad AB = 10.2(1.5141) = 15.4, \text{ or}$$

$$AD/AB = \cos A, \qquad AB = 10.2/0.6604 = 15.4.$$

$$DB/AD = \tan A, \qquad DB = 10.2(1.1369) = 11.6.$$

**9.** Considering the earth as a sphere of radius 3960 miles, find the radius $r$ of the 40th parallel of latitude. Refer to Fig.(a) below.

In the right triangle $OCB$, $\angle OBC = 40°$ and $OB = 3960$.

Then $\cos \angle OBC = CB/OB$ and $r = CB = 3960 \cos 40° = 3960(0.7660) = 3030$ miles.

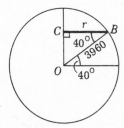     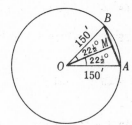     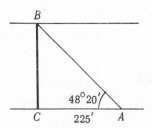

Fig.(a) Prob. 9          Fig.(b) Prob. 10          Fig.(c) Prob. 11

**10.** Find the perimeter of a regular octagon inscribed in a circle of radius 150 feet.

In Fig.(b) above, two consecutive vertices $A$ and $B$ of the octagon are joined to the center $O$ of the circle. The triangle $OAB$ is isosceles with equal sides 150 and $\angle AOB = 360°/8 = 45°$. As in Problem 8, we bisect $\angle AOB$ to form the right triangle $MOB$.

Then $MB = OB \sin \angle MOB = 150 \sin 22°30′ = 150(0.3827) = 57.4$, and the perimeter of the octagon is $16\,MB = 16(57.4) = 918$ ft.

**11.** To find the width of a river, a surveyor set up his transit at $C$ on one bank and sighted across to a point $B$ on the opposite bank; then turning through an angle of 90°, he laid off a distance $CA = 225$ ft. Finally, setting the transit at $A$, he measured $\angle CAB$ as 48°20′. Find the width of the river.

See Fig.(c) above. In the right triangle $ACB$,

$$CB = AC \tan \angle CAB = 225 \tan 48°20′ = 225(1.1237) = 253 \text{ ft.}$$

**12.** In the adjoining figure, the line $AD$ crosses a swamp. In order to locate a point on this line, a surveyor turned through an angle 51°16′ at $A$ and measured 1585 feet to a point $C$. He then turned through an angle of 90° at $C$ and ran a line $CB$. If $B$ is on $AD$, how far must he measure from $C$ to reach $B$?

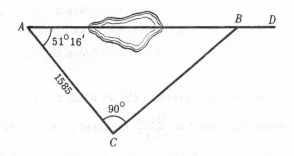

$$CB = AC \tan 51°16′$$

$$= 1585(1.2467) = 1976 \text{ ft.}$$

**13.** From a point $A$ on level ground, the angles of elevation of the top $D$ and bottom $B$ of a flag-pole situated on the top of a hill are measured as $47^\circ 54'$ and $39^\circ 45'$. Find the height of the hill if the height of the flagpole is 115.5 ft. See Fig. $(d)$ below.

Let the line of the pole meet the horizontal through $A$ in $C$.

In the right triangle $ACD$, $AC = DC \cot 47^\circ 54' = (115.5 + BC)(0.9036)$.

In the right triangle $ACB$, $AC = BC \cot 39^\circ 45' = BC(1.2024)$.

Then $$(115.5 + BC)(0.9036) = BC(1.2024)$$

and $$BC = \frac{115.5(0.9036)}{1.2024 - 0.9036} = 349.3 \text{ ft.}$$

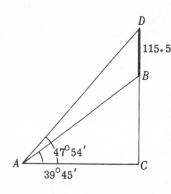

Fig. $(d)$ Prob. 13

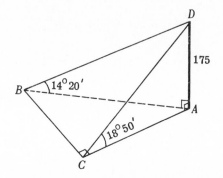

Fig. $(e)$ Prob. 14

**14.** From the top of a lighthouse, 175 ft above the water, the angle of depression of a boat due south is $18^\circ 50'$. Calculate the speed of the boat if, after it moves due west for two minutes, the angle of depression is $14^\circ 20'$.

In Fig. $(e)$ above, $AD$ is the lighthouse, $C$ is the position of the boat when due south of the lighthouse, and $B$ is the position two minutes later.

In the right triangle $CAD$, $AC = AD \cot \angle ACD = 175 \cot 18^\circ 50' = 175(2.9319) = 513.$

In the right triangle $BAD$, $AB = AD \cot \angle ABD = 175 \cot 14^\circ 20' = 175(3.9136) = 685.$

In the right triangle $ABC$, $BC = \sqrt{(AB)^2 - (AC)^2} = \sqrt{(685)^2 - (513)^2} = 454.$

The boat travels 454 ft in 2 min; its speed is 227 ft/min.


# SUPPLEMENTARY PROBLEMS


**15.** Find the natural trigonometric functions of each of the following angles:
   $a)$ $18^\circ 47'$,   $b)$ $32^\circ 13'$,   $c)$ $58^\circ 24'$,   $d)$ $79^\circ 45'$.

| Ans. | sine | cosine | tangent | cotangent | secant | cosecant |
|---|---|---|---|---|---|---|
| $a)$ $18^\circ 47'$ | 0.3220 | 0.9468 | 0.3401 | 2.9403 | 1.0563 | 3.1057 |
| $b)$ $32^\circ 13'$ | 0.5331 | 0.8460 | 0.6301 | 1.5869 | 1.1820 | 1.8757 |
| $c)$ $58^\circ 24'$ | 0.8517 | 0.5240 | 1.6255 | 0.6152 | 1.9084 | 1.1741 |
| $d)$ $79^\circ 45'$ | 0.9840 | 0.1780 | 5.5304 | 0.1808 | 5.6201 | 1.0162 |

**16.** Find (acute) angle $A$, given:

| | | | | | |
|---|---|---|---|---|---|
| $a$) $\sin A = 0.5741$ | *Ans.* $A = 35°\ 2'$ | | $e$) $\cos A = 0.9382$ | *Ans.* $A = 20°15'$ | |
| $b$) $\sin A = 0.9468$ | $A = 71°13'$ | | $f$) $\cos A = 0.6200$ | $A = 51°41'$ | |
| $c$) $\sin A = 0.3510$ | $A = 20°33'$ | | $g$) $\cos A = 0.7120$ | $A = 44°36'$ | |
| $d$) $\sin A = 0.8900$ | $A = 62°52'$ | | $h$) $\cos A = 0.4651$ | $A = 62°17'$ | |

$i$) $\tan A = 0.2725$     $A = 15°15'$          $m$) $\cot A = 0.2315$     $A = 76°58'$
$j$) $\tan A = 1.1652$     $A = 49°22'$          $n$) $\cot A = 2.9715$     $A = 18°36'$
$k$) $\tan A = 0.5200$     $A = 27°28'$          $o$) $\cot A = 0.7148$     $A = 54°27'$
$l$) $\tan A = 2.7775$     $A = 70°12'$          $p$) $\cot A = 1.7040$     $A = 30°24'$

$q$) $\sec A = 1.1161$     $A = 26°22'$          $u$) $\csc A = 3.6882$     $A = 15°44'$
$r$) $\sec A = 1.4382$     $A = 45°57'$          $v$) $\csc A = 1.0547$     $A = 71°28'$
$s$) $\sec A = 1.2618$     $A = 37°35'$          $w$) $\csc A = 1.7631$     $A = 34°33'$
$t$) $\sec A = 2.1584$     $A = 62°24'$          $x$) $\csc A = 1.3436$     $A = 48°\ 6'$

**17.** Solve each of the right triangles $ABC$, given:

$a$) $A = 35°20'$,   $c = 112$          *Ans.*   $B = 54°40'$,   $a = 64.8$,   $b = 91.4$

$b$) $B = 48°40'$,   $c = 225$                   $A = 41°20'$,   $a = 149$,   $b = 169$

$c$) $A = 23°18'$,   $c = 346.4$                 $B = 66°42'$,   $a = 137.0$,   $b = 318.1$

$d$) $B = 54°12'$,   $c = 182.5$                 $A = 35°48'$,   $a = 106.7$,   $b = 148.0$

$e$) $A = 32°10'$,   $a = 75.4$                  $B = 57°50'$,   $b = 120$,   $c = 142$

$f$) $A = 58°40'$,   $b = 38.6$                  $B = 31°20'$,   $a = 63.4$,   $c = 74.2$

$g$) $B = 49°14'$,   $b = 222.2$                 $A = 40°46'$,   $a = 191.6$,   $c = 293.4$

$h$) $A = 66°36'$,   $a = 112.6$                 $B = 23°24'$,   $b = 48.73$,   $c = 122.7$

$i$) $A = 29°48'$,   $b = 458.2$                 $B = 60°12'$,   $a = 262.4$,   $c = 528.0$

$j$) $a = 25.4$,   $b = 38.2$                    $A = 33°37'$,   $B = 56°23'$,   $c = 45.9$

$k$) $a = 45.6$,   $b = 84.8$                    $A = 28°16'$,   $B = 61°44'$,   $c = 96.3$

$l$) $a = 38.64$,   $b = 48.74$                  $A = 38°24'$,   $B = 51°36'$,   $c = 62.21$

$m$) $a = 506.2$,   $c = 984.8$                  $A = 30°56'$,   $B = 59°\ 4'$,   $b = 844.7$

$n$) $b = 672.9$,   $c = 888.1$                  $A = 40°44'$,   $B = 49°16'$,   $a = 579.4$

**18.** Find the base and altitude of an isosceles triangle whose vertical angle is $65°$ and whose equal sides are 415 ft.   *Ans.*   Base = 446 ft.   altitude = 350 ft

**19.** The base of an isosceles triangle is 15.90 in. and the base angles are $54°28'$. Find the equal sides and the altitude.   *Ans.*   Side = 13.68 in.,   altitude = 11.13 in.

**20.** The radius of a circle is 21.4 ft. Find $a$) the length of the chord subtended by a central angle of $110°40'$ and $b$) the distance between two parallel chords on the same side of the center subtended by central angles $118°40'$ and $52°20'$.   *Ans.*   $a$) 35.2 ft,   $b$) 8.29 ft

**21.** Show that the base $b$ of an isosceles triangle whose equal sides are $a$ and whose vertical angle is $\theta$ is given by $b = 2a \sin \frac{1}{2}\theta$.

**22.** Show that the perimeter $P$ of a regular polygon of $n$ sides inscribed in a circle of radius $r$ is given by $P = 2nr \sin(180°/n)$.

**23.** A wheel, 5 ft in diameter, rolls up an incline of $18°20'$. What is the height of the center of the wheel above the base of the incline when the wheel has rolled 5 ft up the incline?   *Ans.*   3.95 ft

**24.** A wall is 15 ft high and 10 ft from a house. Find the length of the shortest ladder which will just touch the top of the wall and reach a window 20.5 ft above the ground.   *Ans.*   **42.5 ft**

# CHAPTER 5

# Practical Applications

THE BEARING OF A POINT B FROM A POINT A, in a horizontal plane, is usually defined as the angle (always acute) made by the half-line drawn from A through B with the north-south line through A. The bearing is then read from the north or south line toward the east or west. For example,

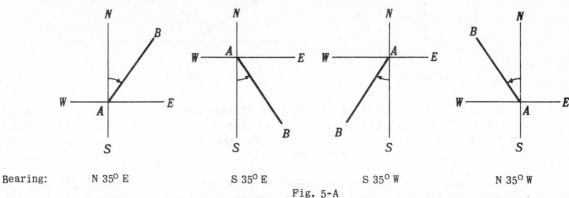

Bearing:  N 35° E  S 35° E  S 35° W  N 35° W

Fig. 5-A

In aeronautics the bearing of B from A is more often given as the angle made by the half-line AB with the north line through A, measured clockwise from the north (i.e., from the north around through the east). For example,

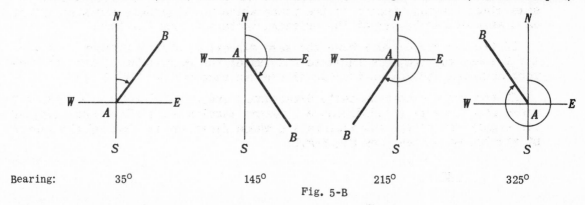

Bearing:  35°  145°  215°  325°

Fig. 5-B

VECTORS. Any physical quantity, as force or velocity, which has both magnitude and direction is called a *vector quantity*. A vector quantity may be represented by a directed line segment (arrow) called a *vector*. The *direction* of the vector is that of the given quantity and the *length* of the vector is proportional to the magnitude of the quantity.

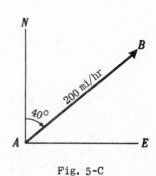

Fig. 5-C

EXAMPLE 1. An airplane is traveling N 40°E at 200 mph. Its velocity is represented by the vector AB of Fig. 5-C.

32

EXAMPLE 2. A motor boat having the speed 12 mph in still water is headed directly across a river whose current is 4 mph. In Fig. 5-D below, the vector *CD* represents the velocity of the current and the vector *AB* represents, to the same scale, the velocity of the boat in still water. Thus, vector *AB* is three times as long as vector *CD*.

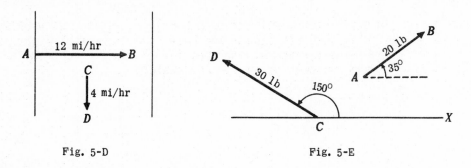

Fig. 5-D                                Fig. 5-E

EXAMPLE 3. In Fig. 5-E above, vector *AB* represents a force of 20 lb making an angle of 35° with the positive direction on the *x*-axis and vector *CD* represents a force of 30 lb at 150° with the positive direction on the *x*-axis. Both vectors are drawn to the same scale.

Two vectors are said to be equal if they have the same magnitude and direction. A vector has no fixed position in a plane and may be moved about in the plane provided only that its magnitude and direction are not changed.

VECTOR ADDITION. The *resultant* or *vector sum* of a number of vectors, all in the same plane, is that vector in the plane which would produce the same effect as that produced by all of the original vectors acting together.

If two vectors $\alpha$ and $\beta$ have the same direction, their resultant is a vector *R* whose magnitude is equal to the sum of the magnitudes of the two vectors and whose direction is that of the two vectors. See Fig. 5-F(*a*).

If two vectors have opposite directions, their resultant is a vector *R* whose magnitude is the difference (greater magnitude – smaller magnitude) of the magnitudes of the two vectors and whose direction is that of the vector of greater magnitude. See Fig. 5-F(*b*).

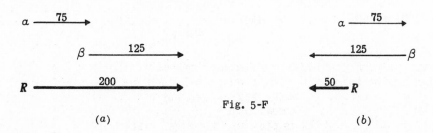

Fig. 5-F

(*a*)                                (*b*)

In all other cases, the magnitude and direction of the resultant of two vectors is obtained by either of the following two methods.

1) PARALLELOGRAM METHOD. Place the tail ends of both vectors at any point *O* in their plane and complete the parallelogram having these vectors as adjacent sides. The directed diagonal issuing from *O* is the resultant or vector sum of the two given vectors. Thus, in Fig. 5-G(*b*) below, the vector *R* is the resultant of the vectors $\alpha$ and $\beta$ of Fig. 5-G(*a*).

2) TRIANGLE METHOD. Choose one of the vectors and label its tail end as *O*.

Place the tail end of the other vector at the arrow end of the first. The
The resultant is then the line segment closing the triangle and directed
from $O$. Thus, in Fig.5-G(c) and 5-G(d) below, $R$ is the resultant of the
vectors $\alpha$ and $\beta$.

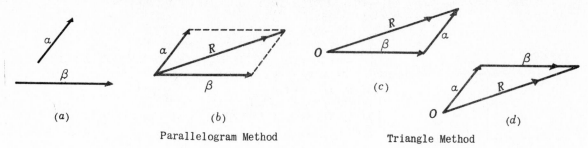

Parallelogram Method                   Triangle Method

Fig. 5-G

EXAMPLE 4. The resultant $R$ of the two vectors of Example 2 represents the
speed and direction in which the boat travels. Fig. 5-H(a) illustrates the
parallelogram method; Fig. 5-H(b) and 5-H(c) illustrate the triangle method.

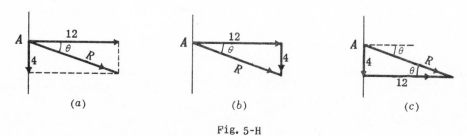

Fig. 5-H

The magnitude of $R = \sqrt{(12)^2 + 4^2} = 12.6$ mph.
From Fig.5-H(a) or 5-H(b), $\tan \theta = 4/12 = 0.3333$ and $\theta = 18°30'$.
Thus, the boat moves down stream in a line making an angle $\theta = 18°30'$ with
the direction in which it is headed or making an angle $90° - \theta = 71°30'$ with
the bank of the river.

THE COMPONENT OF A VECTOR $\alpha$ along a line $L$ is the perpendicular projection of the
vector $\alpha$ on $L$. It is often very useful to resolve
a vector into two components along a pair of perpendicular lines.

EXAMPLE 5. In each of Fig. 5-H(a),(b),(c) the
components of $R$ are 1) 4 mph in the direction of
the current and 2) 12 mph in the direction perpendicular to the current.

EXAMPLE 6. In the adjoining Fig.5-I, the force
$F$ has horizontal component $F_h = F \cos 30°$ and vertical component $F_v = F \sin 30°$. Note that $F$ is the
vector sum or resultant of $F_h$ and $F_v$.

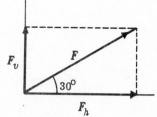

Fig. 5-I

AIR NAVIGATION. The *heading* of an airplane is the direction (determined from a compass reading) in which the airplane is pointed. The heading is measured clockwise from the north.

The *airspeed* (determined from a reading of the airspeed indicator) is the
speed of the airplane in still air.

The *track* (or *course*) of an airplane is the direction in which it moves
relative to the ground. The track is measured clockwise from the north.

The *groundspeed* is the speed of the airplane relative to the ground.

The *drift angle* (or wind correction angle) is the difference (positive) between the heading and the track.

In Fig. 5-J below:  *ON* is the true north line through *O*,
∠*NOA* is the heading
*OA* = the airspeed
*AN* is the true north line through *A*,
∠*NAW* is the wind angle, measured clockwise from north line,
*AB* = the wind speed
∠*NOB* is the track
*OB* = the groundspeed
∠*AOB* is the drift angle.

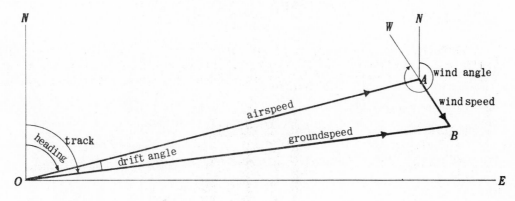

Fig. 5-J

Note that there are three vectors involved:  *OA* representing the **airspeed** and heading, *AB* representing the direction and speed of the wind, and *OB* representing the groundspeed and track. The groundspeed vector is the **resultant** of the airspeed vector and the wind vector.

EXAMPLE 7. Fig. 5-K illustrates an airplane flying at 240 mph on a **heading** of 60° when the wind is 30 mph from 330°.

In constructing the figure put in the airspeed vector at *O* and then **follow** through (note the directions of the arrows) with the wind vector, and **close** the triangle. Note further that the groundspeed vector does not follow **through** from the wind vector.

In the resulting triangle:   Groundspeed $= \sqrt{(240)^2 + (30)^2}$ $= 242$ mph.

$\text{Tan } \theta = 30/240 = 0.1250$ and $\theta = 7°10'$.

$\text{Track} = 60° + \theta = 67°10'$.

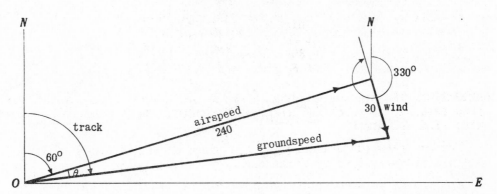

Fig. 5-K

## SOLVED PROBLEMS

1. A motor boat moves in the direction N40° E for 3 hr at 20 mph. How far north and how far east does it travel?

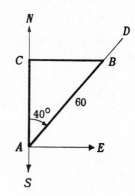

Suppose the boat leaves $A$. Using the north-south line through $A$, draw the half-line $AD$ so that the bearing of $D$ from $A$ is N 40° E. On $AD$ locate $B$ such that $AB$ = 3(20) = 60 miles. Through $B$ pass a line perpendicular to the line $NAS$, meeting it in $C$. In the right triangle $ABC$,

$$AC = AB \cos A = 60 \cos 40° = 60(0.7660) = 45.96$$

and $$CB = AB \sin A = 60 \sin 40° = 60(0.6428) = 38.57.$$

The boat travels 46 miles north and 39 miles east.

2. Three ships are situated as follows: $A$ is 225 miles due north of $C$, and $B$ is 375 miles due east of $C$. What is the bearing a) of $B$ from $A$, b) of $A$ from $B$?

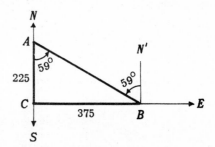

In the right triangle $ABC$,

$$\tan \angle CAB = 375/225 = 1.6667 \quad \text{and} \quad \angle CAB = 59°0'.$$

a) The bearing of $B$ from $A$ (angle $SAB$) is S 59°0' E.

b) The bearing of $A$ from $B$ (angle $N'BA$) is N 59°0' W.

3. Three ships are situated as follows: $A$ is 225 miles west of $C$ while $B$, due south of $C$, bears S 25°10' E from $A$. a) How far is $B$ from $A$? b) How far is $B$ from $C$? c) What is the bearing of $A$ from $B$?

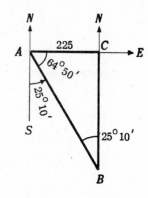

From the figure, $\angle SAB = 25°10'$ and $\angle BAC = 64°50'$. Then

$AB = AC \sec \angle BAC = 225 \sec 64°50' = 225(2.3515) = 529.1$    or

$AB = AC/\cos \angle BAC = 225/\cos 64°50' = 225/0.4253 = 529.0$    and

$CB = AC \tan \angle BAC = 225 \tan 64°50' = 225(2.1283) = 478.9.$

a) $B$ is 529 miles from $A$.      b) $B$ is 479 miles from $C$.

c) Since $\angle CBA = 25°10'$, the bearing of $A$ from $B$ is N 25°10' W.

4. From a boat sailing due north at 16.5 mph, a wrecked ship $K$ and an observation tower $T$ are observed in a line due east. One hour later the wrecked ship and the tower have bearings S 34°40' E and S 65°10' E. Find the distance between the wrecked ship and the tower.

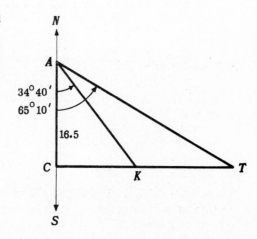

In the figure, $C, K,$ and $T$ represent respectively the boat, the wrecked ship, and the tower when in a line. One hour later the boat is at $A$, 16.5 miles due north of $C$. In the right triangle $ACK$,

$$CK = 16.5 \tan 34°40' = 16.5(0.6916).$$

In the right triangle $ACT$,

$$CT = 16.5 \tan 65°10' = 16.5(2.1609).$$

Then $KT = CT - CK = 16.5(2.1609 - 0.6916) = 24.2$ mi.

5. A ship is sailing due east when a light is observed bearing N 62°10′ E. After the ship has traveled 2250 ft, the light bears N 48°25′ E. If the course is continued, how close will the ship approach the light?

In Fig. (a) below, L is the position of the light, A is the first position of the ship, B is the second position, and C is the position when nearest L.

In the right triangle ACL,    AC = CL cot ∠CAL = CL cot 27°50′  = 1.8940 CL.

In the right triangle BCL,    BC = CL cot ∠CBL = CL cot 41°35′  = 1.1270 CL.

Since  AC = BC + 2250,    1.8940 CL = 1.1270 CL + 2250,  and  CL = $\dfrac{2250}{1.8940 - 1.1270}$ = 2934 ft.

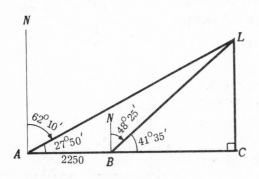

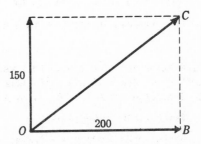

Fig. (a)  Prob. 5                        Fig. (b)  Prob. 6

6. Refer to Fig. (b) above.  A body at O is being acted upon by two forces, one of 150 lb due north and the other of 200 lb due east. Find the magnitude and direction of the resultant.

In the right triangle OBC,    OC = $\sqrt{(OB)^2 + (BC)^2}$ = $\sqrt{(200)^2 + (150)^2}$  = 250 lb,

tan ∠BOC = 150/200 = 0.7500   and   ∠BOC = 36°50′.

The magnitude of the resultant force is 250 lb and its direction is N 53°10′ E.

7. An airplane is moving horizontally at 240 mph when a bullet is shot with speed 2750 ft/sec at right angles to the path of the airplane. Find the resultant speed and direction of the bullet.

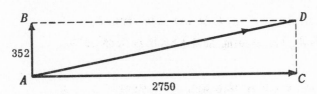

The speed of the airplane is  240 mi/hr = $\dfrac{240(5280)}{60(60)}$ ft/sec = 352 ft/sec.

In the figure, the vector AB represents the velocity of the airplane, the vector AC represents the initial velocity of the bullet, and the vector AD represents the resultant velocity of the bullet.

In the right triangle ACD,    AD = $\sqrt{(352)^2 + (2750)^2}$ = 2770 ft/sec,

tan ∠CAD  = 352/2750 = 0.1280   and   ∠CAD = 7°20′.

Thus, the bullet travels at 2770 ft/sec along a path making an angle of 82°40′ with the path of the airplane.

**8.** A river flows due south at 125 ft/min. A motor boat, moving at 475 ft/min in still water, is headed due east across the river. *a*) Find the direction in which the boat moves and its speed. *b*) In what direction must the boat be headed in order that it move due east and what is its speed in that direction?

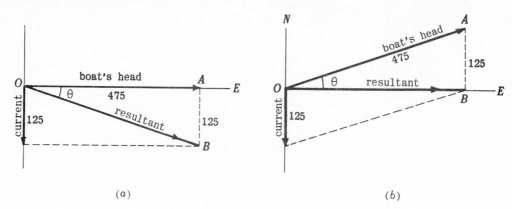

(*a*)                                                    (*b*)

*a*) Refer to Fig.(*a*).  In right triangle *OAB*,  $OB = \sqrt{(475)^2 + (125)^2} = 491$,

$$\tan \theta = 125/475 = 0.2632 \quad \text{and} \quad \theta = 14°40'.$$

Thus the boat moves at 491 ft/min in the direction S 75°20′ E.

*b*) Refer to Fig.(*b*).  In right triangle *OAB*,  $\sin \theta = 125/475 = 0.2632$  and  $\theta = 15°20'$.

Thus the boat must be headed N 74°40′ E and its speed in that direction is

$$OB = \sqrt{(475)^2 - (125)^2} = 458 \text{ ft/min}.$$

**9.** A telegraph pole is kept vertical by a guy wire which makes an angle of 25° with the pole and which exerts a pull of $F = 300$ lb on the top.  Find the horizontal and vertical components $F_h$ and $F_v$ of the pull $F$.

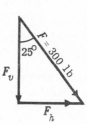 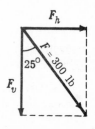

$F_h = 300 \sin 25° = 300(0.4226) = 127$ lb

$F_v = 300 \cos 25° = 300(0.9063) = 272$ lb

**10.** A man pulls a rope attached to a sled with a force of 100 lb. The rope makes an angle of 27° with the ground. *a*) Find the effective pull tending to move the sled along the ground and the effective pull tending to lift the sled vertically. *b*) Find the force which the man must exert in order that the effective force tending to move the sled along the ground is 100 lb.

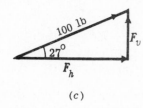   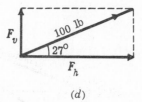   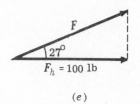

(*c*)                          (*d*)                          (*e*)

*a*) In Fig.(*c*) and (*d*), the 100 lb pull in the rope is resolved into horizontal and vertical components, $F_h$ and $F_v$ respectively.  Then $F_h$ is the force tending to move the sled along the ground and $F_v$ is the force tending to lift the sled.

$F_h = 100 \cos 27° = 100(0.8910) = 89$ lb,     $F_v = 100 \sin 27° = 100(0.4540) = 45$ lb.

*b*) In Fig.(*e*), the horizontal component of the required force $F$ is $F_h = 100$ lb.  Then

$$F = 100/\cos 27° = 100/0.8910 = 112 \text{ lb}.$$

11. A block weighing $W$ = 500 lb rests on a ramp inclined 29° with the horizontal. *a*) Find the force tending to move the block down the ramp and the force of the block on the ramp. *b*) What minimum force must be applied to keep the block from sliding down the ramp? Neglect friction.

*a*) Refer to Fig. (*f*) below. Resolve the weight $W$ of the block into components $F_1$ and $F_2$, respectively parallel and perpendicular to the ramp. $F_1$ is the force tending to move the block down the ramp and $F_2$ is the force of the block on the ramp.

$$F_1 = W \sin 29° = 500(0.4848) = 242 \text{ lb}, \qquad F_2 = W \cos 29° = 500(0.8746) = 437 \text{ lb}.$$

*b*) 242 lb up the ramp.

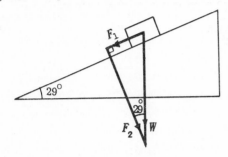

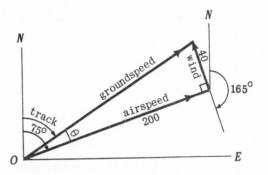

Fig. (*f*)  Prob. 11          Fig. (*g*)  Prob. 12

12. The heading of an airplane is 75° and the airspeed is 200 mph. Find the groundspeed and track is there is a wind of 40 mph from 165°. Refer to Fig. (*g*) above.

Construction. Put in the airspeed vector from $O$, follow through with the wind vector, and close the triangle.

Solution.  Groundspeed = $\sqrt{(200)^2 + (40)^2}$ = 204 mph,

$\tan \theta = 40/200 = 0.2000$ and $\theta = 11°20'$, and track = $75° - \theta = 63°40'$.

13. The airspeed of an airplane is 200 mph. There is a wind of 30 mph from 270°. Find the heading and groundspeed in order to track 0°. Refer to Fig. (*h*) below.

Construction. The groundspeed vector is along $ON$. Lay off the wind vector from $O$, follow through with the airspeed vector (200 units from the head of the wind vector to a point on $ON$), and close the triangle.

Solution.  Groundspeed = $\sqrt{(200)^2 - (30)^2}$ = 198 mph,

$\sin \theta = 30/200 = 0.1500$ and $\theta = 8°40'$, and heading = $360° - \theta = 351°20'$.

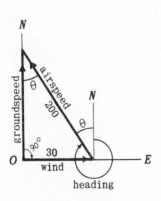

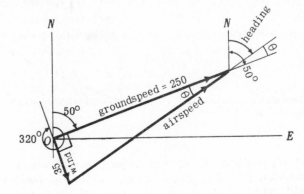

Fig. (*h*)  Prob. 13          Fig. (*i*)  Prob. 14

14. There is a wind of 35 mph from 320°. Find the airspeed and heading in order that the groundspeed and track be 250 mph and 50° respectively. Refer to Fig. (*i*) above.

Construction. Lay off the groundspeed vector from $O$, put in the wind vector at $O$ so that it does not follow through to the groundspeed vector, and close the triangle.

Solution.  Airspeed = $\sqrt{(250)^2 + (35)^2}$ = 252 mph,

tan $\theta$ = 35/250 = 0.1400 and $\theta$ = $8^\circ$,   and   heading = $50^\circ - 8^\circ = 42^\circ$.


# SUPPLEMENTARY PROBLEMS


15. An airplane flies 100 miles in the direction S $38^\circ 10'$ E.  How far south and how far east of the starting point is it?   *Ans.* 78.6 mi south, 61.8 mi east

16. A plane is headed due east with airspeed 240 mph. If a wind at 40 mph from the north is blowing, find the groundspeed and track.   *Ans.* Groundspeed, 243 mph; track, $99^\circ 30'$ or S $80^\circ 30'$ E

17. A body is acted upon by a force of 75 lb, due west, and a force of 125 lb, due north.  Find the magnitude and direction of the resultant force.   *Ans.* 146 lb, N $31^\circ 0'$ W

18. Find the rectangular components of a force of 525.0 lb in a direction $38^\circ 25'$ with the horizontal.   *Ans.* 411.3 lb, 326.2 lb

19. An aviator heads his airplane due west. He finds that due to a wind from the south, the course makes an angle of $20^\circ$ with the heading. If his airspeed is 100 mph, what is his groundspeed and what is the speed of the wind?   *Ans.* Groundspeed, 106 mph; wind, 36 mph

20. An airplane is headed west while a 40 mile wind is blowing from the south.  What is the necessary airspeed to follow a course N $72^\circ$ W and what is the groundspeed? *Ans.* Airspeed, 123 mph; groundspeed, 129 mph

21. A barge is being towed north at the rate 18 mph. A man walks across the deck from west to east at the rate 6 ft/sec. Find the magnitude and direction of the actual velocity. *Ans.* 27 ft/sec, N $12^\circ 50'$ E

22. A ship at $A$ is to sail to $C$, 56 mi north and 258 mi east of $A$.  After sailing N $25^\circ 10'$ E **for** 120 mi to $P$, the ship is headed toward $C$.  Find the distance of $P$ from $C$ and the **required** course to reach $C$.   *Ans.* 214 miles, S $75^\circ 40'$ E

23. A guy wire 78 ft long runs from the top of a telephone pole 56 ft high to the ground and **pulls** on the pole with a force of 290 lb.  What is the horizontal pull on the top of the pole? *Ans.* 202 lb

24. A weight of 200 lb is placed on a smooth plane inclined at an angle of $38^\circ$ with the horizontal and held in place by a rope parallel to the surface and fastened to a peg in the plane. Find the pull on the string.   *Ans.* 123 lb

25. A man wishes to raise a 300 lb weight to the top of a wall 20 ft high by dragging it up an incline.  What is the length of the shortest inclined plane he can use if his pulling strength is 140 lb?   *Ans.* 43 ft

26. A 150 lb shell is dragged up a runway inclined $40^\circ$ to the horizontal.  Find $a$) the force of the shell against the runway  and $b$) the force required to drag the shell. *Ans.* $a$) 115 lb, $b$) 96 lb

# CHAPTER 6

# Logarithms

THE COMMON LOGARITHM of a given positive number $N$ (written, log $N$) is the exponent of the power of 10 which will produce the given number. For example,

$$\log 1 = 0 \text{ since } 10^0 = 1, \qquad \log 100 = 2 \text{ since } 10^2 = 100,$$

$$\log 10 = 1 \text{ since } 10^1 = 10, \qquad \log 0.001 = -3 \text{ since } 10^{-3} = 0.001,$$

while $\qquad \log P = p \text{ if } 10^p = P.$

FUNDAMENTAL LAWS OF LOGARITHMS.

I. The logarithm of a product of two or more positive numbers is equal to the sum of the logarithms of the several numbers, i.e.,

$$\log P \cdot Q = \log P + \log Q,$$

$$\log P \cdot Q \cdot R = \log P + \log Q + \log R, \text{ etc.}$$

II. The logarithm of the quotient of two positive numbers is equal to the logarithm of the dividend minus the logarithm of the divisor, i.e.,

$$\log \frac{P}{Q} = \log P - \log Q.$$

III. The logarithm of a power of a positive number is equal to the logarithm of the number multiplied by the exponent of the power, i.e.,

$$\log (P^n) = n \log P.$$

IV. The logarithm of a root of a positive number is equal to the logarithm of the number divided by the index of the root, i.e.,

$$\log \sqrt[n]{P} = \frac{1}{n} \log P,$$

For proofs of these laws see Problem 1.

The logarithm of an expression involving two or more of the operations in laws I-IV is obtained by combining the results of the several laws, e.g.,

$$\log \frac{P \cdot Q}{R} = \log (P \cdot Q) - \log R = \log P + \log Q - \log R.$$

For other examples see Problems 2-4.

THE COMMON LOGARITHM of a positive number (e.g., log 300 = 2.47712 and log 0.2 = 9.30103-10) consists of two parts: an integral part called the *characteristic*, and a pure decimal part called the *mantissa*.

From Problems 3 and 4 it is seen that the characteristic depends only upon the position of the decimal point in the number. For example,

$$\log 2 = 0.30103 \qquad \text{and} \qquad \log 200 = 2.30103,$$

$$\log 25 = 1.39794 \qquad \text{and} \qquad \log 2.5 = 0.39794.$$

41

The characteristic of the common logarithm of any number greater than 1 is one less than the number of digits to the left of the decimal point in the given number.

The characteristic of the common logarithm of any positive number smaller than 1 is obtained by subtracting the number of zeros immediately following the decimal point from 9 and affixing -10. Thus the characteristic of the common logarithm of 0.2 is 9 −10, of 0.04 is 8 −10, of 0.0005 is 6 −10.

(See also Problem 5.)

The mantissa of the common logarithm of a positive number is usually a continuous decimal. All references here are to a table giving the mantissas to five decimal places.

## TO FIND THE LOGARITHM OF A GIVEN POSITIVE NUMBER:

*a*) Write down the characteristic in accordance with the above rules.

$b_1$) When the given number contains four or fewer significant digits, read the mantissa from the table.

> EXAMPLE 1. Find log 32.86.
>   The characteristic is 1. To find the mantissa locate the entry 51667 in the row opposite 328 and the column headed 6. Then log 32.86 = 1.51667.

> EXAMPLE 2. Find log 5.25.
>   The characteristic is 0. Since 5.25 = 5.250, we find the mantissa by locating the entry 72016 in the row opposite 525 and the column headed 0. Then log 5.25 = 0.72016.

$b_2$) When the given number contains five digits, interpolate using the method of proportional parts.

> EXAMPLE 3. Find log 654.82.
>   The characteristic is 2. For the mantissa, we have

$$
\begin{aligned}
\text{mantissa of log 65480} &= .81611 \\
\text{mantissa of log 65490} &= .81617 \\
\text{tabular difference} &= \overline{.00006} \\
.2 \times \text{tabular difference} &= .000012 \text{ or } .00001 \text{ to five decimal places} \\
\text{mantissa of log 65482} &= .81611 + .00001 = .81612.
\end{aligned}
$$

Then   log 654.82 = 2.81612.

Note that the essential calculation here is 81611 + .2 × 6 = 81612.2 or 81612.

(See also Problems 6-7.)

## TO FIND THE NUMBER CORRESPONDING TO A GIVEN COMMON LOGARITHM:

*a*) When the given mantissa is found in the table, read off the row number and the column heading and then point off using the characteristic rule. The resulting number is called the *antilogarithm* (antilog) of the given logarithm.

> EXAMPLE 4. Antilog 1.88053 = 75.95.
>   The mantissa .88053 is found in the row opposite 759 and the column headed 5. Since the characteristic is 1, there are two digits to the left of the decimal point.

*b*) When the given mantissa is not found in the table, interpolation must be used.

> EXAMPLE 5. Antilog 9.56577-10 = 0.36793.

| Mantissa of log 36790 = .56573 | Given mantissa         = .56577 |
|---|---|
| Mantissa of log 36800 = .56585 | Next smaller mantissa = .56573 |
| Tabular difference = .00012    | Difference = .00004 |

Correction = $\dfrac{.00004}{.00012}$ (.00010) = .000033 or .00003 to five decimal places.

Then antilog 9.56577 – 10 = 0.36790 + .00003 = 0.36793.

Note that the essential operation here is $\dfrac{4 \times 10}{12}$ = 3.3 or 3.

(See also Problem 8.)

THE COLOGARITHM of a positive number $N$ (written, colog $N$) is the logarithm of its reciprocal $\dfrac{1}{N}$. Thus, colog $N = \log \dfrac{1}{N} = \log 1 - \log N = - \log N$.

EXAMPLE 6. Colog 38.386 = 8.41583 – 10.

$$\text{colog } 38.386 = \log \dfrac{1}{38.386} = \log 1 - \log 38.386$$

$$
\begin{array}{ll}
\log 1 & = 10.00000 - 10 \\
(-) \log 38.386 = & \underline{1.58417} \\
& 8.41583 - 10
\end{array}
$$

Note that colog $N$ may be obtained by subtracting each digit (starting at the left) of log $N$ from 9 except the last significant digit, which is subtracted from 10, and affixing –10 when $N$ is greater than 1. For example:
a) log 3163 = 3.50010; colog 3163 = 6.49990 –10.
b) log 0.0399 = 8.60097 –10; colog 0.0399 = 1.39903.

(See also Problems 12-13.)

LOGARITHMS OF TRIGONOMETRIC FUNCTIONS. The table used here is a five place table of the logarithms of the trigonometric functions sine, cosine, tangent, and cotangent of angles from 0° to 90° at intervals of 1 minute.
The procedures for using such a table are essentially those for the table of natural trigonometric functions.

EXAMPLE 7.

a) log sin 22°34′ = 9.58406 – 10

b) log tan 72°18′ = 0.49602

c) log sin 22°34.8′ = 9.58430 – 10

$$
\begin{array}{ll}
\log \sin 22°34' = 9.58406 - 10 \\
\log \sin 22°35' = \underline{9.58436 - 10} \\
\text{Tabular difference} = \quad .00030
\end{array}
$$

Correction = .8 × tabular difference = .00024

Adding the correction, since sine,
log sin 22°34.8′ = 9.58406 –10 + .00024 = 9.58430 –10.

The essential operation here is 58406 + .8(30) = 58406 + 24 = 58430.

d) log cos 66°42.4′ = 9.59708 – 10

$$
\begin{array}{ll}
\log \cos 66°42' = 9.59720 - 10 \\
\text{Tabular difference} = 30
\end{array}
$$

Correction = .4 × tabular difference = .4(30) = 12

Subtracting the correction, since cosine, log cos 66°42.4′ = 9.59708 –10.

(See also Problem 14.)

EXAMPLE 8.

　　*a*) If log sin $A$ = 9.66197 – 10,　then　$A$ = 27°20′.

　　*b*) If log cot $A$ = 0.15262,　then　$A$ = 35°8′.

　　*c*) If log sin $A$ = 9.95472 – 10,　then　$A$ = 64°17.3′.

| | | | | |
|---|---|---|---|---|
| log sin 64°17′ = 9.95470 – 10 | | | log sin 64°17′ = 9.95470 – 10 | |
| log sin 64°18′ = 9.95476 – 10 | | | log sin $A$　　 = 9.95472 – 10 | |
| Tabular difference =　.00006 | | | Difference =　.00002 | |

Correction = $\dfrac{.00002}{.00006}(1') = \dfrac{2}{6}(1') = .3'$. Adding the correction, since sine, $A$ = 64°17.3′.

　　*d*) If log cos $A$ = 9.97888 – 10,　then　$A$ = 17°43.5′.

　　　　log cos 17°44′ = 9.97886 – 10　(next smaller logarithm)
　　　　Tabular difference = 4;　difference = 2.

Correction = $\dfrac{2}{4}(1') = .5'$.　Subtracting the correction, since cosine, $A$ = 17°43.5′.

　　*e*) If log tan $A$ = 0.24372,　then　$A$ = 60°17.6′.

　　　　log tan 60°17′ = 0.24353　(next smaller logarithm)
　　　　Tabular difference = 30 ;　difference = 19.

Correction = $\dfrac{19}{30}(1') = .6'$.　Adding the correction, since tangent, $A$ = 60°17.6′.

　　*f*) If cot $A$ = 9.41640 – 10,　then　$A$ = 75°22.8′.

　　　　log cot 75°23′ = 9.41629 – 10
　　　　Tabular difference = 52;　difference = 11.

Correction = $\dfrac{11}{52}(1') = .2'$.　Subtracting the correction, since cotangent, $A$ = 75°22.8′.

(See also Problem 15.)

## SOLVED PROBLEMS

1. Prove the laws of logarithms.

　　Restricting the proofs to common logarithms,

　let $P$ = $10^p$ and $Q$ = $10^q$ ; then　log $P$ = $p$　and　log $Q$ = $q$.

　　I. Since　$P \cdot Q$ = $10^p \cdot 10^q$ = $10^{p+q}$,　then　log $P \cdot Q$ = $p + q$ = log $P$ + log $Q$.

　　II. Since　$P/Q$ = $10^p/10^q$ = $10^{p-q}$,　then　log $P/Q$ = $p - q$ = log $P$ – log $Q$.

　　III. Since　$P^n$ = $(10^p)^n$ = $10^{np}$,　then　log $P^n$ = $np$ = $n$ log $P$.

　　IV. Since　$\sqrt[n]{P}$ = $(10^p)^{1/n}$ = $10^{p/n}$　then　log $\sqrt[n]{P}$ = $p/n$ = $\dfrac{1}{n}$ log $P$.

2. Express the logarithm of the given expression in terms of the logarithms of the　individual
　letters or numbers involved.

　　*a*) log $\dfrac{P \cdot Q}{R \cdot S}$ = log $(P \cdot Q)$ – log $(R \cdot S)$ = $(\log P + \log Q) - (\log R + \log S)$
　　　　　　　　　　　　　　　　　　　 = log $P$ + log $Q$ – log $R$ – log $S$.

b) $\log \dfrac{\sqrt[3]{P}}{Q^4} = \log \sqrt[3]{P} - \log Q^4 = \dfrac{1}{3} \log P - 4 \log Q$

c) $\log \dfrac{34(104)^2}{(49)^3} = \log 34 + 2 \log 104 - 3 \log 49$

d) $\log \dfrac{(34.2)^2 \sqrt[3]{1.06}}{(9.8)^3 \sqrt{2.33}} = 2 \log 34.2 + \dfrac{1}{3} \log 1.06 - 3 \log 9.8 - \dfrac{1}{2} \log 2.33$

3. Given $\log 2 = 0.30103$ and $\log 3 = 0.47712$, find the logarithm of :

    a) 30,   b) 200,   c) 25,   d) 120,   e) 2.5,   f) $\sqrt{6}$,   g) $\sqrt[3]{24}$ .

a) $30 = 3 \times 10$ ;   $\log 30 = \log 3 + \log 10 = 0.47712 + 1.00000 = 1.47712$

b) $200 = 2 \times 10^2$; $\log 200 = \log 2 + 2 \log 10 = 0.30103 + 2.00000 = 2.30103$

c) $25 = 10^2/2^2$ ; $\log 25 = 2 \log 10 - 2 \log 2 = 2.00000 - 0.60206 = 1.39794$

d) $120 = 2^2 \cdot 3 \cdot 10$ ; $\log 120 = 2 \log 2 + \log 3 + \log 10 = 0.60206 + 0.47712 + 1.00000 = 2.07918$

e) $2.5 = 10/2^2$ ; $\log 2.5 = \log 10 - 2 \log 2 = 1.00000 - 0.60206 = 0.39794$

f) $\sqrt{6} = (2 \times 3)^{\frac{1}{2}}$ ;   $\log \sqrt{6} = \dfrac{1}{2}(\log 2 + \log 3) = \dfrac{1}{2}(0.77815) = 0.38908$

g) $\sqrt[3]{24} = \sqrt[3]{2^3 \times 3} = 2\sqrt[3]{3}$ ; $\log \sqrt[3]{24} = \log 2 + \dfrac{1}{3} \log 3 = 0.30103 + \dfrac{1}{3}(0.47712) = 0.46007$

4. Given $\log 2 = 0.30103$ and $\log 3 = 0.47712$, find the logarithm of :

    a) 0.2,   b) 0.003,   c) 0.5,   d) $(0.02)^3$,   e) $\sqrt[4]{0.006}$

a) $0.2 = 2/10$ ; $\log 0.2 = \log 2 - \log 10 = 0.30103 - 1.00000 = -1 + 0.30103$.
    We shall write this $9.30103 - 10$.

b) $0.003 = 3/10^3$ ; $\log 0.003 = \log 3 - 3 \log 10 = -3 + 0.47712 = 7.47712 - 10$

c) $0.5 = 1/2$ ; $\log 0.5 = \log 1 - \log 2 = 0.00000 - 0.30103$
$$= (10.00000 - 10) - 0.30103 = 9.69897 - 10$$

d) $(0.02)^3 = (2/10^2)^3$ ; $\log (0.02)^3 = 3 \log 2 - 6 \log 10$
$$= 0.90309 - 6.00000$$
$$= (10.90309 - 10) - 6.00000 = 4.90309 - 10$$

e) $\sqrt[4]{0.006} = \sqrt[4]{2 \times 3/10^3}$ ; $\log \sqrt[4]{0.006} = \dfrac{1}{4}(\log 2 + \log 3 - 3 \log 10)$
$$= \dfrac{1}{4}(0.30103 + 0.47712 - 3.00000)$$
$$= \dfrac{1}{4}(7.77815 - 10) = \dfrac{1}{4}(37.77815 - 40) = 9.44454 - 10$$

5. Determine the characteristic of the common logarithm of each of the following numbers:

    a) 3864       c) 8.746      e) 0.3874      g) 0.07295     i) 2.3567     k) 0.44636

    b) 286        d) 982600    f) 0.00826     h) 0.000023    j) 88.725     l) 0.00072358 .

    The characteristics are:

    a) 3          c) 0         e) 9 $-10$      g) 8 $-10$     i) 0       k) 9 $-10$

    b) 2          d) 5         f) 7 $-10$      h) 5 $-10$     j) 1       l) 6 $-10$ .

6. Verify each of the following logarithms.

   a) log 38.64 = 1.58704          e) log 2.3567 = 0.37231          (37218 + 12.6)
   b) log 286 = 2.45637            f) log 88.725 = 1.94804          (94802 + 2.5)
   c) log 0.3874 = 9.58816 − 10    g) log 0.44636 = 9.64968 − 10    (64963 + 5.4)
   d) log 0.00826 = 7.91698 − 10   h) log 0.00072358 = 6.85949 − 10 (85944 + 4.8)

7. Verify each of the following.

   a) log (0.07324 × 0.0006235)  =  log 0.07324 + log 0.0006235
                                 =  8.86475-10 + 6.79484-10  =  15.65959-20  =  5.65959 − 10

   b) log (8.7633 × 0.0074288)  =  log 8.7633 + log 0.0074288
                                =  0.94266 + 7.87092-10  =  8.81358 − 10

   c) log 34.72/5.384  =  log 34.72 − log 5.384
                       =  1.54058 − 0.73111  =  0.80947

   d) log 7218/0.0235  =  log 7218 − log 0.0235
                       =  3.85842 − 8.37107-10  =  13.85842-10 − 8.37107-10  =  5.48735

   e) log (24.56)$^3$  =  3 log 24.56  =  3(1.39023)  =  4.17069

   f) log (0.4893)$^4$  =  4 log 0.4893  =  4(9.68958-10)  =  38.75832 − 40  =  8.75832 − 10

   g) log $\sqrt{876.4}$  =  $\frac{1}{2}$ log 876.4  =  $\frac{1}{2}$(2.94270)  =  1.47135

   h) log $\sqrt[3]{66.75}$  =  $\frac{1}{3}$ log 66.75  =  $\frac{1}{3}$(1.82445)  =  0.60815

   i) log $\sqrt{0.9494}$  =  $\frac{1}{2}$ log 0.9494 =  $\frac{1}{2}$(9.97745-10)  =  $\frac{1}{2}$(19.97745 − 20)  =  9.98872 − 10

8. Verify each of the following.

   a) Antilog 2.56158 = 364.40
   b) Antilog 5.69002 = 489800
   c) Antilog 8.81358-10 = 0.06510.   From Problem 7b), 8.7633 × 0.0074288 = 0.06510.
   d) Antilog 1.43654 = 27.324   (6 × 10/16 = 4)
   e) Antilog 8.69157-10 = 0.049156   (5 × 10/9 = 6)
   f) Antilog 4.17069 = 14814   (13 × 10/29 = 4).   From Problem 7e), (24.56)$^3$ = 14814.
   g) Antilog 1.47135 = 29.604   (6 × 10/15 = 4).   From Problem 7g), $\sqrt{876.4}$ = 29.604.

   Calculate each of the following using logarithms.

9. $N$ = 36.234 × 2.6748 × 0.0071756

$$
\begin{aligned}
\log 36.234 &= 1.55912 \\
(+)\ \log 2.6748 &= 0.42729 \\
(+)\ \log 0.0071756 &= \underline{7.85586 - 10} \\
\log N &= 9.84227 - 10 \\
N &= 0.69546
\end{aligned}
$$

10. $N$ = $\dfrac{47.75 \times 8.643}{6467}$

$$
\begin{aligned}
\log 47.75 &= 1.67897 \\
(+)\ \log 8.643 &= \underline{0.93666} \\
 &\quad\ 12.61563 - 10 \qquad (2.61563 = 12.61563 - 10) \\
(-)\ \log 6467 &= \underline{3.81070} \\
\log N &= 8.80493 - 10 \\
N &= 0.063816
\end{aligned}
$$

11. $N = \sqrt[3]{0.48476}$.

$$\log N = \frac{1}{3} \log 0.48476$$

$$\log 0.48476 = 9.68552 - 10$$
$$= 29.68552 - 30$$

$$\log N = 9.89517 - 10$$
$$N = 0.78554$$

12. Solve Problem 10 using cologarithms. $\quad N = 47.75 \times 8.643 \times \dfrac{1}{6467}$

$$\log 47.75 = 1.67897$$
$$(+) \ \log 8.643 = 0.93666$$
$$(+) \ \text{colog } 6467 = 6.18930 - 10 \qquad (\log 6467 = 3.81070)$$

$$\log N = 8.80493 - 10$$
$$N = 0.063816$$

13. $\dfrac{74.72}{\sqrt{8.394} \ \sqrt[3]{0.002877}} = N.$ $\qquad \log N = \log 74.72 + \dfrac{1}{2} \text{ colog } 8.394 + \dfrac{1}{3} \text{ colog } 0.002877$

$$\log 74.72 = 1.87344$$
$$(+) \ \frac{1}{2} \text{ colog } 8.394 = 9.53802 - 10$$
$$(+) \ \frac{1}{3} \text{ colog } 0.002877 = 0.84702$$

$$\log N = 12.25848 - 10$$
$$= 2.25848$$
$$N = 181.33$$

$$\log 8.394 = 0.92397$$
$$\text{colog } 8.394 = 9.07603 - 10$$

$$\log 0.002877 = 7.45894 - 10$$
$$\text{colog } 0.002877 = 2.54106$$

14. Verify each of the following.

a) $\log \sin 14°28.3' = 9.39777 - 10 \qquad (39762 + .3 \times 39)$
b) $\log \cos 66°44.8' = 9.59638 - 10 \qquad (59661 - .8 \times 29)$
c) $\log \tan 31°26.4' = 9.78630 - 10 \qquad (78618 + .4 \times 29)$
d) $\log \cot 45°54.6' = 9.98620 - 10 \qquad (98635 - .6 \times 25)$
e) $\log \sin 62°29.1' = 9.94787 - 10 \qquad (94786 + .1 \times 7)$
f) $\log \cos 23°33.7' = 9.96220 - 10 \qquad (96223 - .7 \times 5)$
g) $\log \tan 70°20.6' = 0.44709 \qquad (44685 + .6 \times 40)$
h) $\log \cot 11°17.3' = 0.69982 \qquad (70002 - .3 \times 66)$

15. Verify each of the following.

a) $\log \sin A = 9.90020 - 10, \quad$ then $A = 52°37.6' \qquad (\frac{6}{10} \times 1' = .6')$

b) $\log \cos A = 9.93602 - 10, \quad$ then $A = 30°20.6' \qquad (\frac{3}{7} \times 1' = .4')$

c) $\log \tan A = 9.87150 - 10, \quad$ then $A = 36°38.7' \qquad (\frac{18}{26} \times 1' = .7')$

d) $\log \cot A = 0.01245, \quad$ then $A = 44°10.7' \qquad (\frac{7}{25} \times 1' = .3')$

e) $\log \sin A = 9.80172 - 10, \quad$ then $A = 39°18.4' \qquad (\frac{6}{16} \times 1' = .4')$

f) $\log \cos A = 9.55215 - 10, \quad$ then $A = 69°6.6' \qquad (\frac{13}{33} \times 1' = .4')$

g) $\log \tan A = 0.44372, \quad$ then $A = 70°12.1' \qquad (\frac{5}{40} \times 1' = .1')$

h) $\log \cot A = 9.31142 - 10, \quad$ then $A = 78°25.4' \qquad (\frac{38}{64} \times 1' = .6')$

## SUPPLEMENTARY PROBLEMS

16. Find:

a) log 211 = 2.32428
b) log 9.17 = 0.96237
c) log 0.00466 = 7.66839-10
d) log 0.6754 = 9.82956-10
e) log 32.86 = 1.51667
f) log 264.76 = 2.42285
g) log 7.1775 = 0.85597
h) log 0.96634 = 9.98513-10

i) log 4287.6 = 3.63221
j) log 0.0055558 = 7.74474-10
k) log 0.097147 = 8.98743-10
l) log 2.1222 = 0.32679
m) log 66.985 = 1.82598
n) log 781.59 = 2.89298
o) log 2348.9 = 3.37086
p) log 0.091233 = 8.96016-10

17. Find:

a) antilog 1.98646 = 96.930
b) antilog 0.75005 = 5.6240
c) antilog 8.62086-10 = 0.041770
d) antilog 1.09706 = 12.504
e) antilog 2.65612 = 453.02
f) antilog 0.91821 = 8.2834
g) antilog 8.11848-10 = 0.013136
h) antilog 3.66626 = 4637.2

i) antilog 1.12078 = 13.206
j) antilog 2.62821 = 424.83
k) antilog 0.95846 = 9.0878
l) antilog 9.61299-10 = 0.41019
m) antilog 2.23958 = 173.61
n) antilog 1.22251 = 16.692
o) antilog 4.84033 = 69236
p) antilog 2.67183 = 469.71

18. Evaluate:

a) $\dfrac{819(748)}{3670} = 166.9$,   b) $\dfrac{827.6}{518.3} = 1.597$,   c) $\dfrac{48.62}{77.65} = 0.6261$,   d) $787.97(0.0033238) = 2.6190$

e) $\dfrac{(227.3)^2 \sqrt[3]{0.007764}}{(86.35)^3 \sqrt{0.3848}} = 0.02562$,   f) $\sqrt[3]{\dfrac{781.58(3.4342)}{852.74(586.76)}} = 0.17505$

19. Find:

a) log sin 53°18′ = 9.90405-10
b) log cos 18°17′ = 9.97750-10
c) log tan 42°47′ = 9.96636-10
d) log cot 68°14′ = 9.60130-10
e) log sin 71°9.6′ = 9.97608-10
f) log cos 56°44.4′ = 9.73913-10
g) log tan 67°0.3′ = 0.37226
h) log cot 76°9.3′ = 9.39174-10

i) log sin 72°15.4′ = 9.97884-10
j) log cos 20° 9.2′ = 9.97256-10
k) log tan 84°47.1′ = 1.03967
l) log cot 74° 4.2′ = 9.45549-10
m) log sin 22°15.8′ = 9.57849-10
n) log cos 66°17.4′ = 9.60434-10
o) log tan 11°19.8′ = 9.30182-10
p) log cot 25°10.6′ = 0.32784

20. Find acute angle A, given:

a) log sin A = 9.28705-10,   A = 11°10.0′
b) log cos A = 9.48881-10,   A = 72° 3.0′
c) log tan A = 9.82325-10,   A = 33°39.0′
d) log cot A = 9.91765-10,   A = 50°24.0′
e) log sin A = 9.53928-10,   A = 20°15.2′
f) log cos A = 9.89900-10,   A = 37°34.8′
g) log tan A = 9.53042-10,   A = 18°44.1′
h) log cot A = 0.18960,    A = 32°52.4′

i) log sin A = 9.86000-10,   A = 46°25.3′
j) log cos A = 9.75529-10,   A = 55°18.2′
k) log tan A = 9.80888-10,   A = 81°10 4′
l) log cot A = 9.67240-10,   A = 64°48.7′
m) log sin A = 9.80513-10,   A = 39°40.6′
n) log cos A = 9.86892-10,   A = 42°18.8′
o) log tan A = 0.06510,    A = 49°16.7′
p) log cot A = 9.71700-10,   A = 62°28.3′

# CHAPTER 7

# Logarithmic Solution of Right Triangles

**ANY RIGHT TRIANGLE** may be solved and partially checked by using the trigonometric functions sine, cosine, and either tangent or cotangent of one of the acute angles, together with the angle relation $A + B = 90°$. In general, a better check is obtained by using the relation $c^2 = a^2 + b^2$.

**EXAMPLE.** Suppose the sides $a$ and $b$ of the right triangle $ABC$ are given.

1) To find angle $A$, use $\tan A = a/b$; then $B = 90° - A$.

2) To find side $c$, use $c = a/\sin A$.

3) To check, use $a^2 = c^2 - b^2 = (c - b)(c + b)$

or $b^2 = c^2 - a^2 = (c - a)(c + a)$.

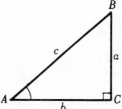

## SOLVED PROBLEMS

**1.** Solve and check the right triangle $ABC$, given $a = 48.620$ and $b = 37.640$. See Fig.(a) below.

| $\tan A = a/b$ | $c = a/\sin A$ | Check: $a^2 = (c-b)(c+b)$ |
|---|---|---|

$\log a = 1.68681$          $\log a = 1.68681$          $c = 61.487$     $\log (c-b) = 1.37744$
$(-)\log b = 1.57565$       $(-)\log \sin A = 9.89803-10$   $b = 37.640$    $(+)\log (c+b) = 1.99620$
$\overline{\log \tan A = 0.11116}$   $\overline{\log c = 1.78878}$   $\overline{c-b = 23.847}$   $\overline{2 \log a = 3.37364}$
$A = 52°15.2'$             $c = 61.487$              $c+b = 99.127$
$B = 37°44.8'$                                                    $\log a = 1.68682$

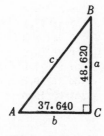

Fig. (a)  Prob. 1

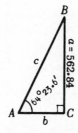

Fig. (b)  Prob. 2

**2.** Solve and check the right triangle $ABC$, given $a = 562.84$ and $A = 64°23.6'$. See Fig.(b) above.

$B = 90° - A = 25°36.4'$.

| $b = a/\tan A$ | $c = a/\sin A$ | Check: $a^2 = (c-b)(c+b)$ |
|---|---|---|

$\log a = 2.75038$          $\log a = 2.75038$          $c = 624.13$     $\log (c-b) = 2.54948$
$(-)\log \tan A = 0.31943$  $(-)\log \sin A = 9.95511-10$   $b = 269.74$    $(+)\log (c+b) = 2.95128$
$\overline{\log b = 2.43095}$   $\overline{\log c = 2.79527}$   $\overline{c-b = 354.39}$   $\overline{2 \log a = 5.50076}$
$b = 269.74$              $c = 624.13$              $c+b = 893.87$
                                                     $\log a = 2.75038$

**3.** Solve and check the right triangle $ABC$, given $b = 583.62$ and $c = 794.86$.  See Fig. $(c)$ below.

| $\cos A = b/c$ | $a = c \sin A$ | Check: $b^2 = (c-a)(c+a)$ |
|---|---|---|

$\log b = 2.76613$    $\log c = 2.90029$    $c = 794.86$    $\log(c-a) = 2.40695$

$(-)\log c = 2.90029$    $(+)\log \sin A = 9.83180-10$    $a = 539.62$    $(+)\log(c+a) = 3.12532$

$\log \cos A = 9.86584-10$    $\log a = 2.73209$    $c-a = 255.24$    $2 \log b = 5.53227$

   $A = 42°45.4'$      $a = 539.62$      $c+a = 1334.48$     $\log b = 2.76614$

   $B = 47°14.6'$               $= 1334.5$

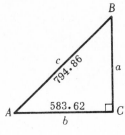

Fig. $(c)$  Prob. 3

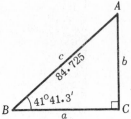

Fig. $(d)$  Prob. 4

**4.** Solve and check the right triangle $ABC$, given $c = 84.725$ and $B = 41°41.3'$.  See Fig. $(d)$ above.

$A = 90° - B = 48°18.7'$.

| $b = c \sin B$ | $a = c \cos B$ | Check: $b^2 = (c-a)(c+a)$ |
|---|---|---|

$\log c = 1.92802$    $\log c = 1.92802$    $c = 84.725$    $\log(c-a) = 1.33151$

$(+)\log \sin B = 9.82287-10$    $(+)\log \cos B = 9.87319-10$    $a = 63.271$    $(+)\log(c+a) = 2.17026$

$\log b = 1.75089$    $\log a = 1.80121$    $c-a = 21.454$    $2 \log b = 3.50177$

   $b = 56.350$      $a = 63.271$     $c+a = 147.996$     $\log b = 1.75088$

                  $= 148.00$

Note that this is a check of $\log b$ and not of $b$.

**5.** At a height of 23,245 ft a pilot of an airplane measures the angle of depression of a light at an airport as $28°45.2'$.  How far is he from the light?

In the adjoining figure, $A$ is the position of the light, $B$ is the position of the pilot, and $c = AB$ is the required distance.  Then

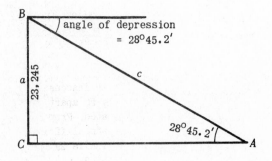

$$c = a/\sin A$$

$\log a = 4.36633$

$(-)\log \sin A = 9.68218-10$

$\log c = 4.68415$

$c = 48,322$

The required distance is 48,322 ft.

**6.** A shell is fired at an angle of elevation $32°14.4'$ with initial velocity 3046.8 ft/sec.  Find the initial horizontal and vertical velocities.

From the figure, $v$ = 3046.8, $\alpha$ = 32°14.4', and

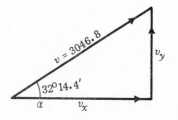

$$v_x = v \cos \alpha \qquad\qquad v_y = v \sin \alpha$$

| | |
|---|---|
| log $v$ = 3.48384 | log $v$ = 3.48384 |
| (+)log cos $\alpha$ = 9.92728-10 | (+)log sin $\alpha$ = 9.72711-10 |
| log $v_x$ = 3.41112 | log $v_y$ = 3.21095 |
| $v_x$ = 2577.1 ft/sec | $v_y$ = 1625.4 ft/sec |

**7.** Two forces of 151.75 lb and 225.80 lb act at right angles. Find the magnitude of the resultant and the angle which it makes with the larger force.

Using the right triangle $ABC$,

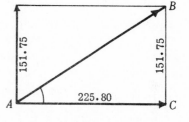

| | |
|---|---|
| tan $A$ = $CB/AC$ | $AB$ = $CB/\sin A$ |
| log $CB$ = 2.18113 | log $CB$ = 2.18113 |
| (−)log $AC$ = 2.35372 | (−)log sin $A$ = 9.74648-10 |
| log tan $A$ = 9.82741-10 | log $AB$ = 2.43465 |
| $A$ = 33°54.2' | $AB$ = 272.05 |

The magnitude of the resultant force is 272.05 lb and it makes an angle of 33°54.2' with the larger force.

**8.** A boat travels N 28°14.6' E for 55.375 miles and then N 61°45.4' W for 94.625 miles. What is its distance and bearing from the starting point?

In the figure, the boat starts at $A$, travels to $C$, and then to $B$. In the right triangle $ABC$,

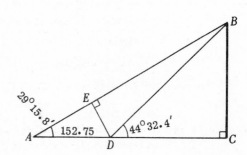

| | |
|---|---|
| tan $\angle CAB$ = $BC/AC$ | $AB$ = $BC/\sin \angle CAB$ |
| log $BC$ = 1.97600 | log $BC$ = 1.97600 |
| (−)log $AC$ = 1.74331 | (−)log sin $\angle CAB$ = 9.93605-10 |
| log tan $\angle CAB$ = 0.23269 | log $AB$ = 2.03995 |
| $\angle CAB$ = 59°39.8' | $AB$ = 109.64 |

The boat is then 109.64 miles from the starting point. Since $\angle NAB = \angle CAB - \angle CAN = 59°39.8' - 28°14.6' = 31°25.2'$, the required bearing is N 31°25.2' W.

**9.** In finding the height of an inaccessible cliff $CB$, two points $A$ and $D$, 152.75 ft apart, on a plain due west of the cliff are located. From $D$ the angle of elevation of the top of the cliff is 44°32.4' and from $A$ the angle of elevation is 29°15.8'. How high is the cliff above the plain?

A solution of a similar problem (Problem 15, Chapter 3) made use of the relation

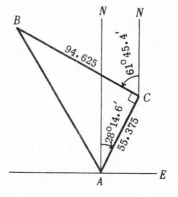

$$CB = \frac{AD}{\cot \angle BAC - \cot \angle BDC} \, .$$

In the solution here, a relation more suitable for logarithmic computation will be used.

In the figure, $DE$ is perpendicular to $AB$. Then

$$\angle DBE = \angle CBA - \angle CBD = (90^\circ - \angle BAC) - (90^\circ - \angle BDC) = \angle BDC - \angle BAC = 15^\circ 16.6'.$$

In the right triangle $AED$,    $DE = AD \sin \angle BAC$.
In the right triangle $BCD$,    $CB = BD \sin \angle BDC$.
In the right triangle $BED$,    $BD = DE/\sin \angle DBE$.

Then $$CB = BD \sin \angle BDC = \frac{DE \sin \angle BDC}{\sin \angle DBE} = \frac{AD \sin \angle BAC \cdot \sin \angle BDC}{\sin \angle DBE}$$

$$= \frac{152.75 \sin 29^\circ 15.8' \cdot \sin 44^\circ 32.4'}{\sin 15^\circ 16.6'}.$$

$$
\begin{array}{rl}
\log 152.75 = & 2.18398 \\
(+) \quad \log \sin 29^\circ 15.8' = & 9.68915\text{-}10 \\
(+) \quad \log \sin 44^\circ 32.4' = & 9.84597\text{-}10 \\
(+) \quad \text{colog} \sin 15^\circ 16.6' = & 0.57925 \\
\hline
\log CB = & 2.29835 \\
CB = & 198.77 \text{ ft}
\end{array}
$$

             $(\log \sin 15^\circ 16.6' = 9.42075 - 10)$

**10.** A tower $AB$ stands on a hill. On level ground at the foot of the hill two points $C$ and $D$, 200.00 ft apart, are located in the same vertical plane with $AB$. From $C$ the angles of elevation of the foot and top of the tower are $25^\circ 10.2'$ and $31^\circ 4.8'$ respectively, while from $D$ they are $13^\circ 50.5'$ and $17^\circ 32.3'$ respectively. Find the height of the tower.

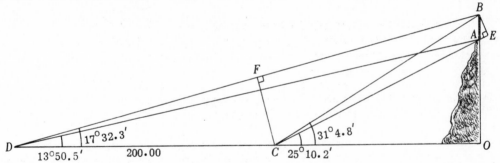

From $C$ drop a perpendicular $CF$ to $BD$ and from $B$ a perpendicular $BE$ to $CA$ extended. Let $AB$ extended meet $CD$ extended in $O$.

In the right triangle $AEB$,    $AB = EB/\sin \angle EAB$.
In the right triangle $CEB$,    $EB = CB \sin \angle ECB$.
In the right triangle $CFB$,    $CB = CF/\sin \angle CBF$.
In the right triangle $CDF$,    $CF = CD \sin \angle CDF$.

Then $$AB = \frac{EB}{\sin \angle EAB} = \frac{CB \sin \angle ECB}{\sin \angle EAB} = \frac{CF \sin \angle ECB}{\sin \angle EAB \cdot \sin \angle CBF} = \frac{CD \sin \angle CDF \cdot \sin \angle ECB}{\sin \angle EAB \cdot \sin \angle CBF}.$$

Now $\angle CDF = 17^\circ 32.3'$, $\angle ECB = \angle OCB - \angle OCA = 5^\circ 54.6'$, $\angle EAB = \angle OAC = 90^\circ - \angle OCA = 64^\circ 49.8'$, and $\angle CBF = \angle OBD - \angle OBC = (90^\circ - \angle ODB) - (90^\circ - \angle OCB) = \angle OCB - \angle ODB = 13^\circ 32.5'$.

$$
\begin{array}{rl}
\log 200.00 = & 2.30103 \\
(+) \quad \log \sin 17^\circ 32.3' = & 9.47906\text{-}10 \\
(+) \quad \log \sin 5^\circ 54.6' = & 9.01269\text{-}10 \\
(+) \quad \text{colog} \sin 64^\circ 49.8' = & 0.04333 \\
(+) \quad \text{colog} \sin 13^\circ 32.5' = & 0.63050 \\
\hline
\log AB = & 1.46661 \\
AB = & 29.283 \text{ ft}
\end{array}
$$

      $(\log \sin 64^\circ 49.8' = 9.95667 - 10)$

      $(\log \sin 13^\circ 32.5' = 9.36950 - 10)$

## SUPPLEMENTARY PROBLEMS

Solve and check each of the following right triangles $ABC$, given:

11. $a = 25.72$,    $A = 36°20'$          Ans.   $B = 53°40'$,   $b = 34.97$,   $c = 43.41$

12. $a = 342.86$,    $A = 55°32.8'$      Ans.   $B = 34°27.2'$,   $b = 235.23$,   $c = 415.81$

13. $a = 574.16$,    $B = 56°20.6'$      Ans.   $A = 33°39.4'$,   $b = 862.32$,   $c = 1036.0$

14. $c = 44.26$,    $A = 56°14'$          Ans.   $B = 33°46'$,   $a = 36.79$,   $b = 24.60$

15. $c = 287.68$,    $A = 38°10.2'$      Ans.   $B = 51°49.8'$,   $a = 177.78$,   $b = 226.17$

16. $c = 67.546$,    $B = 47°25.6'$      Ans.   $A = 42°34.4'$,   $a = 45.697$,   $b = 49.741$

17. $a = 42.420$,    $b = 58.480$       Ans.   $A = 35°57.4'$,   $B = 54°2.6'$,   $c = 72.243$

18. $a = 384.66$,    $b = 254.88$       Ans.   $A = 56°28.3'$,   $B = 33°31.7'$,   $c = 461.44$

19. A straight road is to be constructed joining two towns $A$ and $B$. If $B$ is located 133.75 miles to the east and 256.78 miles to the north of $A$, find the length and direction of the road from $A$.     Ans. 289.53 miles, N 27°30.8' E

20. Two forces of 281.66 lb and 323.54 lb act at right angles. Find the magnitude of the resultant force and the angle which it makes with the larger force.     Ans. 428.97 lb, 41°2.5'

21. Find the base of an isosceles triangle whose vertex angle is 48°27.4' and whose equal legs are 168.14.     Ans. 138.00

22. Given a circle of radius 417.12 ft, find the side and area
    a) of a regular inscribed decagon.     Ans. 257.80 ft, 511,340 ft$^2$
    b) of a regular circumscribed decagon.     Ans. 271.06 ft, 565,320 ft$^2$

23. Given a circle of radius 336.48 ft, find the side and area
    a) of a regular inscribed octagon.     Ans. 257.52 ft, 320,240 ft$^2$
    b) of a regular circumscribed octagon.     Ans. 278.74 ft, 375,170 ft$^2$

24. Two points $A$ and $D$ are in a horizontal line with the foot of a tower $CB$ and on opposite sides. The distance between $A$ and $D$ is 535.4 ft, while the angles of elevation of the top $B$ are 12°46' from $A$ and 18°38' from $D$. Let the perpendicular through $D$ to the line $AB$ produced meet it at $E$ and show that

$$CB = BD \sin \angle BDC = \frac{DE \sin \angle BDC}{\sin \angle DBE} = \frac{AD \sin \angle BAC \, \sin \angle BDC}{\sin \angle DBE} = 72.56 \text{ ft.}$$

# CHAPTER 8

# Reduction to Functions of Positive Acute Angles

**COTERMINAL ANGLES.** Let θ be any angle; then

$$\sin(\theta + n360^\circ) = \sin\theta \qquad \cot(\theta + n360^\circ) = \cot\theta$$
$$\cos(\theta + n360^\circ) = \cos\theta \qquad \sec(\theta + n360^\circ) = \sec\theta$$
$$\tan(\theta + n360^\circ) = \tan\theta \qquad \csc(\theta + n360^\circ) = \csc\theta$$

where $n$ is any positive or negative integer or zero.

Examples. $\sin 400^\circ = \sin(40^\circ + 360^\circ) = \sin 40^\circ$
$\cos 850^\circ = \cos(130^\circ + 2\cdot360^\circ) = \cos 130^\circ$
$\tan(-1000^\circ) = \tan(80^\circ - 3\cdot360^\circ) = \tan 80^\circ$

**FUNCTIONS OF A NEGATIVE ANGLE.** Let θ be any angle; then

$$\sin(-\theta) = -\sin\theta \qquad \cot(-\theta) = -\cot\theta$$
$$\cos(-\theta) = \cos\theta \qquad \sec(-\theta) = \sec\theta$$
$$\tan(-\theta) = -\tan\theta \qquad \csc(-\theta) = -\csc\theta$$

Examples. $\sin(-50^\circ) = -\sin 50^\circ$, $\cos(-30^\circ) = \cos 30^\circ$, $\tan(-200^\circ) = -\tan 200^\circ$.

For a proof of these relations, see Problem 1.

**REDUCTION FORMULAS.** Let θ be any angle; then

$$\sin(90^\circ - \theta) = \cos\theta \qquad \sin(90^\circ + \theta) = \cos\theta$$
$$\cos(90^\circ - \theta) = \sin\theta \qquad \cos(90^\circ + \theta) = -\sin\theta$$
$$\tan(90^\circ - \theta) = \cot\theta \qquad \tan(90^\circ + \theta) = -\cot\theta$$
$$\cot(90^\circ - \theta) = \tan\theta \qquad \cot(90^\circ + \theta) = -\tan\theta$$
$$\sec(90^\circ - \theta) = \csc\theta \qquad \sec(90^\circ + \theta) = -\csc\theta$$
$$\csc(90^\circ - \theta) = \sec\theta \qquad \csc(90^\circ + \theta) = \sec\theta$$

$$\sin(180^\circ - \theta) = \sin\theta \qquad \sin(180^\circ + \theta) = -\sin\theta$$
$$\cos(180^\circ - \theta) = -\cos\theta \qquad \cos(180^\circ + \theta) = -\cos\theta$$
$$\tan(180^\circ - \theta) = -\tan\theta \qquad \tan(180^\circ + \theta) = \tan\theta$$
$$\cot(180^\circ - \theta) = -\cot\theta \qquad \cot(180^\circ + \theta) = \cot\theta$$
$$\sec(180^\circ - \theta) = -\sec\theta \qquad \sec(180^\circ + \theta) = -\sec\theta$$
$$\csc(180^\circ - \theta) = \csc\theta \qquad \csc(180^\circ + \theta) = -\csc\theta$$

For proofs of these relations, see Problems 2,3,4,5.

**GENERAL REDUCTION FORMULA.** Any trigonometric function of $(n\cdot90^\circ \pm \theta)$, where θ is any angle, is *numerically* equal
a) to the same function of θ if $n$ is an even integer,
b) to the corresponding cofunction of θ if $n$ is an odd integer.

The algebraic sign in each case is the same as the sign of the given function for that quadrant in which $n\cdot90^\circ \pm \theta$ lies when θ is a positive acute angle.

For a verification of this formula, see Problem 8.

Examples.

1) $\sin(180° - \theta) = \sin(2 \cdot 90° - \theta) = \sin \theta$ since $180°$ is an even multiple of $90°$ and, when $\theta$ is positive acute, the terminal side of $180° - \theta$ lies in quadrant II.

2) $\cos(180° + \theta) = \cos(2 \cdot 90° + \theta) = -\cos \theta$ since $180°$ is an even multiple of $90°$ and, when $\theta$ is positive acute, the terminal side of $180° + \theta$ lies in quadrant III.

3) $\tan(270° - \theta) = \tan(3 \cdot 90° - \theta) = \cot \theta$ since $270°$ is an odd multiple of $90°$ and, when $\theta$ is positive acute, the terminal side of $270° - \theta$ lies in quadrant III.

4) $\cos(270° + \theta) = \cos(3 \cdot 90° + \theta) = \sin \theta$ since $270°$ is an odd multiple of $90°$ and, when $\theta$ is positive acute, the terminal side of $270° + \theta$ lies in quadrant IV.

## SOLVED PROBLEMS

**1.** Derive formulas for the functions of $(-\theta)$ in terms of the functions of $\theta$.

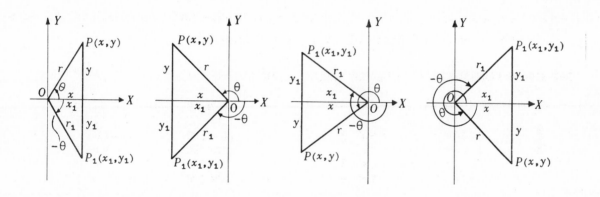

In the figures, $\theta$ and $-\theta$ are constructed in standard position and numerically equal. On their respective terminal sides the points $P(x,y)$ and $P_1(x_1,y_1)$ are located so that $OP = OP_1$. In each of the figures the two triangles are congruent and $r_1 = r$, $x_1 = x$, $y_1 = -y$. Then

$$\sin(-\theta) = \frac{y_1}{r_1} = \frac{-y}{r} = -\frac{y}{r} = -\sin \theta \qquad \cot(-\theta) = \frac{x_1}{y_1} = \frac{x}{-y} = -\frac{x}{y} = -\cot \theta$$

$$\cos(-\theta) = \frac{x_1}{r_1} = \frac{x}{r} = \cos \theta \qquad\qquad \sec(-\theta) = \frac{r_1}{x_1} = \frac{r}{x} = \sec \theta$$

$$\tan(-\theta) = \frac{y_1}{x_1} = \frac{-y}{x} = -\frac{y}{x} = -\tan \theta \qquad \csc(-\theta) = \frac{r_1}{y_1} = \frac{r}{-y} = -\frac{r}{y} = -\csc \theta$$

Except for those cases in which a function is not defined, the above relations are also valid when $\theta$ is a quadrantal angle. This may be verified by making use of the fact that $-0°$ and $0°$, $-90°$ and $270°$, $-180°$ and $180°$, $-270°$ and $90°$ are coterminal.

For example, $\sin(-0°) = \sin 0° = 0 = -\sin 0°$, $\quad \sin(-90°) = \sin 270° = -1 = -\sin 90°$, $\cos(-180°) = \cos 180°$, and $\cot(-270°) = \cot 90° = 0 = -\cot 270°$.

**2.** Derive formulas for the functions of $(90^\circ - \theta)$ in terms of the functions of $\theta$.

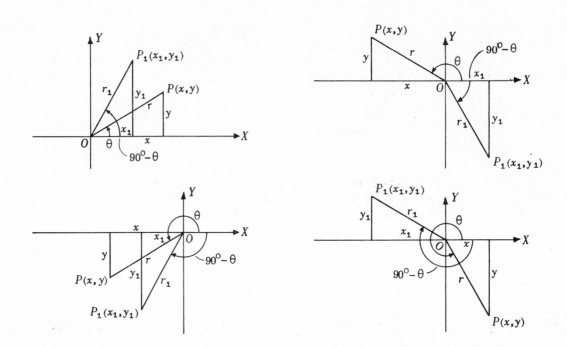

In the figures, $\theta$ and $90^\circ - \theta$ are constructed in standard position and on their respective terminal sides the points $P(x,y)$ and $P_1(x_1 \ y_1)$ are located so that $OP = OP_1$. In each of the figures the two triangles are congruent and $r_1 = r$, $x_1 = y$, $y_1 = x$. Then

$$\sin(90^\circ - \theta) = \frac{y_1}{r_1} = \frac{x}{r} = \cos\theta \qquad\qquad \cot(90^\circ - \theta) = \frac{x_1}{y_1} = \frac{y}{x} = \tan\theta$$

$$\cos(90^\circ - \theta) = \frac{x_1}{r_1} = \frac{y}{r} = \sin\theta \qquad\qquad \sec(90^\circ - \theta) = \frac{r_1}{x_1} = \frac{r}{y} = \csc\theta$$

$$\tan(90^\circ - \theta) = \frac{y_1}{x_1} = \frac{x}{y} = \cot\theta \qquad\qquad \csc(90^\circ - \theta) = \frac{r_1}{y_1} = \frac{r}{x} = \sec\theta$$

As in the case of the formulas of Problem 1, certain of these relations are without meaning when $\theta$ is a quadrantal angle.

**3.** Derive formulas for the functions of $(90^\circ + \theta)$ in terms of the functions of $\theta$.

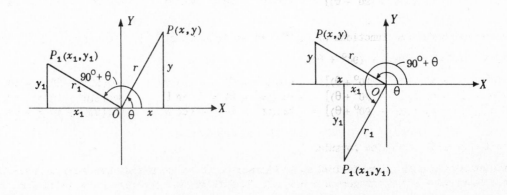

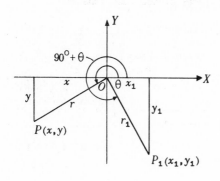 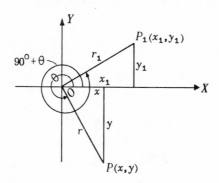

In the figures, $\theta$ and $90^\circ+\theta$ are constructed in standard position and on their respective terminal sides the points $P(x,y)$ and $P_1(x_1\ y_1)$ are located so that $OP = OP_1$. In each of the figures the two triangles are congruent and $r_1 = r$, $x_1 = -y$, $y_1 = x$. Then

$$\sin(90^\circ + \theta) \;=\; \frac{y_1}{r_1} \;=\; \frac{x}{r} \;=\; \cos\theta \qquad\qquad \cot(90^\circ + \theta) \;=\; \frac{x_1}{y_1} \;=\; -\frac{y}{x} \;=\; -\tan\theta$$

$$\cos(90^\circ + \theta) \;=\; \frac{x_1}{r_1} \;=\; -\frac{y}{r} \;=\; -\sin\theta \qquad\qquad \sec(90^\circ + \theta) \;=\; \frac{r_1}{x_1} \;=\; -\frac{r}{y} \;=\; -\csc\theta$$

$$\tan(90^\circ + \theta) \;=\; \frac{y_1}{x_1} \;=\; -\frac{x}{y} \;=\; -\cot\theta \qquad\qquad \csc(90^\circ + \theta) \;=\; \frac{r_1}{y_1} \;=\; \frac{r}{x} \;=\; \sec\theta$$

**4.** Derive formulas for the functions of $(180^\circ - \theta)$ in terms of the functions of $\theta$.

Since $180^\circ - \theta = 90^\circ + (90^\circ - \theta)$,

$$\sin(180^\circ - \theta) \;=\; \sin[90^\circ + (90^\circ - \theta)] \;=\; \cos(90^\circ - \theta) \;=\; \sin\theta$$
$$\cos(180^\circ - \theta) \;=\; \cos[90^\circ + (90^\circ - \theta)] \;=\; -\sin(90^\circ - \theta) \;=\; -\cos\theta, \quad \text{etc.}$$

**5.** Derive formulas for the functions of $(180^\circ + \theta)$ in terms of the functions of $\theta$.

Since $180^\circ + \theta = 90^\circ + (90^\circ + \theta)$,

$$\sin(180^\circ + \theta) \;=\; \sin[90^\circ + (90^\circ + \theta)] \;=\; \cos(90^\circ + \theta) \;=\; -\sin\theta$$
$$\cos(180^\circ + \theta) \;=\; \cos[90^\circ + (90^\circ + \theta)] \;=\; -\sin(90^\circ + \theta) \;=\; -\cos\theta, \quad \text{etc.}$$

**6.** Derive formulas for the functions of $(270^\circ - \theta)$ in terms of the functions of $\theta$.

Since $270^\circ - \theta = 180^\circ + (90^\circ - \theta)$,

$$\sin(270^\circ - \theta) = \sin[180^\circ + (90^\circ - \theta)] = -\sin(90^\circ - \theta) = -\cos\theta \qquad \cot(270^\circ - \theta) = \tan\theta$$
$$\cos(270^\circ - \theta) = \cos[180^\circ + (90^\circ - \theta)] = -\cos(90^\circ - \theta) = -\sin\theta \qquad \sec(270^\circ - \theta) = -\csc\theta$$
$$\tan(270^\circ - \theta) = \tan[180^\circ + (90^\circ - \theta)] = \tan(90^\circ - \theta) = \cot\theta \qquad \csc(270^\circ - \theta) = -\sec\theta.$$

**7.** Derive formulas for the functions of $(270^\circ + \theta)$ in terms of the functions of $\theta$.

Since $270^\circ + \theta = 180^\circ + (90^\circ + \theta)$,

$$\sin(270^\circ + \theta) = \sin[180^\circ + (90^\circ + \theta)] = -\sin(90^\circ + \theta) = -\cos\theta \qquad \cot(270^\circ + \theta) = -\tan\theta$$
$$\cos(270^\circ + \theta) = \cos[180^\circ + (90^\circ + \theta)] = -\cos(90^\circ + \theta) = \sin\theta \qquad \sec(270^\circ + \theta) = \csc\theta$$
$$\tan(270^\circ + \theta) = \tan[180^\circ + (90^\circ + \theta)] = \tan(90^\circ + \theta) = -\cot\theta \qquad \csc(270^\circ + \theta) = -\sec\theta.$$

**8.** Derive the general reduction formula.

By examining the formulas derived in Problems 1-7, it is seen that the general reduction formula is valid for the integers $n = 0,1,2,3$. It follows that the formula is valid for $n$ any integer since $n \cdot 90^\circ \pm \theta$ is coterminal with some one of the angles $\pm\theta$, $90^\circ \pm \theta$, $180^\circ \pm \theta$, $270^\circ \pm \theta$.

9. Express each of the following in terms of a function of $\theta$:

a) $\sin(\theta - 90^\circ)$          d) $\cos(-180^\circ + \theta)$          g) $\sin(540^\circ + \theta)$          j) $\cos(-450^\circ - \theta)$
b) $\cos(\theta - 90^\circ)$          e) $\sin(-270^\circ - \theta)$          h) $\tan(720^\circ - \theta)$          k) $\csc(-900^\circ + \theta)$
c) $\sec(-\theta - 90^\circ)$          f) $\tan(\theta - 360^\circ)$          i) $\tan(720^\circ + \theta)$          l) $\sin(-540^\circ - \theta)$.

a) $\sin(\theta - 90^\circ) = \sin(-90^\circ + \theta) = \sin(-1 \cdot 90^\circ + \theta) = -\cos\theta$,   the sign being negative since, when $\theta$ is positive acute, the terminal side of $\theta - 90^\circ$ lies in quadrant IV.

b) $\cos(\theta - 90^\circ) = \cos(-90^\circ + \theta) = \cos(-1 \cdot 90^\circ + \theta) = \sin\theta$.

c) $\sec(-\theta - 90^\circ) = \sec(-90^\circ - \theta) = \sec(-1 \cdot 90^\circ - \theta) = -\csc\theta$,   the sign being negative since, when $\theta$ is positive acute, the terminal side of $-\theta - 90^\circ$ lies in quadrant III.

d) $\cos(-180^\circ + \theta) = \cos(-2 \cdot 90^\circ + \theta) = -\cos\theta$.     (quadrant III)

e) $\sin(-270^\circ - \theta) = \sin(-3 \cdot 90^\circ - \theta) = \cos\theta$.     (quadrant I)

f) $\tan(\theta - 360^\circ) = \tan(-4 \cdot 90^\circ + \theta) = \tan\theta$.     (quadrant I)

g) $\sin(540^\circ + \theta) = \sin(6 \cdot 90^\circ + \theta) = -\sin\theta$.     (quadrant III)

h) $\tan(720^\circ - \theta) = \tan(8 \cdot 90^\circ - \theta) = -\tan\theta$
   $= \tan(2 \cdot 360^\circ - \theta) = \tan(-\theta) = -\tan\theta$.

i) $\tan(720^\circ + \theta) = \tan(8 \cdot 90^\circ + \theta) = \tan\theta$
   $= \tan(2 \cdot 360^\circ + \theta) = \tan\theta$.

j) $\cos(-450^\circ - \theta) = \cos(-5 \cdot 90^\circ - \theta) = -\sin\theta$.

k) $\csc(-900^\circ + \theta) = \csc(-10 \cdot 90^\circ + \theta) = -\csc\theta$.

l) $\sin(-540^\circ - \theta) = \sin(-6 \cdot 90^\circ - \theta) = \sin\theta$.


10. Express each of the following in terms of functions of a positive acute angle in two **ways**:

a) $\sin 130^\circ$          c) $\sin 200^\circ$          e) $\tan 165^\circ$          g) $\sin 670^\circ$          i) $\csc 865^\circ$          k) $\cos(-680^\circ)$
b) $\tan 325^\circ$          d) $\cos 310^\circ$          f) $\sec 250^\circ$          h) $\cot 930^\circ$          j) $\sin(-100^\circ)$          l) $\tan(-290^\circ)$.

a) $\sin 130^\circ = \sin(2 \cdot 90^\circ - 50^\circ) = \sin 50^\circ$          d) $\cos 310^\circ = \cos(4 \cdot 90^\circ - 50^\circ) = \cos 50^\circ$
   $= \sin(1 \cdot 90^\circ + 40^\circ) = \cos 40^\circ$          $= \cos(3 \cdot 90^\circ + 40^\circ) = \sin 40^\circ$

b) $\tan 325^\circ = \tan(4 \cdot 90^\circ - 35^\circ) = -\tan 35^\circ$          e) $\tan 165^\circ = \tan(2 \cdot 90^\circ - 15^\circ) = -\tan 15^\circ$
   $= \tan(3 \cdot 90^\circ + 55^\circ) = -\cot 55^\circ$          $= \tan(1 \cdot 90^\circ + 75^\circ) = -\cot 75^\circ$

c) $\sin 200^\circ = \sin(2 \cdot 90^\circ + 20^\circ) = -\sin 20^\circ$          f) $\sec 250^\circ = \sec(2 \cdot 90^\circ + 70^\circ) = -\sec 70^\circ$
   $= \sin(3 \cdot 90^\circ - 70^\circ) = -\cos 70^\circ$          $= \sec(3 \cdot 90^\circ - 20^\circ) = -\csc 20^\circ$

g) $\sin 670^\circ = \sin(8 \cdot 90^\circ - 50^\circ) = -\sin 50^\circ$
   $= \sin(7 \cdot 90^\circ + 40^\circ) = -\cos 40^\circ$

or $\sin 670^\circ = \sin(310^\circ + 360^\circ) = \sin 310^\circ = \sin(4 \cdot 90^\circ - 50^\circ) = -\sin 50^\circ$

h) $\cot 930^\circ = \cot(10 \cdot 90^\circ + 30^\circ) = \cot 30^\circ$
   $= \cot(11 \cdot 90^\circ - 60^\circ) = \tan 60^\circ$

or $\cot 930^\circ = \cot(210^\circ + 2 \cdot 360^\circ) = \cot 210^\circ = \cot(2 \cdot 90^\circ + 30^\circ) = \cot 30^\circ$

i) $\csc 865^\circ = \csc(10 \cdot 90^\circ - 35^\circ) = \csc 35^\circ$
   $= \csc(9 \cdot 90^\circ + 55^\circ) = \sec 55^\circ$

or $\csc 865^\circ = \csc(145^\circ + 2 \cdot 360^\circ) = \csc 145^\circ = \csc(2 \cdot 90^\circ - 35^\circ) = \csc 35^\circ$

j) $\sin(-100^\circ) = \sin(-2 \cdot 90^\circ + 80^\circ) = -\sin 80^\circ$
   $= \sin(-1 \cdot 90^\circ - 10^\circ) = -\cos 10^\circ$

or $\sin(-100^\circ) = -\sin 100^\circ = -\sin(2 \cdot 90^\circ - 80^\circ) = -\sin 80^\circ$
or $\sin(-100^\circ) = \sin(-100^\circ + 360^\circ) = \sin 260^\circ = \sin(2 \cdot 90^\circ + 80^\circ) = -\sin 80^\circ$

$k)$ $\cos(-680^{\circ}) = \cos(-8\cdot90^{\circ} + 40^{\circ}) = \cos 40^{\circ}$
$= \cos(-7\cdot90^{\circ} - 50^{\circ}) = \sin 50^{\circ}$

or $\cos(-680^{\circ}) = \cos(-680^{\circ} + 2\cdot360^{\circ}) = \cos 40^{\circ}$

$l)$ $\tan(-290^{\circ}) = \tan(-4\cdot90^{\circ} + 70^{\circ}) = \tan 70^{\circ}$
$= \tan(-3\cdot90^{\circ} - 20^{\circ}) = \cot 20^{\circ}$

or $\tan(-290^{\circ}) = \tan(-290^{\circ} + 360^{\circ}) = \tan 70^{\circ}$

**11.** Find the exact values of the sine, cosine, and tangent of:
   $a)$ $120^{\circ}$,  $b)$ $210^{\circ}$,  $c)$ $315^{\circ}$,  $d)$ $-135^{\circ}$,  $e)$ $-240^{\circ}$,  $f)$ $-330^{\circ}$.

   Call $\theta$, always positive acute, the *related angle* of $\phi$ when $\phi = 180^{\circ} - \theta$, $180^{\circ} + \theta$ or $360^{\circ} - \theta$. Then any function of $\phi$ is numerically equal to the same function of $\theta$. The algebraic sign in each case is that of the function in the quadrant in which the terminal side of $\phi$ lies.

   $a)$ $120^{\circ} = 180^{\circ} - 60^{\circ}$. The related angle is $60^{\circ}$; $120^{\circ}$ is in quadrant II.
   $\sin 120^{\circ} = \sin 60^{\circ} = \sqrt{3}/2$,  $\cos 120^{\circ} = -\cos 60^{\circ} = -1/2$,  $\tan 120^{\circ} = -\tan 60^{\circ} = -\sqrt{3}$.

   $b)$ $210^{\circ} = 180^{\circ} + 30^{\circ}$. The related angle is $30^{\circ}$; $210^{\circ}$ is in quadrant III.
   $\sin 210^{\circ} = -\sin 30^{\circ} = -1/2$,  $\cos 210^{\circ} = -\cos 30^{\circ} = -\sqrt{3}/2$,  $\tan 210^{\circ} = \tan 30^{\circ} = \sqrt{3}/3$.

   $c)$ $315^{\circ} = 360^{\circ} - 45^{\circ}$. The related angle is $45^{\circ}$; $315^{\circ}$ is in quadrant IV.
   $\sin 315^{\circ} = -\sin 45^{\circ} = -\sqrt{2}/2$,  $\cos 315^{\circ} = \cos 45^{\circ} = \sqrt{2}/2$,  $\tan 315^{\circ} = -\tan 45^{\circ} = -1$.

   $d)$ Any function of $-135^{\circ}$ is the same function of $-135^{\circ} + 360^{\circ} = 225^{\circ} = \phi$.
   $225^{\circ} = 180^{\circ} + 45^{\circ}$. The related angle is $45^{\circ}$; $225^{\circ}$ is in quadrant III.
   $\sin(-135^{\circ}) = -\sin 45^{\circ} = -\sqrt{2}/2$,  $\cos(-135^{\circ}) = -\cos 45^{\circ} = -\sqrt{2}/2$,  $\tan(-135^{\circ}) = 1$.

   $e)$ Any function of $-240^{\circ}$ is the same function of $-240^{\circ} + 360^{\circ} = 120^{\circ}$.
   $120^{\circ} = 180^{\circ} - 60^{\circ}$. The related angle is $60^{\circ}$; $120^{\circ}$ is in quadrant II.
   $\sin(-240^{\circ}) = \sin 60^{\circ} = \sqrt{3}/2$,  $\cos(-240^{\circ}) = -\cos 60^{\circ} = -1/2$,  $\tan(-240^{\circ}) = -\tan 60^{\circ} = -\sqrt{3}$.

   $f)$ Any function of $-330^{\circ}$ is the same function of $-330^{\circ} + 360^{\circ} = 30^{\circ}$.
   $\sin(-330^{\circ}) = \sin 30^{\circ} = 1/2$,  $\cos(-330^{\circ}) = \cos 30^{\circ} = \sqrt{3}/2$,  $\tan(-330^{\circ}) = \tan 30^{\circ} = \sqrt{3}/3$.

**12.** Using the table of natural functions, find:
   $a)$ $\sin 125^{\circ}14' = \sin(180^{\circ} - 54^{\circ}46') = \sin 54^{\circ}46' = 0.8168$
   $b)$ $\cos 169^{\circ}40' = \cos(180^{\circ} - 10^{\circ}20') = -\cos 10^{\circ}20' = -0.9838$
   $c)$ $\tan 200^{\circ}23' = \tan(180^{\circ} + 20^{\circ}23') = \tan 20^{\circ}23' = 0.3716$
   $d)$ $\cot 250^{\circ}44' = \cot(180^{\circ} + 70^{\circ}44') = \cot 70^{\circ}44' = 0.3495$
   $e)$ $\cos 313^{\circ}18' = \cos(360^{\circ} - 46^{\circ}42') = \cos 46^{\circ}42' = 0.6858$
   $f)$ $\sin 341^{\circ}52' = \sin(360^{\circ} - 18^{\circ}8') = -\sin 18^{\circ}8' = -0.3112$

**13.** If $\tan 25^{\circ} = a$, find:

   $a)$ $\dfrac{\tan 155^{\circ} - \tan 115^{\circ}}{1 + \tan 155^{\circ}\,\tan 115^{\circ}} = \dfrac{-\tan 25^{\circ} - (-\cot 25^{\circ})}{1 + (-\tan 25^{\circ})(-\cot 25^{\circ})} = \dfrac{-a + 1/a}{1 + a(1/a)} = \dfrac{-a^2 + 1}{a + a} = \dfrac{1 - a^2}{2a}$ .

   $b)$ $\dfrac{\tan 205^{\circ} - \tan 115^{\circ}}{\tan 245^{\circ} + \tan 335^{\circ}} = \dfrac{\tan 25^{\circ} - (-\cot 25^{\circ})}{\cot 25^{\circ} + (-\tan 25^{\circ})} = \dfrac{a + 1/a}{1/a - a} = \dfrac{a^2 + 1}{1 - a^2}$ .

**14.** If $A + B + C = 180^{\circ}$, then
   $a)$ $\sin(B + C) = \sin(180^{\circ} - A) = \sin A$ .
   $b)$ $\sin\frac{1}{2}(B + C) = \sin\frac{1}{2}(180^{\circ} - A) = \sin(90^{\circ} - \frac{1}{2}A) = \cos\frac{1}{2}A$.

15. Show that $\sin\theta$ and $\tan\frac{1}{2}\theta$ have the same sign.

    *a*) Suppose $\theta = n\cdot180°$. If $n$ is even (including zero), say $2m$, then $\sin(2m\cdot180°) = \tan(m\cdot180°) = 0$. The case when $n$ is odd is excluded since then $\tan\frac{1}{2}\theta$ is not defined.

    *b*) Suppose $\theta = n\cdot180° + \phi$, where $0 < \phi < 180°$. If $n$ is even, including zero, $\theta$ is in quadrant I or quadrant II and $\sin\theta$ is positive while $\frac{1}{2}\theta$ is in quadrant I or quadrant III and $\tan\frac{1}{2}\theta$ is positive. If $n$ is odd, $\theta$ is in quadrant III or IV and $\sin\theta$ is negative while $\frac{1}{2}\theta$ is in quadrant II or IV and $\tan\frac{1}{2}\theta$ is negative.

16. Find all positive values of $\theta$ less than $360°$ for which $\sin\theta = -\frac{1}{2}$.

    There will be two angles (see Chapter 2), one in the third quadrant and one in the fourth quadrant. The related angle (see Problem 11) of each has its sine equal to $+\frac{1}{2}$ and is $30°$. Thus the required angles are $\theta = 180° + 30° = 210°$ and $\theta = 360° - 30° = 330°$.

    Note. To obtain *all* values of $\theta$ for which $\sin\theta = -\frac{1}{2}$, add $n\cdot360°$ to each of the above solutions; thus $\theta = 210° + n\cdot360°$ and $\theta = 330° + n\cdot360°$, where $n$ is any integer.

17. Find all positive values of $\theta$ less than $360°$ for which $\cos\theta = 0.9063$.

    There are two solutions, $\theta = 25°$ in the first quadrant and $\theta = 360° - 25° = 335°$ in the fourth quadrant.

18. Find all positive values of $\frac{1}{4}\theta$ less than $360°$, given $\sin\theta = 0.6428$.

    The two positive angles less than $360°$ for which $\sin\theta = 0.6428$ are $\theta = 40°$ and $\theta = 180° - 40° = 140°$. But if $\frac{1}{4}\theta$ is to include all values less than $360°$, $\theta$ must include all values less than $4\cdot360° = 1440°$. Hence, for $\theta$ we take the two angles above and all coterminal angles less than $1440°$, that is,

$$\theta = 40°, \ 400°, \ 760°, \ 1120°; \ 140°, \ 500°, \ 860°, \ 1220° \quad \text{and}$$
$$\tfrac{1}{4}\theta = 10°, \ 100°, \ 190°, \ 280°; \ 35°, \ 125°, \ 215°, \ 305°.$$

19. Find all positive values of $\theta$ less than $360°$ which satisfy $\sin2\theta = \cos\frac{1}{2}\theta$.

    Since $\cos\frac{1}{2}\theta = \sin(90° - \frac{1}{2}\theta) = \sin2\theta$, $\ 2\theta = 90° - \frac{1}{2}\theta, \ 450° - \frac{1}{2}\theta, \ 810° - \frac{1}{2}\theta, \ 1170° - \frac{1}{2}\theta, \ \dots$
Then $\frac{5}{2}\theta = 90°, \ 450°, \ 810°, \ 1170°, \ \dots$ and $\theta = 36°, \ 180°, \ 324°, \ 468°, \ \dots$

    Since $\cos\frac{1}{2}\theta = \sin(90° + \frac{1}{2}\theta) = \sin2\theta$, $\ 2\theta = 90° + \frac{1}{2}\theta, \ 450° + \frac{1}{2}\theta, \ 810° + \frac{1}{2}\theta, \ \dots$
Then $\frac{3}{2}\theta = 90°, \ 450°, \ 810°, \ \dots$ and $\theta = 60°, \ 300°, \ 540°, \ \dots$

    The required solutions are: $36°, \ 180°, \ 324°; \ 60°, \ 300°$.

## SUPPLEMENTARY PROBLEMS

**20.** Express each of the following in terms of functions of a positive acute angle.

a) sin 145°       d) cot 155°       g) sin (−200°)       j) cot 610°

b) cos 215°       e) sec 325°       h) cos (−760°)       k) sec 455°

c) tan 440°       f) csc 190°       i) tan (−1385°)       l) csc 825°

*Ans.*  a) sin 35° or cos 55°            g) sin 20° or cos 70°

       b) − cos 35° or − sin 55°         h) cos 40° or sin 50°

       c) tan 80° or cot 10°            i) tan 55° or cot 35°

       d) − cot 25° or − tan 65°        j) cot 70° or tan 20°

       e) sec 35° or csc 55°            k) − sec 85° or − csc 5°

       f) − csc 10° or − sec 80°         l) csc 75° or sec 15°

**21.** Find the exact values of the sine, cosine, and tangent of:

a) 150°,  b) 225°,  c) 300°,  d) −120°,  e) −210°,  f) −315°.

*Ans.*  a) $1/2$, $-\sqrt{3}/2$, $-1/\sqrt{3}$           d) $-\sqrt{3}/2$, $-1/2$, $\sqrt{3}$

       b) $-\sqrt{2}/2$, $-\sqrt{2}/2$, $1$          e) $1/2$, $-\sqrt{3}/2$, $-1/\sqrt{3}$

       c) $-\sqrt{3}/2$, $1/2$, $-\sqrt{3}$           f) $\sqrt{2}/2$, $\sqrt{2}/2$, $1$

**22.** Using appropriate tables, find:

a) sin 155°13′ = 0.4192           f) log sin 129°44.8′ = 9.88586−10

b) cos 104°38′ = −0.2526         g) log sin 110°32.7′ = 9.97146−10

c) tan 305°24′ = −1.4071        h) log sin 162°35.6′ = 9.47589−10

d) sin 114°18′ = 0.9114           i) log sin 138°30.5′ = 9.82119−10

e) cos 166°51′ = −0.9738         j) log sin 174°22.7′ = 8.99104−10

**23.** Find all angles, $0 \leqq \theta < 360°$, for which:

a) sin θ = $\sqrt{2}/2$,  b) cos θ = −1,  c) sin θ = −0.6180,  d) cos θ = 0.5125,  e) tan θ = −1.5301

*Ans.*  a) 45°, 135°       c) 218°10′, 321°50′         e) 123°10′, 303°10′

       b) 180°           d) 59°10′, 300°50′

**24.** When θ is a second quadrant angle for which tan θ = −2/3, show that

a) $\dfrac{\sin(90° - \theta) - \cos(180° - \theta)}{\tan(270° + \theta) + \cot(360° - \theta)} = -\dfrac{2}{\sqrt{13}}$,       b) $\dfrac{\tan(90° + \theta) + \cos(180° + \theta)}{\sin(270° - \theta) - \cot(-\theta)} = \dfrac{2 + \sqrt{13}}{2 - \sqrt{13}}$

# CHAPTER 9

# Variations and Graphs of the Trigonometric Functions

LINE REPRESENTATIONS OF THE TRIGONOMETRIC FUNCTIONS. Let θ be any given angle in standard position. (See the figures below for θ in each of the quadrants.) With the vertex $O$ as center describe a circle of radius one unit cutting the initial side $OX$ of θ at $A$, the positive $y$-axis at $B$, and the terminal side of θ at $P$. Draw $MP$ perpendicular to $OX$; draw also the tangents to the circle at $A$ and $B$ meeting the terminal side of θ or its extension through $O$ in the points $Q$ and $R$ respectively.

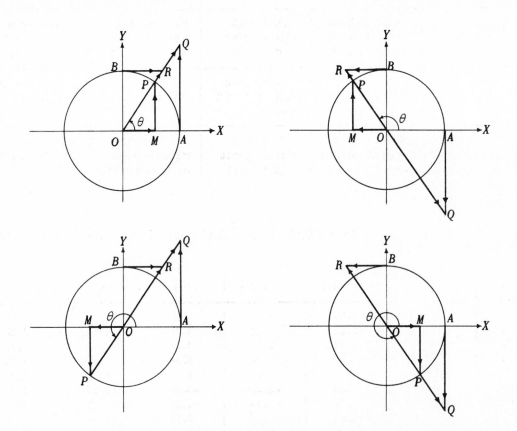

In each of the figures, the right triangles $OMP$, $OAQ$, and $OBR$ are similar, and

$$\sin θ = MP/OP = MP \qquad \cot θ = OM/MP = BR/OB = BR$$
$$\cos θ = OM/OP = OM \qquad \sec θ = OP/OM = OQ/OA = OQ$$
$$\tan θ = MP/OM = AQ/OA = AQ \qquad \csc θ = OP/MP = OR/OB = OR.$$

The segments $MP$, $OM$, $AQ$, etc., are directed line segments, the magnitude of a function being given by the length of the corresponding segment and the sign being given by the indicated direction. The directed segments $OQ$ and $OR$ are to be considered positive when measured on the terminal side of the angle and negative when measured on the terminal side extended.

62

**VARIATIONS OF THE TRIGONOMETRIC FUNCTIONS.** Let $P$ move counterclockwise about the unit circle, starting at $A$, so that $\theta = \angle XOP$ varies continuously from $0°$ to $360°$. Using the figures above, it is seen that ($I.$ = increases, $D.$ = decreases):

| As $\theta$ increases from | $0°$ to $90°$ | $90°$ to $180°$ | $180°$ to $270°$ | $270°$ to $360°$ |
|---|---|---|---|---|
| $\sin \theta$ | $I.$ from 0 to 1 | $D.$ from 1 to 0 | $D.$ from 0 to –1 | $I.$ from –1 to 0 |
| $\cos \theta$ | $D.$ from 1 to 0 | $D.$ from 0 to –1 | $I.$ from –1 to 0 | $I.$ from 0 to 1 |
| $\tan \theta$ | $I.$ from 0 without limit (0 to $+\infty$) | $I.$ from large negative values to 0. ($-\infty$ to 0) | $I.$ from 0 without limit (0 to $+\infty$) | $I.$ from large negative values to 0. ($-\infty$ to 0) |
| $\cot \theta$ | $D.$ from large positive values to 0. ($+\infty$ to 0) | $D.$ from 0 without limit (0 to $-\infty$) | $D.$ from large positive values to 0. ($+\infty$ to 0) | $D.$ from 0 without limit (0 to $-\infty$) |
| $\sec \theta$ | $I.$ from 1 without limit (1 to $+\infty$) | $I.$ from large negative values to –1. ($-\infty$ to –1) | $D.$ from –1 without limit (–1 to $-\infty$) | $D.$ from large positive values to 1. ($+\infty$ to 1) |
| $\csc \theta$ | $D.$ from large positive values to 1. ($+\infty$ to 1) | $I.$ from 1 without limit (1 to $+\infty$) | $I.$ from large negative values to –1. ($-\infty$ to –1) | $D.$ from –1 without limit (–1 to $-\infty$) |

**GRAPHS OF THE TRIGONOMETRIC FUNCTIONS.** In the following table, values of the angle $x$ are given in radians.

| $x$ | $y = \sin x$ | $y = \cos x$ | $y = \tan x$ | $y = \cot x$ | $y = \sec x$ | $y = \csc x$ |
|---|---|---|---|---|---|---|
| 0 | 0 | 1.00 | 0 | $\pm\infty$ | 1.00 | $\pm\infty$ |
| $\pi/6$ | 0.50 | 0.87 | 0.58 | 1.73 | 1.15 | 2.00 |
| $\pi/4$ | 0.71 | 0.71 | 1.00 | 1.00 | 1.41 | 1.41 |
| $\pi/3$ | 0.87 | 0.50 | 1.73 | 0.58 | 2.00 | 1.15 |
| $\pi/2$ | 1.00 | 0 | $\pm\infty$ | 0 | $\pm\infty$ | 1.00 |
| $2\pi/3$ | 0.87 | –0.50 | –1.73 | –0.58 | –2.00 | 1.15 |
| $3\pi/4$ | 0.71 | –0.71 | –1.00 | –1.00 | –1.41 | 1.41 |
| $5\pi/6$ | 0.50 | –0.87 | –0.58 | –1.73 | –1.15 | 2.00 |
| $\pi$ | 0 | –1.00 | 0 | $\pm\infty$ | –1.00 | $\pm\infty$ |
| $7\pi/6$ | –0.50 | –0.87 | 0.58 | 1.73 | –1.15 | –2.00 |
| $5\pi/4$ | –0.71 | –0.71 | 1.00 | 1.00 | –1.41 | –1.41 |
| $4\pi/3$ | –0.87 | –0.50 | 1.73 | 0.58 | –2.00 | –1.15 |
| $3\pi/2$ | –1.00 | 0 | $\pm\infty$ | 0 | $\pm\infty$ | –1.00 |
| $5\pi/3$ | –0.87 | 0.50 | –1.73 | –0.58 | 2.00 | –1.15 |
| $7\pi/4$ | –0.71 | 0.71 | –1.00 | –1.00 | 1.41 | –1.41 |
| $11\pi/6$ | –0.50 | 0.87 | –0.58 | –1.73 | 1.15 | –2.00 |
| $2\pi$ | 0 | 1.00 | 0 | $\pm\infty$ | 1.00 | $\pm\infty$ |

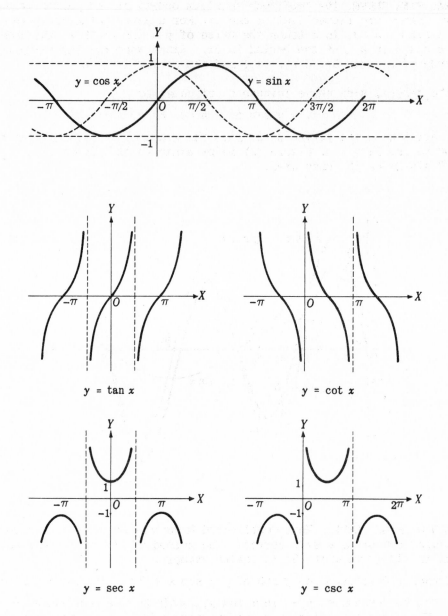

Note 1. Since $\sin(\tfrac{1}{2}\pi + x) = \cos x$, the graph of $y = \cos x$ may be obtained most easily by shifting the graph of $y = \sin x$ a distance $\tfrac{1}{2}\pi$ to the left.

Note 2. Since $\csc(\tfrac{1}{2}\pi + x) = \sec x$, the graph of $y = \csc x$ may be obtained by shifting the graph of $y = \sec x$ a distance $\tfrac{1}{2}\pi$ to the right.

PERIODIC FUNCTIONS.    Any function of a variable $x$, $f(x)$, which repeats its values in definite cycles, is called *periodic*.    The smallest range of values of $x$ which corresponds to a complete cycle of values of the function is called the period of the function.    It is evident from the graphs of the trigonometric functions that the sine, cosine, secant, and cosecant are of period $2\pi$ while the tangent and cotangent are of period $\pi$.

**THE GENERAL SINE CURVE.** The *amplitude* (maximum ordinate) and period (wave length) of $y = \sin x$ are respectively 1 and $2\pi$. For a given value of $x$, the value of $y = a \sin x$, $a > 0$, is $a$ times the value of $y = \sin x$. Thus, the amplitude of $y = a \sin x$ is $a$ and the period is $2\pi$. Since when $bx = 2\pi$, $x = 2\pi/b$, the amplitude of $y = \sin bx$, $b > 0$, is 1 and the period is $2\pi/b$.

The general sine curve (sinusoid) of equation

$$y = a \sin bx, \quad a > 0, \quad b > 0,$$

has amplitude $a$ and period $2\pi/b$. Thus the graph of $y = 3 \sin 2x$ has amplitude 3 and period $2\pi/2 = \pi$. Figure (a) below exhibits the graphs of $y = \sin x$ and $y = 3 \sin 2x$ on the same axes.

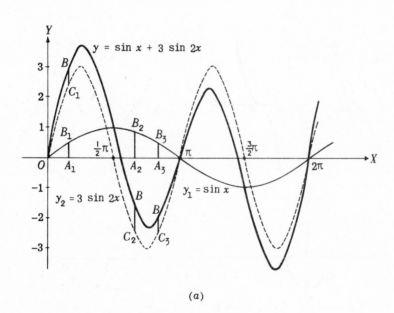

(a)

**COMPOSITION OF SINE CURVES.** More complicated forms of wave motions are obtained by combining two or more sine curves. The method of adding corresponding ordinates is illustrated in the following example.

EXAMPLE. Construct the graph of $y = \sin x + 3 \sin 2x$.    See Fig. (a) above.

First the graphs of $y_1 = \sin x$ and $y_2 = 3 \sin 2x$ are constructed on the same axes. Then, corresponding to a given value $x = OA_1$, the ordinate $A_1B$ of $y = \sin x + 3 \sin 2x$ is the *algebraic* sum of the ordinates $A_1B_1$ of $y_1 = \sin x$ and $A_1C_1$ of $y_2 = 3 \sin 2x$. Also, $A_2B = A_2B_2 + A_2C_2$, $A_3B = A_3B_3 + A_3C_3$, etc.

## SOLVED PROBLEMS

1. Sketch the graphs of the following for one wave length.

   a) $y = 4 \sin x$       c) $y = 3 \sin \frac{1}{2}x$       e) $y = 3 \cos \frac{1}{2}x = 3 \sin (\frac{1}{2}x + \frac{1}{2}\pi)$

   b) $y = \sin 3x$       d) $y = 2 \cos x = 2 \sin (x + \frac{1}{2}\pi)$

   In each case we use the same curve and then put in the y-axis and choose the units on each axis to satisfy the requirements of amplitude and period of each curve.

   a) $y = 4 \sin x$ has amplitude = 4 and period = $2\pi$.

   b) $y = \sin 3x$ has amplitude = 1 and period = $2\pi/3$.

   c) $y = 3 \sin \frac{1}{2}x$ has amplitude = 3 and period = $2\pi/\frac{1}{2} = 4\pi$.

   d) $y = 2 \cos x$ has amplitude = 2 and period = $2\pi$.  Note the position of the y-axis.

   e) $y = 3 \cos \frac{1}{2}x$ has amplitude = 3 and period = $4\pi$.

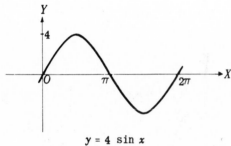

$y = 4 \sin x$

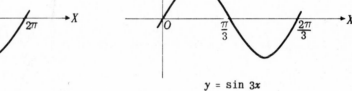

$y = \sin 3x$

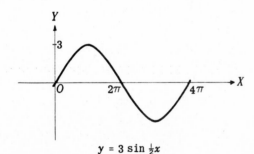

$y = 3 \sin \frac{1}{2}x$

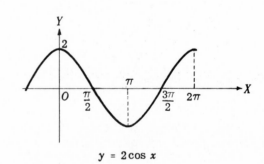

$y = 2\cos x$

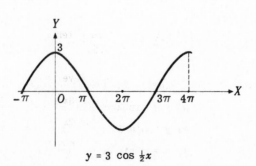

$y = 3 \cos \frac{1}{2}x$

**2.** Construct the graph of each of the following.

a) $y = \sin x + \cos x$     c) $y = \sin 2x - \cos 3x$

b) $y = \sin 2x + \cos 3x$     d) $y = 3 \sin 2x + 2 \cos 3x$

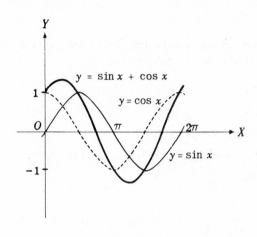

(a)

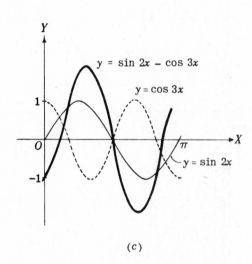

(c)

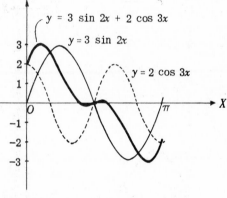

(b)

(d)

## SUPPLEMENTARY PROBLEMS

**3.** Sketch the graph of each of the following for one wave length: a) $y = 3 \sin x$, b) $y = \sin 2x$, c) $y = 4 \sin x/2$, d) $y = 4 \cos x$, e) $y = 2 \cos x/3$.

**4.** Construct the graph of each of the following for one wave length.

a) $y = \sin x + 2 \cos x$          d) $y = \sin 2x + \sin 3x$

b) $y = \sin 3x + \cos 2x$          e) $y = \sin 3x - \cos 2x$

c) $y = \sin x + \sin 2x$           f) $y = 2 \sin 3x + 3 \cos 2x$

# CHAPTER 10

# Fundamental Relations and Identities

**FUNDAMENTAL RELATIONS.**

| Reciprocal Relations | Quotient Relations | Pythagorean Relations |
|---|---|---|
| $\csc\theta = 1/\sin\theta$ | $\tan\theta = \sin\theta/\cos\theta$ | $\sin^2\theta + \cos^2\theta = 1$ |
| $\sec\theta = 1/\cos\theta$ | $\cot\theta = \cos\theta/\sin\theta$ | $1 + \tan^2\theta = \sec^2\theta$ |
| $\cot\theta = 1/\tan\theta$ | | $1 + \cot^2\theta = \csc^2\theta$ |

The above relations hold for every value of $\theta$ for which the functions involved are defined.

Thus, $\sin^2\theta + \cos^2\theta = 1$ holds for every value of $\theta$ while $\tan\theta = \sin\theta/\cos\theta$ holds for all values of $\theta$ for which $\tan\theta$ is defined, i.e., for all $\theta \neq n \cdot 90°$ where $n$ is odd. Note that for the excluded values of $\theta$, $\cos\theta = 0$ and $\sin\theta \neq 0$.

For proofs of the quotient and Pythagorean relations, see Problems 1, 2. The reciprocal relations were treated in Chapter 2. (See also Problems 3-6.)

**SIMPLIFICATION OF TRIGONOMETRIC EXPRESSIONS.** It is frequently desirable to transform or reduce a given expression involving trigonometric functions to a simpler form.

EXAMPLE 1. *a*) Using $\csc\theta = \dfrac{1}{\sin\theta}$, $\cos\theta \csc\theta = \cos\theta\,\dfrac{1}{\sin\theta} = \dfrac{\cos\theta}{\sin\theta} = \cot\theta.$

*b*) Using $\tan\theta = \dfrac{\sin\theta}{\cos\theta}$, $\cos\theta \tan\theta = \cos\theta\,\dfrac{\sin\theta}{\cos\theta} = \sin\theta.$

EXAMPLE 2. Using the relation $\sin^2\theta + \cos^2\theta = 1$,

*a*) $\sin^3\theta + \sin\theta\cos^2\theta = (\sin^2\theta + \cos^2\theta)\sin\theta = (1)\sin\theta = \sin\theta.$

*b*) $\dfrac{\cos^2\theta}{1 - \sin\theta} = \dfrac{1 - \sin^2\theta}{1 - \sin\theta} = \dfrac{(1 - \sin\theta)(1 + \sin\theta)}{1 - \sin\theta} = 1 + \sin\theta.$

Note. The relation $\sin^2\theta + \cos^2\theta = 1$ may be written as $\sin^2\theta = 1 - \cos^2\theta$ and as $\cos^2\theta = 1 - \sin^2\theta$. Each form is equally useful.

(See Problems 7-9.)

**TRIGONOMETRIC IDENTITIES.** A relation involving the trigonometric functions which is valid for all values of the angle for which the functions are defined is called a trigonometric identity. The eight fundamental relations above are trigonometric identities; so also are

$$\cos\theta \csc\theta = \cot\theta \qquad \text{and} \qquad \cos\theta \tan\theta = \sin\theta$$

of Example 1 above.

A trigonometric identity is verified by transforming one member (your choice) into the other. In general, one begins with the more complicated side.

Success in verifying identities requires:
*a)* complete familiarity with the fundamental relations,
*b)* complete familiarity with the processes of factoring, adding fractions, etc.,
*c)* practice.

(See Problems 10 – 18.)

## SOLVED PROBLEMS

**1.** Prove the quotient relations: $\tan\theta = \dfrac{\sin\theta}{\cos\theta}$, $\cot\theta = \dfrac{\cos\theta}{\sin\theta}$.

For any angle $\theta$, $\sin\theta = y/r$, $\cos\theta = x/r$, $\tan\theta = y/x$, and $\cot\theta = x/y$, where $P(x,y)$ is any point on the terminal side of $\theta$ at a distance $r$ from the origin.

Then $\tan\theta = \dfrac{y}{x} = \dfrac{y/r}{x/r} = \dfrac{\sin\theta}{\cos\theta}$ and $\cot\theta = \dfrac{x}{y} = \dfrac{x/r}{y/r} = \dfrac{\cos\theta}{\sin\theta}$. (Also, $\cot\theta = \dfrac{1}{\tan\theta} = \dfrac{\cos\theta}{\sin\theta}$.)

**2.** Prove the Pythagorean relations: *a)* $\sin^2\theta + \cos^2\theta = 1$, *b)* $1 + \tan^2\theta = \sec^2\theta$, *c)* $1 + \cot^2\theta = \csc^2\theta$.

For $P(x,y)$ defined as in Problem 1, we have *A)* $x^2 + y^2 = r^2$.

*a)* Dividing *A)* by $r^2$, $(x/r)^2 + (y/r)^2 = 1$ and $\sin^2\theta + \cos^2\theta = 1$.

*b)* Dividing *A)* by $x^2$, $1 + (y/x)^2 = (r/x)^2$ and $1 + \tan^2\theta = \sec^2\theta$.

Also, dividing $\sin^2\theta + \cos^2\theta = 1$ by $\cos^2\theta$, $\left(\dfrac{\sin\theta}{\cos\theta}\right)^2 + 1 = \left(\dfrac{1}{\cos\theta}\right)^2$ or $\tan^2\theta + 1 = \sec^2\theta$.

*c)* Dividing *A)* by $y^2$, $(x/y)^2 + 1 = (r/y)^2$ and $\cot^2\theta + 1 = \csc^2\theta$.

Also, dividing $\sin^2\theta + \cos^2\theta = 1$ by $\sin^2\theta$, $1 + \left(\dfrac{\cos\theta}{\sin\theta}\right)^2 = \left(\dfrac{1}{\sin\theta}\right)^2$ or $1 + \cot^2\theta = \csc^2\theta$.

**3.** Express each of the other functions of $\theta$ in terms of $\sin\theta$.

$$\cos^2\theta = 1 - \sin^2\theta \quad\text{and}\quad \cos\theta = \pm\sqrt{1 - \sin^2\theta},$$

$$\tan\theta = \frac{\sin\theta}{\cos\theta} = \frac{\sin\theta}{\pm\sqrt{1 - \sin^2\theta}}, \qquad \cot\theta = \frac{1}{\tan\theta} = \frac{\pm\sqrt{1 - \sin^2\theta}}{\sin\theta},$$

$$\sec\theta = \frac{1}{\cos\theta} = \frac{1}{\pm\sqrt{1 - \sin^2\theta}}, \qquad \csc\theta = \frac{1}{\sin\theta}.$$

Note that $\cos\theta = \pm\sqrt{1 - \sin^2\theta}$. Writing $\cos\theta = \sqrt{1 - \sin^2\theta}$ limits angle $\theta$ to those quadrants (first and fourth) in which the cosine is positive.

**4.** Express each of the other functions of $\theta$ in terms of $\tan\theta$.

$$\sec^2\theta = 1 + \tan^2\theta \quad\text{and}\quad \sec\theta = \pm\sqrt{1 + \tan^2\theta}, \qquad \cos\theta = \frac{1}{\sec\theta} = \frac{1}{\pm\sqrt{1 + \tan^2\theta}},$$

$$\frac{\sin\theta}{\cos\theta} = \tan\theta \quad\text{and}\quad \sin\theta = \tan\theta\cos\theta = \tan\theta\,\frac{1}{\pm\sqrt{1 + \tan^2\theta}} = \frac{\tan\theta}{\pm\sqrt{1 + \tan^2\theta}},$$

$$\csc\theta = \frac{1}{\sin\theta} = \frac{\pm\sqrt{1 + \tan^2\theta}}{\tan\theta}, \qquad \cot\theta = \frac{1}{\tan\theta}.$$

**5.** Using the fundamental relations, find the values of the functions of $\theta$, given $\sin \theta = 3/5$.

From $\cos^2\theta = 1 - \sin^2\theta$, $\cos \theta = \pm \sqrt{1 - \sin^2\theta} = \pm \sqrt{1 - (3/5)^2} = \pm \sqrt{16/25} = \pm 4/5$.

Now $\sin \theta$ and $\cos \theta$ are both positive when $\theta$ is a first quadrant angle while $\sin \theta = +$ and $\cos \theta = -$ when $\theta$ is a second quadrant angle. Thus,

| first quadrant | | second quadrant | |
|---|---|---|---|
| $\sin \theta = 3/5$ | $\cot \theta = 4/3$ | $\sin \theta = 3/5$ | $\cot \theta = -4/3$ |
| $\cos \theta = 4/5$ | $\sec \theta = 5/4$ | $\cos \theta = -4/5$ | $\sec \theta = -5/4$ |
| $\tan \theta = \dfrac{3/5}{4/5} = 3/4$ | $\csc \theta = 5/3$ | $\tan \theta = -3/4$ | $\csc \theta = 5/3.$ |

**6.** Using the fundamental relations, find the values of the functions of $\theta$, given $\tan \theta = -5/12$.

Since $\tan \theta = -$, $\theta$ is either a second or fourth quadrant angle.

| second quadrant | fourth quadrant |
|---|---|
| $\tan \theta = -5/12$ | $\tan \theta = -5/12$ |
| $\cot \theta = 1/\tan \theta = -12/5$ | $\cot \theta = -12/5$ |
| $\sec \theta = -\sqrt{1 + \tan^2\theta} = -13/12$ | $\sec \theta = 13/12$ |
| $\cos \theta = 1/\sec \theta = -12/13$ | $\cos \theta = 12/13$ |
| $\csc \theta = \sqrt{1 + \cot^2\theta} = 13/5$ | $\csc \theta = -13/5$ |
| $\sin \theta = 1/\csc \theta = 5/13$ | $\sin \theta = -5/13$ |

**7.** Perform the indicated operations.

*a)* $(\sin \theta - \cos \theta)(\sin \theta + \cos \theta) = \sin^2\theta - \cos^2\theta$

*b)* $(\sin A + \cos A)^2 = \sin^2 A + 2 \sin A \cos A + \cos^2 A$

*c)* $(\sin x + \cos y)(\sin y - \cos x) = \sin x \sin y - \sin x \cos x + \sin y \cos y - \cos x \cos y$

*d)* $(\tan^2 A - \cot A)^2 = \tan^4 A - 2 \tan^2 A \cot A + \cot^2 A$

*e)* $1 + \dfrac{\cos \theta}{\sin \theta} = \dfrac{\sin \theta + \cos \theta}{\sin \theta}$

*f)* $1 - \dfrac{\sin \theta}{\cos \theta} + \dfrac{2}{\cos^2\theta} = \dfrac{\cos^2\theta - \sin \theta \cos \theta + 2}{\cos^2\theta}$

**8.** Factor.

*a)* $\sin^2\theta - \sin \theta \cos \theta = \sin \theta (\sin \theta - \cos \theta)$

*b)* $\sin^2\theta + \sin^2\theta \cos^2\theta = \sin^2\theta (1 + \cos^2\theta)$

*c)* $\sin^2\theta + \sin \theta \sec \theta - 6 \sec^2\theta = (\sin \theta + 3 \sec \theta)(\sin \theta - 2 \sec \theta)$

*d)* $\sin^3\theta \cos^2\theta - \sin^2\theta \cos^3\theta + \sin \theta \cos^2\theta = \sin \theta \cos^2\theta (\sin^2\theta - \sin \theta \cos \theta + 1)$

*e)* $\sin^4\theta - \cos^4\theta = (\sin^2\theta + \cos^2\theta)(\sin^2\theta - \cos^2\theta) = (\sin^2\theta + \cos^2\theta)(\sin \theta - \cos \theta)(\sin \theta + \cos \theta)$

**9.** Simplify each of the following.

*a)* $\sec \theta - \sec \theta \sin^2\theta = \sec \theta (1 - \sin^2\theta) = \sec \theta \cos^2\theta = \dfrac{1}{\cos \theta} \cos^2\theta = \cos \theta$

b) $\sin\theta \sec\theta \cot\theta = \sin\theta \dfrac{1}{\cos\theta} \dfrac{\cos\theta}{\sin\theta} = \dfrac{\sin\theta\cos\theta}{\cos\theta\sin\theta} = 1$

c) $\sin^2\theta(1 + \cot^2\theta) = \sin^2\theta \csc^2\theta = \sin^2\theta \dfrac{1}{\sin^2\theta} = 1$

d) $\sin^2\theta \sec^2\theta - \sec^2\theta = (\sin^2\theta - 1)\sec^2\theta = -\cos^2\theta \sec^2\theta = -\cos^2\theta \dfrac{1}{\cos^2\theta} = -1$

e) $(\sin\theta + \cos\theta)^2 + (\sin\theta - \cos\theta)^2 = \sin^2\theta + 2\sin\theta\cos\theta + \cos^2\theta + \sin^2\theta$
$$- 2\sin\theta\cos\theta + \cos^2\theta = 2(\sin^2\theta + \cos^2\theta) = 2$$

f) $\tan^2\theta \cos^2\theta + \cot^2\theta \sin^2\theta = \dfrac{\sin^2\theta}{\cos^2\theta}\cos^2\theta + \dfrac{\cos^2\theta}{\sin^2\theta}\sin^2\theta = \sin^2\theta + \cos^2\theta = 1$

g) $\tan\theta + \dfrac{\cos\theta}{1 + \sin\theta} = \dfrac{\sin\theta}{\cos\theta} + \dfrac{\cos\theta}{1 + \sin\theta} = \dfrac{\sin\theta(1 + \sin\theta) + \cos^2\theta}{\cos\theta(1 + \sin\theta)}$

$$= \dfrac{\sin\theta + \sin^2\theta + \cos^2\theta}{\cos\theta(1 + \sin\theta)} = \dfrac{\sin\theta + 1}{\cos\theta(1 + \sin\theta)} = \dfrac{1}{\cos\theta} = \sec\theta$$

Verify the following identities.

**10.** $\sec^2\theta \csc^2\theta = \sec^2\theta + \csc^2\theta$

$\sec^2\theta + \csc^2\theta = \dfrac{1}{\cos^2\theta} + \dfrac{1}{\sin^2\theta} = \dfrac{\sin^2\theta + \cos^2\theta}{\sin^2\theta \cos^2\theta} = \dfrac{1}{\sin^2\theta \cos^2\theta} = \dfrac{1}{\sin^2\theta}\dfrac{1}{\cos^2\theta} = \csc^2\theta \sec^2\theta$

**11.** $\sec^4\theta - \sec^2\theta = \tan^4\theta + \tan^2\theta$

$\tan^4\theta + \tan^2\theta = \tan^2\theta(\tan^2\theta + 1) = \tan^2\theta \sec^2\theta = (\sec^2\theta - 1)\sec^2\theta = \sec^4\theta - \sec^2\theta$ **or**

$\sec^4\theta - \sec^2\theta = \sec^2\theta(\sec^2\theta - 1) = \sec^2\theta \tan^2\theta = (1 + \tan^2\theta)\tan^2\theta = \tan^2\theta + \tan^4\theta$

**12.** $2\csc x = \dfrac{\sin x}{1 + \cos x} + \dfrac{1 + \cos x}{\sin x}$

$\dfrac{\sin x}{1 + \cos x} + \dfrac{1 + \cos x}{\sin x} = \dfrac{\sin^2 x + (1 + \cos x)^2}{\sin x(1 + \cos x)} = \dfrac{\sin^2 x + 1 + 2\cos x + \cos^2 x}{\sin x(1 + \cos x)}$

$$= \dfrac{2 + 2\cos x}{\sin x(1 + \cos x)} = \dfrac{2(1 + \cos x)}{\sin x(1 + \cos x)} = \dfrac{2}{\sin x} = 2\csc x$$

**13.** $\dfrac{1 - \sin x}{\cos x} = \dfrac{\cos x}{1 + \sin x}$

$\dfrac{\cos x}{1 + \sin x} = \dfrac{\cos^2 x}{\cos x(1 + \sin x)} = \dfrac{1 - \sin^2 x}{\cos x(1 + \sin x)} = \dfrac{(1 - \sin x)(1 + \sin x)}{\cos x(1 + \sin x)} = \dfrac{1 - \sin x}{\cos x}$

**14.** $\dfrac{\sec A - \csc A}{\sec A + \csc A} = \dfrac{\tan A - 1}{\tan A + 1}$

$\dfrac{\sec A - \csc A}{\sec A + \csc A} = \dfrac{\dfrac{1}{\cos A} - \dfrac{1}{\sin A}}{\dfrac{1}{\cos A} + \dfrac{1}{\sin A}} = \dfrac{\dfrac{\sin A}{\cos A} - 1}{\dfrac{\sin A}{\cos A} + 1} = \dfrac{\tan A - 1}{\tan A + 1}$

**15.** $\dfrac{\tan x - \sin x}{\sin^3 x} = \dfrac{\sec x}{1 + \cos x}$

$$\frac{\tan x - \sin x}{\sin^3 x} = \frac{\dfrac{\sin x}{\cos x} - \sin x}{\sin^3 x} = \frac{\sin x - \sin x \cos x}{\cos x \, \sin^3 x} = \frac{\sin x (1 - \cos x)}{\cos x \, \sin^3 x}$$

$$= \frac{1 - \cos x}{\cos x \, \sin^2 x} = \frac{1 - \cos x}{\cos x \,(1 - \cos^2 x)} = \frac{1}{\cos x \,(1 + \cos x)} = \frac{\sec x}{1 + \cos x}$$

**16.** $\dfrac{\cos A \cot A - \sin A \tan A}{\csc A - \sec A} = 1 + \sin A \cos A$

$$\frac{\cos A \cot A - \sin A \tan A}{\csc A - \sec A} = \frac{\cos A \dfrac{\cos A}{\sin A} - \sin A \dfrac{\sin A}{\cos A}}{\dfrac{1}{\sin A} - \dfrac{1}{\cos A}} = \frac{\cos^3 A - \sin^3 A}{\cos A - \sin A}$$

$$= \frac{(\cos A - \sin A)(\cos^2 A + \cos A \sin A + \sin^2 A)}{\cos A - \sin A} = \cos^2 A + \cos A \sin A + \sin^2 A = 1 + \cos A \sin A$$

**17.** $\dfrac{\sin \theta - \cos \theta + 1}{\sin \theta + \cos \theta - 1} = \dfrac{\sin \theta + 1}{\cos \theta}$

$$\frac{\sin \theta + 1}{\cos \theta} = \frac{(\sin \theta + 1)(\sin \theta + \cos \theta - 1)}{\cos \theta \,(\sin \theta + \cos \theta - 1)} = \frac{\sin^2 \theta + \sin \theta \cos \theta + \cos \theta - 1}{\cos \theta \,(\sin \theta + \cos \theta - 1)}$$

$$= \frac{-\cos^2 \theta + \sin \theta \cos \theta + \cos \theta}{\cos \theta \,(\sin \theta + \cos \theta - 1)} = \frac{\cos \theta \,(\sin \theta - \cos \theta + 1)}{\cos \theta \,(\sin \theta + \cos \theta - 1)} = \frac{\sin \theta - \cos \theta + 1}{\sin \theta + \cos \theta - 1}$$

**18.** $\dfrac{\tan \theta + \sec \theta - 1}{\tan \theta - \sec \theta + 1} = \tan \theta + \sec \theta$

$$\frac{\tan \theta + \sec \theta - 1}{\tan \theta - \sec \theta + 1} = \frac{\tan \theta + \sec \theta + \tan^2 \theta - \sec^2 \theta}{\tan \theta - \sec \theta + 1} = \frac{(\tan \theta + \sec \theta)(1 + \tan \theta - \sec \theta)}{\tan \theta - \sec \theta + 1}$$

$$= \tan \theta + \sec \theta$$

or

$$\tan \theta + \sec \theta = (\tan \theta + \sec \theta) \frac{\tan \theta - \sec \theta + 1}{\tan \theta - \sec \theta + 1} = \frac{\tan^2 \theta - \sec^2 \theta + \tan \theta + \sec \theta}{\tan \theta - \sec \theta + 1}$$

$$= \frac{-1 + \tan \theta + \sec \theta}{\tan \theta - \sec \theta + 1}$$

Note. When expressed in terms of sin θ and cos θ, this identity becomes that of Problem 17.

## SUPPLEMENTARY PROBLEMS

19. Find the values of the trigonometric functions of $\theta$, given $\sin \theta = 2/3$.

   *Ans.*  Quad I :   $2/3$, $\sqrt{5}/3$, $2/\sqrt{5}$, $\sqrt{5}/2$, $3/\sqrt{5}$, $3/2$
   Quad II:  $2/3$, $-\sqrt{5}/3$, $-2/\sqrt{5}$, $-\sqrt{5}/2$, $-3/\sqrt{5}$, $3/2$

20. Find the values of the trigonometric functions of $\theta$, given $\cos \theta = -5/6$.

   *Ans.*  Quad II :  $\sqrt{11}/6$, $-5/6$, $-\sqrt{11}/5$, $-5/\sqrt{11}$, $-6/5$, $6/\sqrt{11}$
   Quad III:  $-\sqrt{11}/6$, $-5/6$, $\sqrt{11}/5$, $5/\sqrt{11}$, $-6/5$, $-6/\sqrt{11}$

21. Find the values of the trigonometric functions of $\theta$, given $\tan \theta = 5/4$.

   *Ans.*  Quad I:   $5/\sqrt{41}$, $4/\sqrt{41}$, $5/4$, $4/5$, $\sqrt{41}/4$, $\sqrt{41}/5$
   Quad III:  $-5/\sqrt{41}$, $-4/\sqrt{41}$, $5/4$, $4/5$, $-\sqrt{41}/4$, $-\sqrt{41}/5$

22. Find the values of the trigonometric functions of $\theta$, given $\cot \theta = -\sqrt{3}$.

   *Ans.*  Quad II:  $1/2$, $-\sqrt{3}/2$, $-1/\sqrt{3}$, $-\sqrt{3}$, $-2/\sqrt{3}$, $2$
   Quad IV:  $-1/2$, $\sqrt{3}/2$, $-1/\sqrt{3}$, $-\sqrt{3}$, $2/\sqrt{3}$, $-2$

23. Find the value of $\dfrac{\sin \theta + \cos \theta - \tan \theta}{\sec \theta + \csc \theta - \cot \theta}$ when $\tan \theta = -4/3$.

   *Ans.*  Quad II: $23/5$;  Quad IV: $34/35$

Verify the following identities.

24. $\sin \theta \sec \theta = \tan \theta$

25. $(1 - \sin^2 A)(1 + \tan^2 A) = 1$

26. $(1 - \cos \theta)(1 + \sec \theta) \cot \theta = \sin \theta$

27. $\csc^2 x \, (1 - \cos^2 x) = 1$

28. $\dfrac{\sin \theta}{\csc \theta} + \dfrac{\cos \theta}{\sec \theta} = 1$

29. $\dfrac{1 - 2\cos^2 A}{\sin A \cos A} = \tan A - \cot A$

30. $\tan^2 x \, \csc^2 x \, \cot^2 x \, \sin^2 x = 1$

31. $\sin A \cos A \, (\tan A + \cot A) = 1$

32. $1 - \dfrac{\cos^2 \theta}{1 + \sin \theta} = \sin \theta$

33. $\dfrac{1}{\sec \theta + \tan \theta} = \sec \theta - \tan \theta$

34. $\dfrac{1}{1 - \sin A} + \dfrac{1}{1 + \sin A} = 2\sec^2 A$

35. $\dfrac{1 - \cos x}{1 + \cos x} = \dfrac{\sec x - 1}{\sec x + 1} = (\cot x - \csc x)^2$

36. $\tan \theta \sin \theta + \cos \theta = \sec \theta$

37. $\tan \theta - \csc \theta \sec \theta \, (1 - 2\cos^2 \theta) = \cot \theta$

38. $\dfrac{\sin \theta}{\sin \theta + \cos \theta} = \dfrac{\sec \theta}{\sec \theta + \csc \theta}$

39. $\dfrac{\sin x + \tan x}{\cot x + \csc x} = \sin x \tan x$

40. $\dfrac{\sec x + \csc x}{\tan x + \cot x} = \sin x + \cos x$

41. $\dfrac{\sin^3 \theta + \cos^3 \theta}{\sin \theta + \cos \theta} = 1 - \sin \theta \cos \theta$

42. $\cot \theta + \dfrac{\sin \theta}{1 + \cos \theta} = \csc \theta$

43. $\dfrac{\sin \theta \cos \theta}{\cos^2 \theta - \sin^2 \theta} = \dfrac{\tan \theta}{1 - \tan^2 \theta}$

44. $(\tan x + \tan y)(1 - \cot x \cot y) + (\cot x + \cot y)(1 - \tan x \tan y) = 0$

45. $(x \sin \theta - y \cos \theta)^2 + (x \cos \theta + y \sin \theta)^2 = x^2 + y^2$

46. $(2r \sin \theta \cos \theta)^2 + r^2(\cos^2 \theta - \sin^2 \theta)^2 = r^2$

47. $(r \sin \theta \cos \phi)^2 + (r \sin \theta \sin \phi)^2 + (r \cos \theta)^2 = r^2$

# CHAPTER 11

# Trigonometric Functions of Two Angles

**ADDITION FORMULAS.**

$$\sin(\alpha + \beta) = \sin \alpha \cos \beta + \cos \alpha \sin \beta$$

$$\cos(\alpha + \beta) = \cos \alpha \cos \beta - \sin \alpha \sin \beta$$

$$\tan(\alpha + \beta) = \frac{\tan \alpha + \tan \beta}{1 - \tan \alpha \tan \beta}$$

For a proof of these formulas, see Problems 1-3.

**SUBTRACTION FORMULAS.**

$$\sin(\alpha - \beta) = \sin \alpha \cos \beta - \cos \alpha \sin \beta$$

$$\cos(\alpha - \beta) = \cos \alpha \cos \beta + \sin \alpha \sin \beta$$

$$\tan(\alpha - \beta) = \frac{\tan \alpha - \tan \beta}{1 + \tan \alpha \tan \beta}$$

For a proof of these formulas, see Problem 4.

**DOUBLE ANGLE FORMULAS.**

$$\sin 2\alpha = 2 \sin \alpha \cos \alpha$$

$$\cos 2\alpha = \cos^2\alpha - \sin^2\alpha = 1 - 2 \sin^2\alpha = 2 \cos^2\alpha - 1$$

$$\tan 2\alpha = \frac{2 \tan \alpha}{1 - \tan^2\alpha}$$

For a proof of these formulas, see Problem 10.

**HALF ANGLE FORMULAS.**

$$\sin \tfrac{1}{2}\theta = \pm \sqrt{\frac{1 - \cos \theta}{2}}$$

$$\cos \tfrac{1}{2}\theta = \pm \sqrt{\frac{1 + \cos \theta}{2}}$$

$$\tan \tfrac{1}{2}\theta = \pm \sqrt{\frac{1 - \cos \theta}{1 + \cos \theta}} = \frac{\sin \theta}{1 + \cos \theta} = \frac{1 - \cos \theta}{\sin \theta}$$

For a proof of these formulas, see Problem 11.

74

## SOLVED PROBLEMS

**1.** Prove  1)  $\sin(\alpha + \beta) = \sin\alpha\cos\beta + \cos\alpha\sin\beta$

and  2)  $\cos(\alpha + \beta) = \cos\alpha\cos\beta - \sin\alpha\sin\beta$   when $\alpha$ and $\beta$ are positive acute angles.

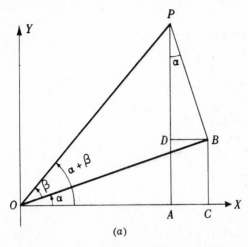

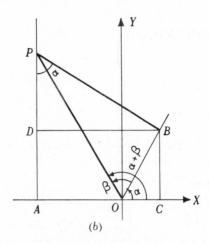

(a)                                    (b)

Let $\alpha$ and $\beta$ be positive acute angles such that  $\alpha + \beta < 90^\circ$  (Fig. $a$)  and  $\alpha + \beta > 90^\circ$ (Fig. $b$).

To construct these figures, place angle $\alpha$ in standard position and then place angle $\beta$ with its vertex at $O$ and with its initial side along the terminal side of angle $\alpha$. Let $P$ be any point on the terminal side of angle $(\alpha + \beta)$. Draw $PA$ perpendicular to $OX$, $PB$ perpendicular to the terminal side of angle $\alpha$, $BC$ perpendicular to $OX$, and $BD$ perpendicular to $AP$.

Now $\angle APB = \alpha$  since corresponding sides $(OA$ and $AP$, $OB$ and $BP)$ are perpendicular.  Then

$$\sin(\alpha + \beta) = \frac{AP}{OP} = \frac{AD + DP}{OP} = \frac{CB + DP}{OP} = \frac{CB}{OP} + \frac{DP}{OP} = \frac{CB}{OB}\cdot\frac{OB}{OP} + \frac{DP}{BP}\cdot\frac{BP}{OP}$$

$$= \sin\alpha\cos\beta + \cos\alpha\sin\beta$$

and  $$\cos(\alpha + \beta) = \frac{OA}{OP} = \frac{OC - AC}{OP} = \frac{OC - DB}{OP} = \frac{OC}{OP} - \frac{DB}{OP} = \frac{OC}{OB}\cdot\frac{OB}{OP} - \frac{DB}{BP}\cdot\frac{BP}{OP}$$

$$= \cos\alpha\cos\beta - \sin\alpha\sin\beta.$$

**2.** Show that 1) and 2) of Problem 1 are valid when $\alpha$ and $\beta$ are any angles.

First check the formulas for the case $\alpha = 0^\circ$, $\beta = 0^\circ$.  Since

$$\sin(0^\circ + 0^\circ) = \sin 0^\circ \cos 0^\circ + \cos 0^\circ \sin 0^\circ = 0\cdot 1 + 1\cdot 0 = 0 = \sin 0^\circ$$

and  $$\cos(0^\circ + 0^\circ) = \cos 0^\circ \cos 0^\circ - \sin 0^\circ \sin 0^\circ = 1\cdot 1 - 0\cdot 0 = 1 = \cos 0^\circ,$$

the formulas are valid for this case.

Next, it will be shown that if 1) and 2) are valid for any two given angles $\alpha$ and $\beta$, the formulas are also valid when, say, $\alpha$ is increased by $90^\circ$. Let $\alpha$ and $\beta$ be two angles for which 1) and 2) hold and consider

$a)$  $\sin(\alpha + \beta + 90^\circ) = \sin(\alpha + 90^\circ)\cos\beta + \cos(\alpha + 90^\circ)\sin\beta$

and  $b)$  $\cos(\alpha + \beta + 90^\circ) = \cos(\alpha + 90^\circ)\cos\beta - \sin(\alpha + 90^\circ)\sin\beta.$

By the reduction formulas of Chapter 8, $\sin(\theta + 90^\circ) = \cos\theta$, $\cos(\theta + 90^\circ) = -\sin\theta$, it follows that  $\sin(\alpha + \beta + 90^\circ) = \cos(\alpha + \beta)$, $\cos(\alpha + \beta + 90^\circ) = -\sin(\alpha + \beta)$.  Then $a)$ and $b)$ reduce to

$a'$)   $\cos(\alpha + \beta) \;=\; \cos\alpha\cos\beta + (-\sin\alpha)\sin\beta \;=\; \cos\alpha\cos\beta - \sin\alpha\sin\beta$

and   $b'$) $- \sin(\alpha + \beta) \;=\; -\sin\alpha\cos\beta - \cos\alpha\sin\beta$    or

$\sin(\alpha + \beta) \;=\; \sin\alpha\cos\beta + \cos\alpha\sin\beta$

which, by assumption, are valid relations. Thus, $a$) and $b$) are valid relations.

The same argument may be made to show that if 1) and 2) are valid for two angles $\alpha$ and $\beta$, they are also valid when $\beta$ is increased by $90^{\circ}$. Thus, the formulas are valid when both $\alpha$ and $\beta$ are increased by $90^{\circ}$. Now any positive angle can be expressed as a multiple of $90^{\circ}$ plus $\theta$, where $\theta$ is either $0^{\circ}$ or an acute angle. Thus, by a finite number of repetitions of the argument we show that the formulas are valid for any two given positive angles.

It will be left for the reader to carry through the argument when, instead of an increase, there is a decrease of $90^{\circ}$ and thus to show that 1) and 2) are valid when one angle is positive and the other negative, and when both are negative.

3. Prove:   $\tan(\alpha + \beta) \;=\; \dfrac{\tan\alpha + \tan\beta}{1 - \tan\alpha\tan\beta}$ .

$$\tan(\alpha + \beta) \;=\; \frac{\sin(\alpha + \beta)}{\cos(\alpha + \beta)} \;=\; \frac{\sin\alpha\cos\beta + \cos\alpha\sin\beta}{\cos\alpha\cos\beta - \sin\alpha\sin\beta}$$

$$=\; \frac{\dfrac{\sin\alpha\cos\beta}{\cos\alpha\cos\beta} + \dfrac{\cos\alpha\sin\beta}{\cos\alpha\cos\beta}}{\dfrac{\cos\alpha\cos\beta}{\cos\alpha\cos\beta} - \dfrac{\sin\alpha\sin\beta}{\cos\alpha\cos\beta}} \;=\; \frac{\tan\alpha + \tan\beta}{1 - \tan\alpha\tan\beta}$$

4. Prove the Subtraction formulas.

$\sin(\alpha - \beta) \;=\; \sin[\alpha + (-\beta)] \;=\; \sin\alpha\cos(-\beta) + \cos\alpha\sin(-\beta)$
            $=\; \sin\alpha\,(\cos\beta) + \cos\alpha\,(-\sin\beta) \;=\; \sin\alpha\cos\beta - \cos\alpha\sin\beta$

$\cos(\alpha - \beta) \;=\; \cos[\alpha + (-\beta)] \;=\; \cos\alpha\cos(-\beta) - \sin\alpha\sin(-\beta)$
            $=\; \cos\alpha\,(\cos\beta) - \sin\alpha\,(-\sin\beta) \;=\; \cos\alpha\cos\beta + \sin\alpha\sin\beta$

$\tan(\alpha - \beta) \;=\; \tan[\alpha + (-\beta)] \;=\; \dfrac{\tan\alpha + \tan(-\beta)}{1 - \tan\alpha\tan(-\beta)}$

$$=\; \frac{\tan\alpha + (-\tan\beta)}{1 - \tan\alpha\,(-\tan\beta)} \;=\; \frac{\tan\alpha - \tan\beta}{1 + \tan\alpha\tan\beta}$$

5. Find the values of sine, cosine, and tangent of $15^{\circ}$, using   ($a$) $15^{\circ} = 45^{\circ} - 30^{\circ}$   and   ($b$) $15^{\circ} = 60^{\circ} - 45^{\circ}$.

$a$)    $\sin 15^{\circ} \;=\; \sin(45^{\circ} - 30^{\circ}) \;=\; \sin 45^{\circ}\cos 30^{\circ} - \cos 45^{\circ}\sin 30^{\circ}$

$$=\; \frac{1}{\sqrt{2}}\cdot\frac{\sqrt{3}}{2} - \frac{1}{\sqrt{2}}\cdot\frac{1}{2} \;=\; \frac{\sqrt{3} - 1}{2\sqrt{2}} \;=\; \frac{\sqrt{2}}{4}(\sqrt{3} - 1)$$

$\cos 15^{\circ} \;=\; \cos(45^{\circ} - 30^{\circ}) \;=\; \cos 45^{\circ}\cos 30^{\circ} + \sin 45^{\circ}\sin 30^{\circ}$

$$=\; \frac{1}{\sqrt{2}}\cdot\frac{\sqrt{3}}{2} + \frac{1}{\sqrt{2}}\cdot\frac{1}{2} \;=\; \frac{\sqrt{2}}{4}(\sqrt{3} + 1)$$

$\tan 15^{\circ} \;=\; \tan(45^{\circ} - 30^{\circ}) \;=\; \dfrac{\tan 45^{\circ} - \tan 30^{\circ}}{1 + \tan 45^{\circ}\tan 30^{\circ}} \;=\; \dfrac{1 - 1/\sqrt{3}}{1 + 1(1/\sqrt{3})} \;=\; \dfrac{\sqrt{3} - 1}{\sqrt{3} + 1} \;=\; 2 - \sqrt{3}$

*b)* $\sin 15^{\circ} = \sin(60^{\circ} - 45^{\circ}) = \sin 60^{\circ} \cos 45^{\circ} - \cos 60^{\circ} \sin 45^{\circ} = \dfrac{\sqrt{3}}{2} \cdot \dfrac{1}{\sqrt{2}} - \dfrac{1}{2} \cdot \dfrac{1}{\sqrt{2}} = \dfrac{\sqrt{2}}{4}(\sqrt{3} - 1)$

$\cos 15^{\circ} = \cos(60^{\circ} - 45^{\circ}) = \cos 60^{\circ} \cos 45^{\circ} + \sin 60^{\circ} \sin 45^{\circ} = \dfrac{1}{2} \cdot \dfrac{1}{\sqrt{2}} + \dfrac{\sqrt{3}}{2} \cdot \dfrac{1}{\sqrt{2}} = \dfrac{\sqrt{2}}{4}(\sqrt{3} + 1)$

$\tan 15^{\circ} = \tan(60^{\circ} - 45^{\circ}) = \dfrac{\tan 60^{\circ} - \tan 45^{\circ}}{1 + \tan 60^{\circ} \tan 45^{\circ}} = \dfrac{\sqrt{3} - 1}{\sqrt{3} + 1} = 2 - \sqrt{3}$

**6.** Prove:    *a)* $\sin(45^{\circ} + \theta) - \sin(45^{\circ} - \theta) = \sqrt{2} \sin \theta$,    *b)* $\sin(30^{\circ} + \theta) + \cos(60^{\circ} + \theta) = \cos \theta$.

*a)* $\sin(45^{\circ} + \theta) - \sin(45^{\circ} - \theta) = (\sin 45^{\circ} \cos \theta + \cos 45^{\circ} \sin \theta) - (\sin 45^{\circ} \cos \theta - \cos 45^{\circ} \sin \theta)$
$$= 2 \cos 45^{\circ} \sin \theta = 2 \dfrac{1}{\sqrt{2}} \sin \theta = \sqrt{2} \sin \theta$$

*b)* $\sin(30^{\circ} + \theta) + \cos(60^{\circ} + \theta) = (\sin 30^{\circ} \cos \theta + \cos 30^{\circ} \sin \theta) + (\cos 60^{\circ} \cos \theta - \sin 60^{\circ} \sin \theta)$
$$= (\dfrac{1}{2} \cos \theta + \dfrac{\sqrt{3}}{2} \sin \theta) + (\dfrac{1}{2} \cos \theta - \dfrac{\sqrt{3}}{2} \sin \theta) = \cos \theta$$

**7.** Simplify:    *a)* $\sin(\alpha + \beta) + \sin(\alpha - \beta)$,    *b)* $\cos(\alpha + \beta) - \cos(\alpha - \beta)$,    *c)* $\dfrac{\tan(\alpha + \beta) - \tan \alpha}{1 + \tan(\alpha + \beta) \tan \alpha}$,
    *d)* $(\sin \alpha \cos \beta - \cos \alpha \sin \beta)^{2} + (\cos \alpha \cos \beta + \sin \alpha \sin \beta)^{2}$.

*a)* $\sin(\alpha + \beta) + \sin(\alpha - \beta) = (\sin \alpha \cos \beta + \cos \alpha \sin \beta) + (\sin \alpha \cos \beta - \cos \alpha \sin \beta)$
    $= 2 \sin \alpha \cos \beta$

*b)* $\cos(\alpha + \beta) - \cos(\alpha - \beta) = (\cos \alpha \cos \beta - \sin \alpha \sin \beta) - (\cos \alpha \cos \beta + \sin \alpha \sin \beta)$
    $= -2 \sin \alpha \sin \beta$

*c)* $\dfrac{\tan(\alpha + \beta) - \tan \alpha}{1 + \tan(\alpha + \beta) \tan \alpha} = \tan[(\alpha + \beta) - \alpha] = \tan \beta$

*d)* $(\sin \alpha \cos \beta - \cos \alpha \sin \beta)^{2} + (\cos \alpha \cos \beta + \sin \alpha \sin \beta)^{2} = \sin^{2}(\alpha - \beta) + \cos^{2}(\alpha - \beta) = 1$

**8.** Find $\sin(\alpha + \beta)$, $\cos(\alpha + \beta)$, $\sin(\alpha - \beta)$, $\cos(\alpha - \beta)$ and determine the quadrants in which $(\alpha + \beta)$ and $(\alpha - \beta)$ terminate, given
*a)* $\sin \alpha = 4/5$, $\cos \beta = 5/13$; $\alpha$ and $\beta$ in quadrant I.
*b)* $\sin \alpha = 2/3$, $\cos \beta = 3/4$; $\alpha$ in quadrant II, $\beta$ in quadrant IV.

*a)* $\cos \alpha = 3/5$ and $\sin \beta = 12/13$.

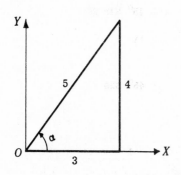
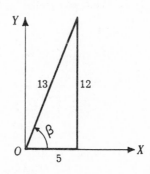

$$\sin(\alpha+\beta) \;=\; \sin\alpha\cos\beta + \cos\alpha\sin\beta \;=\; \frac{4}{5}\cdot\frac{5}{13} + \frac{3}{5}\cdot\frac{12}{13} \;=\; \frac{56}{65}$$

$(\alpha+\beta)$ in quadrant II

$$\cos(\alpha+\beta) \;=\; \cos\alpha\cos\beta - \sin\alpha\sin\beta \;=\; \frac{3}{5}\cdot\frac{5}{13} - \frac{4}{5}\cdot\frac{12}{13} \;=\; -\frac{33}{65}$$

$$\sin(\alpha-\beta) \;=\; \sin\alpha\cos\beta - \cos\alpha\sin\beta \;=\; \frac{4}{5}\cdot\frac{5}{13} - \frac{3}{5}\cdot\frac{12}{13} \;=\; -\frac{16}{65}$$

$(\alpha-\beta)$ in quadrant IV

$$\cos(\alpha-\beta) \;=\; \cos\alpha\cos\beta + \sin\alpha\sin\beta \;=\; \frac{3}{5}\cdot\frac{5}{13} + \frac{4}{5}\cdot\frac{12}{13} \;=\; \frac{63}{65}$$

*b*) $\cos\alpha = -\sqrt{5}/3$ and $\sin\beta = -\sqrt{7}/4$.

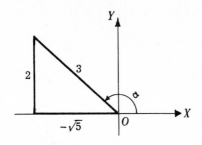
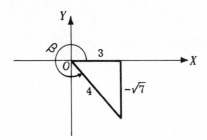

$$\sin(\alpha+\beta) \;=\; \sin\alpha\cos\beta + \cos\alpha\sin\beta \;=\; \frac{2}{3}\cdot\frac{3}{4} + \left(-\frac{\sqrt{5}}{3}\right)\left(-\frac{\sqrt{7}}{4}\right) \;=\; \frac{6+\sqrt{35}}{12}$$

$(\alpha+\beta)$ in quadrant II

$$\cos(\alpha+\beta) \;=\; \cos\alpha\cos\beta - \sin\alpha\sin\beta \;=\; \left(-\frac{\sqrt{5}}{3}\right)\frac{3}{4} - \frac{2}{3}\left(-\frac{\sqrt{7}}{4}\right) \;=\; \frac{-3\sqrt{5}+2\sqrt{7}}{12}$$

$$\sin(\alpha-\beta) \;=\; \sin\alpha\cos\beta - \cos\alpha\sin\beta \;=\; \frac{2}{3}\cdot\frac{3}{4} - \left(-\frac{\sqrt{5}}{3}\right)\left(-\frac{\sqrt{7}}{4}\right) \;=\; \frac{6-\sqrt{35}}{12}$$

$(\alpha-\beta)$ in quadrant II

$$\cos(\alpha-\beta) \;=\; \cos\alpha\cos\beta + \sin\alpha\sin\beta \;=\; \left(-\frac{\sqrt{5}}{3}\right)\frac{3}{4} + \frac{2}{3}\left(-\frac{\sqrt{7}}{4}\right) \;=\; \frac{-3\sqrt{5}-2\sqrt{7}}{12}$$

**9.** Prove:   *a*) $\cot(\alpha+\beta) = \dfrac{\cot\alpha\cot\beta - 1}{\cot\beta + \cot\alpha}$,    *b*) $\cot(\alpha-\beta) = \dfrac{\cot\alpha\cot\beta + 1}{\cot\beta - \cot\alpha}$.

*a*) $\cot(\alpha+\beta) \;=\; \dfrac{1}{\tan(\alpha+\beta)} \;=\; \dfrac{1-\tan\alpha\tan\beta}{\tan\alpha+\tan\beta} \;=\; \dfrac{1-\dfrac{1}{\cot\alpha\cot\beta}}{\dfrac{1}{\cot\alpha}+\dfrac{1}{\cot\beta}} \;=\; \dfrac{\cot\alpha\cot\beta - 1}{\cot\beta + \cot\alpha}$

*b*) $\cot(\alpha-\beta) \;=\; \cot[\alpha + (-\beta)] \;=\; \dfrac{\cot\alpha\cot(-\beta) - 1}{\cot(-\beta) + \cot\alpha} \;=\; \dfrac{-\cot\alpha\cot\beta - 1}{-\cot\beta + \cot\alpha} \;=\; \dfrac{\cot\alpha\cot\beta + 1}{\cot\beta - \cot\alpha}$

**10.** Prove the double angle formulas.

In   $\sin(\alpha+\beta) = \sin\alpha\cos\beta + \cos\alpha\sin\beta$,   $\cos(\alpha+\beta) = \cos\alpha\cos\beta - \sin\alpha\sin\beta$,   and   $\tan(\alpha+\beta) = \dfrac{\tan\alpha+\tan\beta}{1-\tan\alpha\tan\beta}$   put $\beta = \alpha$.   Then

$$\sin 2\alpha = \sin\alpha\cos\alpha + \cos\alpha\sin\alpha = 2\sin\alpha\cos\alpha,$$

$$\cos 2\alpha = \cos\alpha\cos\alpha - \sin\alpha\sin\alpha$$
$$= \cos^2\alpha - \sin^2\alpha \;=\; (1-\sin^2\alpha) - \sin^2\alpha \;=\; 1 - 2\sin^2\alpha$$
$$= \cos^2\alpha - (1-\cos^2\alpha) \;=\; 2\cos^2\alpha - 1,$$

$$\tan 2\alpha = \frac{\tan \alpha + \tan \alpha}{1 - \tan \alpha \tan \alpha} = \frac{2 \tan \alpha}{1 - \tan^2 \alpha}.$$

11. Prove the half angle formulas.

In $\cos 2\alpha = 1 - 2 \sin^2 \alpha$, put $\alpha = \frac{1}{2}\theta$. Then

$$\cos \theta = 1 - 2 \sin^2 \tfrac{1}{2}\theta, \qquad \sin^2 \tfrac{1}{2}\theta = \frac{1 - \cos \theta}{2} \qquad \text{and} \qquad \sin \tfrac{1}{2}\theta = \pm \sqrt{\frac{1 - \cos \theta}{2}}.$$

In $\cos 2\alpha = 2 \cos^2 \alpha - 1$, put $\alpha = \frac{1}{2}\theta$. Then

$$\cos \theta = 2 \cos^2 \tfrac{1}{2}\theta - 1, \qquad \cos^2 \tfrac{1}{2}\theta = \frac{1 + \cos \theta}{2} \qquad \text{and} \qquad \cos \tfrac{1}{2}\theta = \pm \sqrt{\frac{1 + \cos \theta}{2}}.$$

Finally, $\quad \tan \tfrac{1}{2}\theta = \dfrac{\sin \tfrac{1}{2}\theta}{\cos \tfrac{1}{2}\theta} = \pm \sqrt{\dfrac{1 - \cos \theta}{1 + \cos \theta}}$

$$= \pm \sqrt{\frac{(1 - \cos \theta)(1 + \cos \theta)}{(1 + \cos \theta)(1 + \cos \theta)}} = \pm \sqrt{\frac{1 - \cos^2 \theta}{(1 + \cos \theta)^2}} = \frac{\sin \theta}{1 + \cos \theta}$$

$$= \pm \sqrt{\frac{(1 - \cos \theta)(1 - \cos \theta)}{(1 + \cos \theta)(1 - \cos \theta)}} = \pm \sqrt{\frac{(1 - \cos \theta)^2}{1 - \cos^2 \theta}} = \frac{1 - \cos \theta}{\sin \theta}.$$

The signs $\pm$ are not needed here since $\tan \tfrac{1}{2}\theta$ and $\sin \theta$ always have the same sign (Problem 15, Chapter 8) and $1 - \cos \theta$ is always positive.

12. Using the half angle formulas, find the exact values of $a)$ $\sin 15^\circ$, $b)$ $\sin 292\tfrac{1}{2}^\circ$.

$a)$ $\sin 15^\circ = \sqrt{\dfrac{1 - \cos 30^\circ}{2}} = \sqrt{\dfrac{1 - \sqrt{3}/2}{2}} = \tfrac{1}{2}\sqrt{2 - \sqrt{3}}$

$b)$ $\sin 292\tfrac{1}{2}^\circ = -\sqrt{\dfrac{1 - \cos 585^\circ}{2}} = -\sqrt{\dfrac{1 - \cos 225^\circ}{2}} = -\sqrt{\dfrac{1 + 1/\sqrt{2}}{2}} = -\tfrac{1}{2}\sqrt{2 + \sqrt{2}}$

13. Find the values of sine, cosine, and tangent of $\tfrac{1}{2}\theta$, given $a)$ $\sin \theta = 5/13$, $\theta$ in quadrant II and $b)$ $\cos \theta = 3/7$, $\theta$ in quadrant IV.

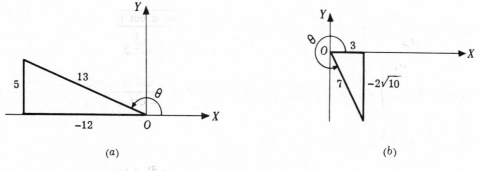

$(a)$ $\qquad\qquad\qquad\qquad\qquad\qquad (b)$

$a)$ $\sin \theta = 5/13$, $\cos \theta = -12/13$, and $\tfrac{1}{2}\theta$ in quadrant I.

$$\sin \tfrac{1}{2}\theta = \sqrt{\frac{1 - \cos \theta}{2}} = \sqrt{\frac{1 + 12/13}{2}} = \sqrt{\frac{25}{26}} = \frac{5\sqrt{26}}{26}$$

$$\cos \tfrac{1}{2}\theta = \sqrt{\frac{1 + \cos \theta}{2}} = \sqrt{\frac{1 - 12/13}{2}} = \sqrt{\frac{1}{26}} = \frac{\sqrt{26}}{26}$$

$$\tan \tfrac{1}{2}\theta \;=\; \frac{1 - \cos \theta}{\sin \theta} \;=\; \frac{1 + 12/13}{5/13} \;=\; 5$$

*b*) $\sin \theta = -2\sqrt{10}/7$,   $\cos \theta = 3/7$,   and   $\tfrac{1}{2}\theta$ in quadrant II.

$$\sin \tfrac{1}{2}\theta \;=\; \sqrt{\frac{1 - \cos \theta}{2}} \;=\; \sqrt{\frac{1 - 3/7}{2}} \;=\; \frac{\sqrt{14}}{7}$$

$$\cos \tfrac{1}{2}\theta \;=\; -\sqrt{\frac{1 + \cos \theta}{2}} \;=\; -\sqrt{\frac{1 + 3/7}{2}} \;=\; -\frac{\sqrt{35}}{7}, \qquad \tan \tfrac{1}{2}\theta \;=\; \frac{1 - \cos \theta}{\sin \theta} \;=\; \frac{1 - 3/7}{-2\sqrt{10}/7} \;=\; -\frac{\sqrt{10}}{5}$$

**14.** Show that:   *a*) $\sin \theta = 2 \sin \tfrac{1}{2}\theta \cos \tfrac{1}{2}\theta$      *d*) $\cos 6\theta = 1 - 2 \sin^2 3\theta$

              *b*) $\sin A = \pm \sqrt{\dfrac{1 - \cos 2A}{2}}$      *e*) $\sin^2 \tfrac{1}{2}\theta = \tfrac{1}{2}(1 - \cos \theta)$,   $\cos^2 \tfrac{1}{2}\theta = \tfrac{1}{2}(1 + \cos \theta)$.

              *c*) $\tan 4x = \dfrac{\sin 8x}{1 + \cos 8x}$

*a*) This is obtained from   $\sin 2\alpha = 2 \sin \alpha \cos \alpha$   by putting $\alpha = \tfrac{1}{2}\theta$.

*b*) This is obtained from   $\sin \tfrac{1}{2}\theta = \pm \sqrt{\dfrac{1 - \cos \theta}{2}}$   by putting $\theta = 2A$.

*c*) This is obtained from   $\tan \tfrac{1}{2}\theta = \dfrac{\sin \theta}{1 + \cos \theta}$   by putting $\theta = 8x$.

*d*) This is obtained from   $\cos 2\alpha = 1 - 2 \sin^2\alpha$   by putting $\alpha = 3\theta$.

*e*) These formulas are obtained by squaring   $\sin \tfrac{1}{2}\theta = \pm\sqrt{\dfrac{1 - \cos \theta}{2}}$   and   $\cos \tfrac{1}{2}\theta = \pm\sqrt{\dfrac{1 + \cos \theta}{2}}$.

**15.** Express   *a*) $\sin 3\alpha$ in terms of $\sin \alpha$,   *b*) $\cos 4\alpha$ in terms of $\cos \alpha$.

*a*) $\sin 3\alpha = \sin(2\alpha + \alpha) = \sin 2\alpha \cos \alpha + \cos 2\alpha \sin \alpha$

         $= (2 \sin \alpha \cos \alpha)\cos \alpha + (1 - 2 \sin^2\alpha)\sin \alpha \; = \; 2 \sin \alpha \cos^2\alpha + (1 - 2 \sin^2\alpha)\sin \alpha$

         $= 2 \sin \alpha (1 - \sin^2\alpha) + (1 - 2 \sin^2\alpha)\sin \alpha \; = \; 3 \sin \alpha - 4 \sin^3\alpha$.

*b*) $\cos 4\alpha = \cos 2(2\alpha) = 2 \cos^2 2\alpha - 1 = 2(2 \cos^2\alpha - 1)^2 - 1 = 8 \cos^4\alpha - 8 \cos^2\alpha + 1$.

**16.** Prove   $\cos 2x = \cos^4 x - \sin^4 x$.

     $\cos^4 x - \sin^4 x = (\cos^2 x + \sin^2 x)(\cos^2 x - \sin^2 x) = \cos^2 x - \sin^2 x = \cos 2x$

**17.** Prove   $1 - \tfrac{1}{2} \sin 2x = \dfrac{\sin^3 x + \cos^3 x}{\sin x + \cos x}$.

$$\frac{\sin^3 x + \cos^3 x}{\sin x + \cos x} \;=\; \frac{(\sin x + \cos x)(\sin^2 x - \sin x \cos x + \cos^2 x)}{\sin x + \cos x}$$

         $= 1 - \sin x \cos x = 1 - \tfrac{1}{2}(2 \sin x \cos x) = 1 - \tfrac{1}{2} \sin 2x$

**18.** Prove   $\cos \theta = \sin(\theta + 30^\circ) + \cos(\theta + 60^\circ)$.

     $\sin(\theta + 30^\circ) + \cos(\theta + 60^\circ) = (\sin \theta \cos 30^\circ + \cos \theta \sin 30^\circ) + (\cos \theta \cos 60^\circ - \sin \theta \sin 60^\circ)$

$$= \frac{\sqrt{3}}{2} \sin \theta + \frac{1}{2} \cos \theta + \frac{1}{2} \cos \theta - \frac{\sqrt{3}}{2} \sin \theta = \cos \theta$$

**19.** Prove $\cos x = \dfrac{1 - \tan^2 \frac{1}{2}x}{1 + \tan^2 \frac{1}{2}x}$.

$$\frac{1 - \tan^2 \frac{1}{2}x}{1 + \tan^2 \frac{1}{2}x} = \frac{1 - \dfrac{\sin^2 \frac{1}{2}x}{\cos^2 \frac{1}{2}x}}{\sec^2 \frac{1}{2}x} = \frac{\left(1 - \dfrac{\sin^2 \frac{1}{2}x}{\cos^2 \frac{1}{2}x}\right)\cos^2 \frac{1}{2}x}{\sec^2 \frac{1}{2}x \cos^2 \frac{1}{2}x} = \cos^2 \tfrac{1}{2}x - \sin^2 \tfrac{1}{2}x = \cos x$$

**20.** Prove $2 \tan 2x = \dfrac{\cos x + \sin x}{\cos x - \sin x} - \dfrac{\cos x - \sin x}{\cos x + \sin x}$.

$$\frac{\cos x + \sin x}{\cos x - \sin x} - \frac{\cos x - \sin x}{\cos x + \sin x} = \frac{(\cos x + \sin x)^2 - (\cos x - \sin x)^2}{(\cos x - \sin x)(\cos x + \sin x)}$$

$$= \frac{(\cos^2 x + 2 \sin x \cos x + \sin^2 x) - (\cos^2 x - 2 \sin x \cos x + \sin^2 x)}{\cos^2 x - \sin^2 x}$$

$$= \frac{4 \sin x \cos x}{\cos^2 x - \sin^2 x} = \frac{2 \sin 2x}{\cos 2x} = 2 \tan 2x$$

**21.** Prove $\sin^4 A = \dfrac{3}{8} - \dfrac{1}{2} \cos 2A + \dfrac{1}{8} \cos 4A$.

$$\sin^4 A = (\sin^2 A)^2 = \left(\frac{1 - \cos 2A}{2}\right)^2 = \frac{1 - 2 \cos 2A + \cos^2 2A}{4}$$

$$= \frac{1}{4}\left(1 - 2 \cos 2A + \frac{1 + \cos 4A}{2}\right) = \frac{3}{8} - \frac{1}{2} \cos 2A + \frac{1}{8} \cos 4A$$

**22.** Prove $\tan^6 x = \tan^4 x \sec^2 x - \tan^2 x \sec^2 x + \sec^2 x - 1$.

$$\tan^6 x = \tan^4 x \tan^2 x = \tan^4 x(\sec^2 x - 1) = \tan^4 x \sec^2 x - \tan^2 x \tan^2 x$$

$$= \tan^4 x \sec^2 x - \tan^2 x(\sec^2 x - 1) = \tan^4 x \sec^2 x - \tan^2 x \sec^2 x + \tan^2 x$$

$$= \tan^4 x \sec^2 x - \tan^2 x \sec^2 x + \sec^2 x - 1$$

**23.** When $A + B + C = 180°$, show that $\sin 2A + \sin 2B + \sin 2C = 4 \sin A \sin B \sin C$.

Since $C = 180° - (A + B)$,

$$\begin{aligned}
\sin 2A + \sin 2B + \sin 2C &= \sin 2A + \sin 2B + \sin[360° - 2(A + B)] \\
&= \sin 2A + \sin 2B - \sin 2(A + B) \\
&= \sin 2A + \sin 2B - \sin 2A \cos 2B - \cos 2A \sin 2B \\
&= (\sin 2A)(1 - \cos 2B) + (\sin 2B)(1 - \cos 2A) \\
&= 2 \sin 2A \sin^2 B + 2 \sin 2B \sin^2 A \\
&= 4 \sin A \cos A \sin^2 B + 4 \sin B \cos B \sin^2 A \\
&= 4 \sin A \sin B [\sin A \cos B + \cos A \sin B] \\
&= 4 \sin A \sin B \sin(A + B) \\
&= 4 \sin A \sin B \sin[180° - (A + B)] = 4 \sin A \sin B \sin C.
\end{aligned}$$

**24.** When $A + B + C = 180^\circ$, show that $\tan A + \tan B + \tan C = \tan A \tan B \tan C$.

Since $C = 180^\circ - (A + B)$,

$$\tan A + \tan B + \tan C = \tan A + \tan B + \tan[180^\circ - (A+B)] = \tan A + \tan B - \tan(A+B)$$

$$= \tan A + \tan B - \frac{\tan A + \tan B}{1 - \tan A \tan B} = (\tan A + \tan B)(1 - \frac{1}{1 - \tan A \tan B})$$

$$= (\tan A + \tan B)(- \frac{\tan A \tan B}{1 - \tan A \tan B}) = - \tan A \tan B \frac{\tan A + \tan B}{1 - \tan A \tan B}$$

$$= - \tan A \tan B \tan(A+B) = \tan A \tan B \tan[180^\circ - (A+B)] = \tan A \tan B \tan C.$$

## SUPPLEMENTARY PROBLEMS

**25.** Find the values of sine, cosine, and tangent of a) $75^\circ$, b) $255^\circ$.

Ans.  a) $\frac{\sqrt{2}}{4}(\sqrt{3}+1)$, $\frac{\sqrt{2}}{4}(\sqrt{3}-1)$, $2+\sqrt{3}$      b) $-\frac{\sqrt{2}}{4}(\sqrt{3}+1)$, $-\frac{\sqrt{2}}{4}(\sqrt{3}-1)$, $2+\sqrt{3}$

**26.** Find the values of $\sin(\alpha + \beta)$, $\cos(\alpha + \beta)$, and $\tan(\alpha + \beta)$, given:

a) $\sin \alpha = 3/5$, $\cos \beta = 5/13$, $\alpha$ and $\beta$ in Quadrant I.     Ans. $63/65$, $-16/65$, $-63/16$

b) $\sin \alpha = 8/17$, $\tan \beta = 5/12$, $\alpha$ and $\beta$ in Quadrant I.     Ans. $171/221$, $140/221$, $171/140$

c) $\cos \alpha = -12/13$, $\cot \beta = 24/7$, $\alpha$ in Quadrant II, $\beta$ in Quadrant III.

Ans. $-36/325$, $323/325$, $-36/323$

d) $\sin \alpha = 1/3$, $\sin \beta = 2/5$, $\alpha$ in Quadrant I, $\beta$ in Quadrant II.

Ans. $\dfrac{4\sqrt{2} - \sqrt{21}}{15}$, $-\dfrac{2 + 2\sqrt{42}}{15}$, $-\dfrac{4\sqrt{2} - \sqrt{21}}{2 + 2\sqrt{42}}$

**27.** Find the values of $\sin(\alpha - \beta)$, $\cos(\alpha - \beta)$, and $\tan(\alpha - \beta)$, given:

a) $\sin \alpha = 3/5$, $\sin \beta = 5/13$, $\alpha$ and $\beta$ in Quadrant I.     Ans. $16/65$, $63/65$, $16/63$

b) $\sin \alpha = 8/17$, $\tan \beta = 5/12$, $\alpha$ and $\beta$ in Quadrant I.     Ans. $21/221$, $220/221$, $21/220$

c) $\cos \alpha = -12/13$, $\cot \beta = 24/7$, $\alpha$ in Quadrant II, $\beta$ in Quadrant I.

Ans. $204/325$, $-253/325$, $-204/253$

d) $\sin \alpha = 1/3$, $\sin \beta = 2/5$, $\alpha$ in Quadrant II, $\beta$ in Quadrant I.

Ans. $\dfrac{4\sqrt{2} + \sqrt{21}}{15}$, $-\dfrac{2\sqrt{42} - 2}{15}$, $-\dfrac{4\sqrt{2} + \sqrt{21}}{2\sqrt{42} - 2}$

**28.** Prove:

a) $\sin(\alpha + \beta) - \sin(\alpha - \beta) = 2 \cos \alpha \sin \beta$

b) $\cos(\alpha + \beta) + \cos(\alpha - \beta) = 2 \cos \alpha \cos \beta$

f) $\dfrac{\sin(x + y)}{\cos(x - y)} = \dfrac{\tan x + \tan y}{1 + \tan x \tan y}$

c) $\tan(45^\circ - \theta) = \dfrac{1 - \tan \theta}{1 + \tan \theta}$

g) $\tan(45^\circ + \theta) = \dfrac{\cos \theta + \sin \theta}{\cos \theta - \sin \theta}$

d) $\dfrac{\tan(\alpha + \beta)}{\cot(\alpha - \beta)} = \dfrac{\tan^2\alpha - \tan^2\beta}{1 - \tan^2\alpha \tan^2\beta}$

h) $\sin(\alpha + \beta) \sin(\alpha - \beta) = \sin^2\alpha - \sin^2\beta$

e) $\tan(\alpha + \beta + \gamma) = \tan[(\alpha+\beta) + \gamma] = \dfrac{\tan \alpha + \tan \beta + \tan \gamma - \tan \alpha \tan \beta \tan \gamma}{1 - \tan \alpha \tan \beta - \tan \beta \tan \gamma - \tan \gamma \tan \alpha}$

**29.** If $A$ and $B$ are acute angles, find $A+B$ given:

    $a)$ $\tan A = 1/4$, $\tan B = 3/5$. Hint: $\tan(A+B) = 1$.     *Ans.* $45^{\circ}$

    $b)$ $\tan A = 5/3$, $\tan B = 4$.     *Ans.* $135^{\circ}$

**30.** If $\tan(x+y) = 33$ and $\tan x = 3$, show that $\tan y = 0.3$.

**31.** Find the values of $\sin 2\theta$, $\cos 2\theta$, and $\tan 2\theta$, given:

    $a)$ $\sin \theta = 3/5$, $\theta$ in Quadrant I.     *Ans.* $24/25$, $7/25$, $24/7$

    $b)$ $\sin \theta = 3/5$, $\theta$ in Quadrant II.     *Ans.* $-24/25$, $7/25$, $-24/7$

    $c)$ $\sin \theta = -1/2$, $\theta$ in Quadrant IV.     *Ans.* $-\sqrt{3}/2$, $1/2$, $-\sqrt{3}$

    $d)$ $\tan \theta = -1/5$, $\theta$ in Quadrant II.     *Ans.* $-5/13$, $12/13$, $-5/12$

    $e)$ $\tan \theta = u$, $\theta$ in Quadrant I.     *Ans.* $\dfrac{2u}{1+u^2}$, $\dfrac{1-u^2}{1+u^2}$, $\dfrac{2u}{1-u^2}$

**32.** Prove:

    $a)$ $\tan \theta \sin 2\theta = 2 \sin^2\theta$

    $b)$ $\cot \theta \sin 2\theta = 1 + \cos 2\theta$

    $c)$ $\dfrac{\sin^3 x - \cos^3 x}{\sin x - \cos x} = 1 + \frac{1}{2}\sin 2x$

    $d)$ $\dfrac{1 - \sin 2A}{\cos 2A} = \dfrac{1 - \tan A}{1 + \tan A}$

    $e)$ $\cos 2\theta = \dfrac{1 - \tan^2\theta}{1 + \tan^2\theta}$

    $f)$ $\dfrac{1 + \cos 2\theta}{\sin 2\theta} = \cot \theta$

    $g)$ $\cos 3\theta = 4 \cos^3\theta - 3 \cos \theta$

    $h)$ $\cos^4 x = \dfrac{3}{8} + \dfrac{1}{2}\cos 2x + \dfrac{1}{8}\cos 4x$

**33.** Find the values of sine, cosine, and tangent of

    $a)$ $30^{\circ}$, given $\cos 60^{\circ} = 1/2$.     *Ans.* $1/2$, $\sqrt{3}/2$, $1/\sqrt{3}$

    $b)$ $105^{\circ}$, given $\cos 210^{\circ} = -\sqrt{3}/2$     *Ans.* $\frac{1}{2}\sqrt{2+\sqrt{3}}$, $-\frac{1}{2}\sqrt{2-\sqrt{3}}$, $-(2+\sqrt{3})$

    $c)$ $\frac{1}{2}\theta$, given $\sin \theta = 3/5$, $\theta$ in Quadrant I.     *Ans.* $1/\sqrt{10}$, $3/\sqrt{10}$, $1/3$

    $d)$ $\theta$, given $\cot 2\theta = 7/24$, $2\theta$ in Quadrant I.     *Ans.* $3/5$, $4/5$, $3/4$

    $e)$ $\theta$, given $\cot 2\theta = -5/12$, $2\theta$ in Quadrant II.     *Ans.* $3/\sqrt{13}$, $2/\sqrt{13}$, $3/2$

**34.** Prove:

    $a)$ $\cos x = 2\cos^2\frac{1}{2}x - 1 = 1 - 2\sin^2\frac{1}{2}x$

    $b)$ $\sin x = 2\sin\frac{1}{2}x\cos\frac{1}{2}x$

    $c)$ $(\sin\frac{1}{2}\theta - \cos\frac{1}{2}\theta)^2 = 1 - \sin \theta$

    $d)$ $\tan\frac{1}{2}\theta = \csc \theta - \cot \theta$

    $e)$ $\dfrac{1 - \tan\frac{1}{2}\theta}{1 + \tan\frac{1}{2}\theta} = \dfrac{1 - \sin \theta}{\cos \theta} = \dfrac{\cos \theta}{1 + \sin \theta}$

    $f)$ $\dfrac{2\tan\frac{1}{2}x}{1 + \tan^2\frac{1}{2}x} = \sin x$

**35.** In the right triangle $ABC$, in which $C$ is the right angle, prove:

$$\sin 2A = \frac{2ab}{c^2}, \quad \cos 2A = \frac{b^2 - a^2}{c^2}, \quad \sin\tfrac{1}{2}A = \sqrt{\frac{c - b}{2c}}, \quad \cos\tfrac{1}{2}A = \sqrt{\frac{c + b}{2c}}.$$

**36.** Prove: $a)$ $\dfrac{\sin 3x}{\sin x} - \dfrac{\cos 3x}{\cos x} = 2$,     $b)$ $\tan 50^{\circ} - \tan 40^{\circ} = 2\tan 10^{\circ}$.

**37.** If $A+B+C = 180^{\circ}$, prove:

    $a)$ $\sin A + \sin B + \sin C = 4\cos\frac{1}{2}A\cos\frac{1}{2}B\cos\frac{1}{2}C$

    $b)$ $\cos A + \cos B + \cos C = 1 + 4\sin\frac{1}{2}A\sin\frac{1}{2}B\sin\frac{1}{2}C$

    $c)$ $\sin^2 A + \sin^2 B - \sin^2 C = 2\sin A\sin B\cos C$

    $d)$ $\tan\frac{1}{2}A\tan\frac{1}{2}B + \tan\frac{1}{2}B\tan\frac{1}{2}C + \tan\frac{1}{2}C\tan\frac{1}{2}A = 1$.

# CHAPTER 12

# Sum, Difference, and Product Formulas

**PRODUCTS OF SINES AND COSINES.**

$$\sin \alpha \cos \beta = \tfrac{1}{2}[\sin(\alpha+\beta) + \sin(\alpha-\beta)]$$
$$\cos \alpha \sin \beta = \tfrac{1}{2}[\sin(\alpha+\beta) - \sin(\alpha-\beta)]$$
$$\cos \alpha \cos \beta = \tfrac{1}{2}[\cos(\alpha+\beta) + \cos(\alpha-\beta)]$$
$$\sin \alpha \sin \beta = -\tfrac{1}{2}[\cos(\alpha+\beta) - \cos(\alpha-\beta)]$$

For proofs of these formulas, see Problem 1.

**SUM AND DIFFERENCE OF SINES AND COSINES.**

$$\sin A + \sin B = 2 \sin \tfrac{1}{2}(A+B) \cos \tfrac{1}{2}(A-B)$$
$$\sin A - \sin B = 2 \cos \tfrac{1}{2}(A+B) \sin \tfrac{1}{2}(A-B)$$
$$\cos A + \cos B = 2 \cos \tfrac{1}{2}(A+B) \cos \tfrac{1}{2}(A-B)$$
$$\cos A - \cos B = -2 \sin \tfrac{1}{2}(A+B) \sin \tfrac{1}{2}(A-B)$$

For proofs of these formulas, see Problem 2.

## SOLVED PROBLEMS

1. Derive the product formulas.

Since $\sin(\alpha+\beta) + \sin(\alpha-\beta) = (\sin \alpha \cos \beta + \cos \alpha \cos \beta) + (\sin \alpha \cos \beta - \cos \alpha \sin \beta)$
$= 2 \sin \alpha \cos \beta,$

$$\sin \alpha \cos \beta = \tfrac{1}{2}[\sin(\alpha+\beta) + \sin(\alpha-\beta)].$$

Since $\sin(\alpha+\beta) - \sin(\alpha-\beta) = 2 \cos \alpha \sin \beta,$

$$\cos \alpha \sin \beta = \tfrac{1}{2}[\sin(\alpha+\beta) - \sin(\alpha-\beta)].$$

Since $\cos(\alpha+\beta) + \cos(\alpha-\beta) = (\cos \alpha \cos \beta - \sin \alpha \sin \beta) + (\cos \alpha \cos \beta + \sin \alpha \sin \beta)$
$= 2 \cos \alpha \cos \beta,$

$$\cos \alpha \cos \beta = \tfrac{1}{2}[\cos(\alpha+\beta) + \cos(\alpha-\beta)].$$

Since $\cos(\alpha+\beta) - \cos(\alpha-\beta) = -2 \sin \alpha \sin \beta,$

$$\sin \alpha \sin \beta = -\tfrac{1}{2}[\cos(\alpha+\beta) - \cos(\alpha-\beta)].$$

2. Derive the sum and difference formulas.

Let $\alpha+\beta = A$ and $\alpha-\beta = B$ so that $\alpha = \tfrac{1}{2}(A+B)$ and $\beta = \tfrac{1}{2}(A-B)$. Then (see Problem 1)

$\sin(\alpha+\beta) + \sin(\alpha-\beta) = 2 \sin \alpha \cos \beta$ becomes $\sin A + \sin B = 2 \sin \tfrac{1}{2}(A+B) \cos \tfrac{1}{2}(A-B),$

$\sin(\alpha+\beta) - \sin(\alpha-\beta) = 2 \cos \alpha \sin \beta$    becomes    $\sin A - \sin B = 2 \cos \frac{1}{2}(A+B) \sin \frac{1}{2}(A-B)$,

$\cos(\alpha+\beta) + \cos(\alpha-\beta) = 2 \cos \alpha \cos \beta$    becomes    $\cos A + \cos B = 2 \cos \frac{1}{2}(A+B) \cos \frac{1}{2}(A-B)$,

$\cos(\alpha+\beta) - \cos(\alpha-\beta) = -2 \sin \alpha \cos \beta$    becomes    $\cos A - \cos B = -2 \sin \frac{1}{2}(A+B) \sin \frac{1}{2}(A-B)$.

**3.** Express each of the following as a sum or difference:
    a) $\sin 40^\circ \cos 30^\circ$,   b) $\cos 110^\circ \sin 55^\circ$,   c) $\cos 50^\circ \cos 35^\circ$,   d) $\sin 55^\circ \sin 40^\circ$.

a) $\sin 40^\circ \cos 30^\circ = \frac{1}{2}[\sin(40^\circ + 30^\circ) + \sin(40^\circ - 30^\circ)] = \frac{1}{2}(\sin 70^\circ + \sin 10^\circ)$

b) $\cos 110^\circ \sin 55^\circ = \frac{1}{2}[\sin(110^\circ + 55^\circ) - \sin(110^\circ - 55^\circ)] = \frac{1}{2}(\sin 165^\circ - \sin 55^\circ)$

c) $\cos 50^\circ \cos 35^\circ = \frac{1}{2}[\cos(50^\circ + 35^\circ) + \cos(50^\circ - 35^\circ)] = \frac{1}{2}(\cos 85^\circ + \cos 15^\circ)$

d) $\sin 55^\circ \sin 40^\circ = -\frac{1}{2}[\cos(55^\circ + 40^\circ) - \cos(55^\circ - 40^\circ)] = -\frac{1}{2}(\cos 95^\circ - \cos 15^\circ)$

**4.** Express each of the following as a product:
    a) $\sin 50^\circ + \sin 40^\circ$,   b) $\sin 70^\circ - \sin 20^\circ$,   c) $\cos 55^\circ + \cos 25^\circ$,   d) $\cos 35^\circ - \cos 75^\circ$.

a) $\sin 50^\circ + \sin 40^\circ = 2 \sin \frac{1}{2}(50^\circ + 40^\circ) \cos \frac{1}{2}(50^\circ - 40^\circ) = 2 \sin 45^\circ \cos 5^\circ$

b) $\sin 70^\circ - \sin 20^\circ = 2 \cos \frac{1}{2}(70^\circ + 20^\circ) \sin \frac{1}{2}(70^\circ - 20^\circ) = 2 \cos 45^\circ \sin 25^\circ$

c) $\cos 55^\circ + \cos 25^\circ = 2 \cos \frac{1}{2}(55^\circ + 25^\circ) \cos \frac{1}{2}(55^\circ - 25^\circ) = 2 \cos 40^\circ \cos 15^\circ$

d) $\cos 35^\circ - \cos 75^\circ = -2 \sin \frac{1}{2}(35^\circ + 75^\circ) \sin \frac{1}{2}(35^\circ - 75^\circ) = -2 \sin 55^\circ \sin (-20^\circ)$
$$= 2 \sin 55^\circ \sin 20^\circ$$

**5.** Prove   $\dfrac{\sin 4A + \sin 2A}{\cos 4A + \cos 2A} = \tan 3A$.

$$\frac{\sin 4A + \sin 2A}{\cos 4A + \cos 2A} = \frac{2 \sin \frac{1}{2}(4A + 2A) \cos \frac{1}{2}(4A - 2A)}{2 \cos \frac{1}{2}(4A + 2A) \cos \frac{1}{2}(4A - 2A)} = \frac{\sin 3A}{\cos 3A} = \tan 3A$$

**6.** Prove   $\dfrac{\sin A - \sin B}{\sin A + \sin B} = \dfrac{\tan \frac{1}{2}(A - B)}{\tan \frac{1}{2}(A + B)}$.

$$\frac{\sin A - \sin B}{\sin A + \sin B} = \frac{2 \cos \frac{1}{2}(A+B) \sin \frac{1}{2}(A-B)}{2 \sin \frac{1}{2}(A+B) \cos \frac{1}{2}(A-B)} = \cot \frac{1}{2}(A+B) \tan \frac{1}{2}(A-B) = \frac{\tan \frac{1}{2}(A-B)}{\tan \frac{1}{2}(A+B)}$$

**7.** Prove   $\cos^3 x \sin^2 x = \dfrac{1}{16}(2 \cos x - \cos 3x - \cos 5x)$.

$$\cos^3 x \sin^2 x = (\sin x \cos x)^2 \cos x = \frac{1}{4} \sin^2 2x \cos x = \frac{1}{4}(\sin 2x)(\sin 2x \cos x)$$

$$= \frac{1}{4}(\sin 2x)[\frac{1}{2}(\sin 3x + \sin x)] = \frac{1}{8}(\sin 3x \sin 2x + \sin 2x \sin x)$$

$$= \frac{1}{8}\{-\frac{1}{2}(\cos 5x - \cos x) + [-\frac{1}{2}(\cos 3x - \cos x)]\}$$

$$= \frac{1}{16}(2 \cos x - \cos 3x - \cos 5x)$$

**8.** Prove   $1 + \cos 2x + \cos 4x + \cos 6x = 4 \cos x \cos 2x \cos 3x$.

$$1 + (\cos 2x + \cos 4x) + \cos 6x = 1 + 2 \cos 3x \cos x + \cos 6x = (1 + \cos 6x) + 2 \cos 3x \cos x$$
$$= 2 \cos^2 3x + 2 \cos 3x \cos x = 2 \cos 3x (\cos 3x + \cos x)$$
$$= 2 \cos 3x (2 \cos 2x \cos x) = 4 \cos x \cos 2x \cos 3x$$

**9.** Transform $4 \cos x + 3 \sin x$ into the form $c \cos(x - \alpha)$.

Since $c \cos(x - \alpha) = c(\cos x \cos \alpha + \sin x \sin \alpha)$, set $c \cos \alpha = 4$ and $c \sin \alpha = 3$.

Then $\cos \alpha = 4/c$ and $\sin \alpha = 3/c$. Since $\sin^2 \alpha + \cos^2 \alpha = 1$, $c = 5$ and $-5$.

Using $c = 5$, $\cos \alpha = 4/5$, $\sin \alpha = 3/5$, and $\alpha = 36°52'$. Thus,
$$4 \cos x + 3 \sin x = 5 \cos(x - 36°52').$$

Using $c = -5$, $\alpha = 216°52'$ and
$$4 \cos x + 3 \sin x = -5 \cos(x - 216°52').$$

**10.** Find the maximum and minimum values of $4 \cos x + 3 \sin x$ on the interval $0 \leq x \leq 2\pi$.

From Problem 9, $4 \cos x + 3 \sin x = 5 \cos(x - 36°52')$.

Now on the prescribed interval, $\cos \theta$ attains its maximum value 1 when $\theta = 0$ and its minimum value $-1$ when $\theta = \pi$. Thus, the maximum value of $4 \cos x + 3 \sin x$ is 5 which occurs when $x - 36°52' = 0$ or when $x = 36°52'$ while the minimum value is $-5$ which occurs when $x - 36°52' = \pi$ or when $x = 216°52'$.

# SUPPLEMENTARY PROBLEMS

**11.** Express each of the following products as a sum or difference of sines or of cosines.

a) $\sin 35° \cos 25° = \frac{1}{2}(\sin 60° + \sin 10°)$

b) $\sin 25° \cos 75° = \frac{1}{2}(\sin 100° - \sin 50°)$

c) $\cos 50° \cos 70° = \frac{1}{2}(\cos 120° + \cos 20°)$

d) $\sin 130° \sin 55° = -\frac{1}{2}(\cos 185° - \cos 75°)$

e) $\sin 4x \cos 2x = \frac{1}{2}(\sin 6x + \sin 2x)$

f) $\sin x/2 \cos 3x/2 = \frac{1}{2}(\sin 2x - \sin x)$

g) $\cos 7x \cos 4x = \frac{1}{2}(\cos 11x + \cos 3x)$

h) $\sin 5x \sin 4x = -\frac{1}{2}(\cos 9x - \cos x)$

**12.** Show that

a) $2 \sin 45° \cos 15° = \frac{1}{2}(\sqrt{3} + 1)$ and $\cos 15° = \frac{\sqrt{2}}{4}(\sqrt{3} + 1)$,

b) $2 \sin 82\frac{1}{2}° \cos 37\frac{1}{2}° = \frac{1}{2}(\sqrt{3} + \sqrt{2})$,    c) $2 \sin 127\frac{1}{2}° \sin 97\frac{1}{2}° = \frac{1}{2}(\sqrt{3} + \sqrt{2})$.

**13.** Express each of the following as a product.

a) $\sin 50° + \sin 20° = 2 \sin 35° \cos 15°$    e) $\sin 4x + \sin 2x = 2 \sin 3x \cos x$

b) $\sin 75° - \sin 35° = 2 \cos 55° \sin 20°$    f) $\sin 7\theta - \sin 3\theta = 2 \cos 5\theta \sin 2\theta$

c) $\cos 65° + \cos 15° = 2 \cos 40° \cos 25°$    g) $\cos 6\theta + \cos 2\theta = 2 \cos 4\theta \cos 2\theta$

d) $\cos 80° - \cos 70° = -2 \sin 75° \sin 5°$    h) $\cos 3x/2 - \cos 9x/2 = 2 \sin 3x \sin 3x/2$

**14.** Show that

a) $\sin 40° + \sin 20° = \cos 10°$,           c) $\cos 465° + \cos 165° = -\sqrt{6}/2$,

b) $\sin 105° + \sin 15° = \sqrt{6}/2$,       d) $\dfrac{\sin 75° - \sin 15°}{\cos 75° + \cos 15°} = 1/\sqrt{3}$.

15. Prove:

a) $\dfrac{\sin A + \sin 3A}{\cos A + \cos 3A} = \tan 2A$

b) $\dfrac{\sin 2A + \sin 4A}{\cos 2A + \cos 4A} = \tan 3A$

c) $\dfrac{\sin A + \sin B}{\sin A - \sin B} = \dfrac{\tan \frac{1}{2}(A + B)}{\tan \frac{1}{2}(A - B)}$

d) $\dfrac{\cos A + \cos B}{\cos A - \cos B} = - \cot \frac{1}{2}(A - B) \cot \frac{1}{2}(A + B)$

e) $\sin \theta + \sin 2\theta + \sin 3\theta = \sin 2\theta + (\sin \theta + \sin 3\theta) = \sin 2\theta (1 + 2 \cos \theta)$

f) $\cos \theta + \cos 2\theta + \cos 3\theta = \cos 2\theta (1 + 2 \cos \theta)$

g) $\sin 2\theta + \sin 4\theta + \sin 6\theta = (\sin 2\theta + \sin 4\theta) + 2 \sin 3\theta \cos 3\theta$
$$= 4 \cos \theta \cos 2\theta \sin 3\theta$$

h) $\dfrac{\sin 3x + \sin 5x + \sin 7x + \sin 9x}{\cos 3x + \cos 5x + \cos 7x + \cos 9x} = \dfrac{(\sin 3x + \sin 9x) + (\sin 5x + \sin 7x)}{(\cos 3x + \cos 9x) + (\cos 5x + \cos 7x)} = \tan 6x$

16. Prove:

a) $\cos 130^\circ + \cos 110^\circ + \cos 10^\circ = 0$,

b) $\cos 220^\circ + \cos 100^\circ + \cos 20^\circ = 0$.

17. Prove:

a) $\cos^2\theta \sin^3\theta = \dfrac{1}{16}(2 \sin \theta + \sin 3\theta - \sin 5\theta)$

b) $\cos^2\theta \sin^4\theta = \dfrac{1}{32}(2 - \cos 2\theta - 2 \cos 4\theta + \cos 6\theta)$

c) $\cos^5\theta = \dfrac{1}{16}(10 \cos \theta + 5 \cos 3\theta + \cos 5\theta)$

d) $\sin^5\theta = \dfrac{1}{16}(10 \sin \theta - 5 \sin 3\theta + \sin 5\theta)$

18. Transform:

a) $4 \cos x + 3 \sin x$ into the form $c \sin(x + \alpha)$  *Ans.* $5 \sin(x + 53^\circ 8')$

b) $4 \cos x + 3 \sin x$ into the form $c \sin(x - \alpha)$.  *Ans.* $5 \sin(x - 306^\circ 52')$

c) $\sin x - \cos x$ into the form $c \sin(x - \alpha)$.  *Ans.* $\sqrt{2} \sin(x - 45^\circ)$

d) $5 \cos 3t + 12 \sin 3t$ into the form $c \cos(3t - \alpha)$.  *Ans.* $13 \cos(3t - 67^\circ 23')$

19. Find the maximum and minimum values of each sum of Problem 18 and a value of $x$ or $t$ between 0 and $2\pi$ at which each occurs.

*Ans.* a) Maximum = 5, when $x = 36^\circ 52'$ (i.e., when $x + 53^\circ 8' = 90^\circ$); minimum = $-5$, when $x = 216^\circ 52'$.

b) Same as a).

c) Maximum = $\sqrt{2}$, when $x = 135^\circ$; minimum = $-\sqrt{2}$, when $x = 315^\circ$.

d) Maximum = 13, when $t = 22^\circ 28'$; minimum = $-13$, when $t = 82^\circ 28'$.

# CHAPTER 13

# Oblique Triangles. Non-logarithmic Solution

AN OBLIQUE TRIANGLE is one which does not contain a right angle. Such a triangle contains either three acute angles or two acute angles and one obtuse angle.

The convention of denoting the angles by $A, B, C$ and the lengths of the corresponding opposite sides by $a, b, c$ will be used here.

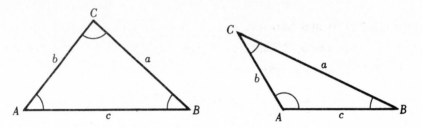

When three parts, not all angles, are known, the triangle is uniquely determined, except in one case to be noted below. The four cases of oblique triangles are:

Case I.   Given one side and two angles.
Case II.  Given two sides and the angle opposite one of them.
Case III. Given two sides and the included angle.
Case IV.  Given the three sides.

THE LAW OF SINES. In any triangle, the sides are proportional to the sines of the opposite angles, i.e.,

$$\frac{a}{\sin A} = \frac{b}{\sin B} = \frac{c}{\sin C}.$$

The following relations follow readily:

$$\frac{a}{b} = \frac{\sin A}{\sin B}, \qquad \frac{b}{c} = \frac{\sin B}{\sin C}, \qquad \frac{c}{a} = \frac{\sin C}{\sin A}.$$

For a proof of the law of sines, see Problem 1.

MOLLWEIDE'S FORMULAS. In any triangle $ABC$,

$$\frac{a+b}{c} = \frac{\cos \frac{1}{2}(A-B)}{\sin \frac{1}{2}C}, \qquad \frac{a-b}{c} = \frac{\sin \frac{1}{2}(A-B)}{\cos \frac{1}{2}C}$$

together with those obtained by cyclic changes of the letters, i.e.,

$$\frac{b+c}{a} = \frac{\cos \frac{1}{2}(B-C)}{\sin \frac{1}{2}A}, \qquad \frac{b-c}{a} = \frac{\sin \frac{1}{2}(B-C)}{\cos \frac{1}{2}A}$$

$$\frac{c+a}{b} = \frac{\cos \frac{1}{2}(C-A)}{\sin \frac{1}{2}B}, \qquad \frac{c-a}{b} = \frac{\sin \frac{1}{2}(C-A)}{\cos \frac{1}{2}B}$$

and those obtained by interchanging the two letters (small and capital) in the numerators of each relation.

For derivations of these formulas, see Problem 2.

**PROJECTION FORMULAS.** In any triangle $ABC$,

$$a = b \cos C + c \cos B$$

$$b = c \cos A + a \cos C$$

$$c = a \cos B + b \cos A .$$

For the derivation of these formulas, see Problem 3.

**CASE I. Given one side and two angles.**

EXAMPLE.   Suppose $a$, $B$, and $C$ are given.

To find $A$, use $A = 180^\circ - (B + C)$.

To find $b$, use $\dfrac{b}{a} = \dfrac{\sin B}{\sin A}$   whence   $b = \dfrac{a \sin B}{\sin A}$.

To find $c$, use $\dfrac{c}{a} = \dfrac{\sin C}{\sin A}$   whence   $c = \dfrac{a \sin C}{\sin A}$.

To check, use one of the Mollweide formulas or one of the projection formulas.

See Problems 4-8.

**CASE II. Given two sides and the angle opposite one of them.**

EXAMPLE.   Suppose $b$, $c$, and $B$ are given.

From $\dfrac{\sin C}{\sin B} = \dfrac{c}{b}$,   $\sin C = \dfrac{c \sin B}{b}$.

If $\sin C > 1$,   no angle $C$ is determined.
If $\sin C = 1$,   $C = 90^\circ$ and a right triangle is determined.
If $\sin C < 1$,   two angles are determined: an acute angle $C$ and an obtuse angle $C' = 180^\circ - C$.   Thus, there may be one or two triangles determined.

This case is discussed geometrically in Problem 9.   The results obtained may be summarized as follows:

When the given angle is *acute*, there will be
a) *one* solution if the side opposite the given angle is equal to or greater than the other given side,
b) *no* solution, *one* solution (right triangle) or *two* solutions if the side opposite the given angle is less than the other given side.

When the given angle is *obtuse*, there will be
c) *no* solution when the side opposite the given angle is less than or equal to the other given side,
d) *one* solution if the side opposite the given angle is greater than the other given side.

EXAMPLE.   1) When $b = 30$, $c = 20$, and $B = 40^\circ$, there is one solution since $B$ is acute and $b > c$.

2) When $b = 20$, $c = 30$, and $B = 40^\circ$, there is either no solution, one solution, or two solutions.   The particular subcase is determined after computing $\sin C = \dfrac{c \sin B}{b}$.

3) When $b = 30$, $c = 20$, and $B = 140^\circ$, there is one solution.
4) When $b = 20$, $c = 30$, and $B = 140^\circ$, there is no solution.

This, the so-called ambiguous case, is solved by the law of sines and may be checked by either the Mollweide formulas or the projection formulas.

See Problems 10-13.

**THE LAW OF COSINES.** In any triangle $ABC$, the square of any side is equal to the sum of the squares of the other two sides diminished by twice the product of these sides and the cosine of their included angle, i.e.,

$$a^2 = b^2 + c^2 - 2bc \cos A$$

$$b^2 = c^2 + a^2 - 2ca \cos B$$

$$c^2 = a^2 + b^2 - 2ab \cos C$$

For the derivation of these formulas, see Problem 14.

**CASE III.** Given two sides and the included angle.

    **EXAMPLE.** Suppose $a$, $b$, and $C$ are given.

      To find $c$, use $c^2 = a^2 + b^2 - 2ab \cos C$.

      To find $A$, use $\sin A = \dfrac{a \sin C}{c}$.    To find $B$, use $\sin B = \dfrac{b \sin C}{c}$.

      To check, use $A + B + C = 180^\circ$.

See Problems 15-18.

**CASE IV.** Given the three sides.

    **EXAMPLE.** With $a$, $b$, and $c$ given, solve the law of cosines for each of the angles.

      To find the angles, use $\cos A = \dfrac{b^2 + c^2 - a^2}{2bc}$, $\cos B = \dfrac{c^2 + a^2 - b^2}{2ca}$, $\cos C = \dfrac{a^2 + b^2 - c^2}{2ab}$.

      To check, use $A + B + C = 180^\circ$.

See Problems 19-21.

## SOLVED PROBLEMS

1. Derive the law of sines.

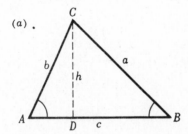 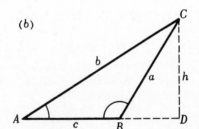

Let $ABC$ be any oblique triangle. In Fig. (a), angles $A$ and $B$ are acute while in Fig. (b), angle $B$ is obtuse. Draw $CD$ perpendicular to $AB$ or $AB$ extended and denote its length by $h$.

In the right triangle $ACD$ of either figure, $h = b \sin A$ while in the right triangle $BCD$, $h = a \sin B$ since in Fig. (b), $h = a \sin \angle DBC = a \sin(180^\circ - B) = a \sin B$. Thus,

$$a \sin B = b \sin A \qquad \text{or} \qquad \frac{a}{\sin A} = \frac{b}{\sin B}.$$

In a similar manner (by drawing a perpendicular from $B$ to $AC$ or a perpendicular from $A$ to $BC$), we obtain

$$\frac{a}{\sin A} = \frac{c}{\sin C} \quad \text{or} \quad \frac{b}{\sin B} = \frac{c}{\sin C}.$$

Thus, finally, $\quad \dfrac{a}{\sin A} = \dfrac{b}{\sin B} = \dfrac{c}{\sin C}.$

**2.** Derive a pair of Mollweide's formulas.

By the law of sines, $\quad \dfrac{a}{c} = \dfrac{\sin A}{\sin C}\quad$ and $\quad \dfrac{b}{c} = \dfrac{\sin B}{\sin C}.$

Then $\quad \dfrac{a+b}{c} = \dfrac{\sin A + \sin B}{\sin C} = \dfrac{2 \sin \frac{1}{2}(A+B) \cos \frac{1}{2}(A-B)}{2 \sin \frac{1}{2}C \cos \frac{1}{2}C} = \dfrac{\cos \frac{1}{2}(A-B)}{\sin \frac{1}{2}C},$

since $\sin \frac{1}{2}(A+B) = \sin \frac{1}{2}(180^\circ - C) = \sin(90^\circ - \frac{1}{2}C) = \cos \frac{1}{2}C.$

Similarly, $\quad \dfrac{a-b}{c} = \dfrac{\sin A - \sin B}{\sin C} = \dfrac{2 \cos \frac{1}{2}(A+B) \sin \frac{1}{2}(A-B)}{2 \sin \frac{1}{2}C \cos \frac{1}{2}C} = \dfrac{\sin \frac{1}{2}(A-B)}{\cos \frac{1}{2}C},$

since $\cos \frac{1}{2}(A+B) = \cos(90^\circ - \frac{1}{2}C) = \sin \frac{1}{2}C.$

**3.** Derive one of the projection formulas.

Refer to the figures of Problem 1. In the right triangle $ACD$ of either figure, $AD = b \cos A.$

In the right triangle $BCD$ of Fig. (a), $DB = a \cos B.$ Thus, in Fig. (a),

$$c = AB = AD + DB = b \cos A + a \cos B = a \cos B + b \cos A.$$

In the right triangle $BCD$ of Fig. (b), $BD = a \cos \angle DBC = a \cos(180^\circ - B) = -a \cos B.$ Thus, in Fig. (b),

$$c = AB = AD - BD = b \cos A - (-a \cos B) = a \cos B + b \cos A.$$

## CASE I.

**4.** Solve the triangle $ABC$, given $c = 25$, $A = 35^\circ$, and $B = 68^\circ$.

To find $C$: $\quad C = 180^\circ - (A + B) = 180^\circ - 103^\circ = 77^\circ.$

To find $a$: $\quad a = \dfrac{c \sin A}{\sin C} = \dfrac{25 \sin 35^\circ}{\sin 77^\circ} = \dfrac{25(0.5736)}{0.9744} = 15.$

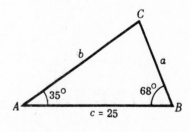

To find $b$: $\quad b = \dfrac{c \sin B}{\sin C} = \dfrac{25 \sin 68^\circ}{\sin 77^\circ} = \dfrac{25(0.9272)}{0.9744} = 24.$

To check by Mollweide's formula:

$$\frac{b+a}{c} = \frac{\cos \frac{1}{2}(B-A)}{\sin \frac{1}{2}C} \quad \text{or} \quad (b+a) \sin \tfrac{1}{2}C = c \cos \tfrac{1}{2}(B-A)$$

$$
\begin{aligned}
(b+a) \sin \tfrac{1}{2}C &= 39 \sin 38^\circ 30' = 39(0.6225) = 24.3 \\
c \cos \tfrac{1}{2}(B-A) &= 25 \cos 16^\circ 30' = 25(0.9588) = 24.0
\end{aligned}
$$

To check by projection formula: $\quad c = a \cos B + b \cos A = 15 \cos 68^\circ + 24 \cos 35^\circ$
$$= 15(0.3746) + 24(0.8192) = 25.3.$$

The required parts are $a = 15$, $b = 24$, and $C = 77^\circ.$

**5.** Solve the triangle $ABC$, given $a = 62.5$, $A = 112^\circ 20'$, and $C = 42^\circ 10'$.

For $B$ :    $B = 180^\circ - (C + A) = 180^\circ - 154^\circ 30' = 25^\circ 30'$.

For $b$ :    $b = \dfrac{a \sin B}{\sin A} = \dfrac{62.5 \sin 25^\circ 30'}{\sin 112^\circ 20'} = \dfrac{62.5(0.4305)}{0.9250} = 29.1.$

$(\sin 112^\circ 20' = \sin(180^\circ - 112^\circ 20') = \sin 67^\circ 40')$

For $c$ :    $c = \dfrac{a \sin C}{\sin A} = \dfrac{62.5 \sin 42^\circ 10'}{\sin 112^\circ 20'} = \dfrac{62.5(0.6713)}{0.9250} = 45.4.$

Check :    $(c + b) \sin \tfrac{1}{2}A = a \cos \tfrac{1}{2}(C - B)$

$(c + b) \sin \tfrac{1}{2}A = 74.5 \sin 56^\circ 10' = 74.5(0.8307) = 61.89$
$a \cos \tfrac{1}{2}(C - B) = 62.5 \cos 8^\circ 20' = 62.5(0.9894) = 61.84 ;$

or    $a = b \cos C + c \cos B = 29.1(0.7412) + 45.4(0.9026) = 62.55.$

The required parts are    $b = 29.1$, $c = 45.4$, and $B = 25^\circ 30'$.

**6.** $A$ and $B$ are two points on opposite banks of a river. From $A$ a line $AC = 275$ ft is laid off and the angles $CAB = 125^\circ 40'$ and $ACB = 48^\circ 50'$ are measured.  Find the length of $AB$.

In the triangle $ABC$ of Fig. $(a)$ below,    $B = 180^\circ - (C + A) = 5^\circ 30'$    and

$$AB = c = \frac{b \sin C}{\sin B} = \frac{275 \sin 48^\circ 50'}{\sin 5^\circ 30'} = \frac{275(0.7528)}{0.0958} = 2160 \text{ ft.}$$

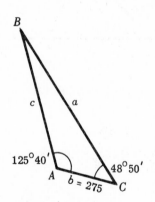

Fig. $(a)$  Prob. 6

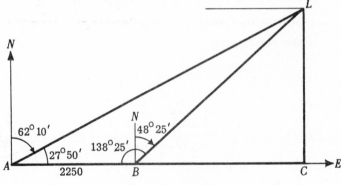

Fig. $(b)$  Prob. 7

**7.** A ship is sailing due east when a light is observed bearing N $62^\circ 10'$ E. After the ship has traveled 2250 ft, the light bears N $48^\circ 25'$ E.  If the course is continued, how close will  the ship approach the light? (See Problem 5, Chapter 5.)

Refer to Fig. $(b)$ above.

In the oblique triangle $ABL$ :    $AB = 2250$, $\angle BAL = 27^\circ 50'$,  and $\angle ABL = 138^\circ 25'$.

$\angle ALB = 180^\circ - (\angle BAL + \angle ABL) = 13^\circ 45'$.

$$BL = \frac{AB \sin \angle BAL}{\sin \angle ALB} = \frac{2250 \sin 27^\circ 50'}{\sin 13^\circ 45'} = \frac{2250(0.4669)}{0.2377} = 4420.$$

In the right triangle $BLC$ :    $BL = 4420$ and $\angle CBL = 90^\circ - 48^\circ 25' = 41^\circ 35'$.

$CL = BL \sin \angle CBL = 4420 \sin 41^\circ 35' = 4420(0.6637) = 2934$ ft.

For an alternate solution, find $AL$ in the oblique triangle $ABL$ and then $CL$ in the right triangle $ALC$.

8. A tower 125 ft high is on a cliff on the bank of a river. From the top of the tower the angle of depression of a point on the opposite shore is $28°40'$ and from the base of the tower the angle of depression of the same point is $18°20'$. Find the width of the river and the height of the cliff.

In the figure $BC$ represents the tower, $DB$ represents the cliff, and $A$ is the point on the opposite shore.

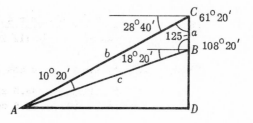

In triangle $ABC$,  $C = 90° - 28°40' = 61°20'$,
$$B = 90° + 18°20' = 108°20',$$
$$A = 180° - (B + C) = 10°20'.$$

$$c = \frac{a \sin C}{\sin A} = \frac{125 \sin 61°20'}{\sin 10°20'} = \frac{125(0.8774)}{0.1794} = 611.$$

In right triangle $ABD$,

$$DB = c \sin 18°20' = 611(0.3145) = 192,$$
$$AD = c \cos 18°20' = 611(0.9492) = 580.$$

The river is 580 ft wide and the cliff is 192 ft high.

9. Discuss the several special cases when two sides and the angle opposite one of them are given.

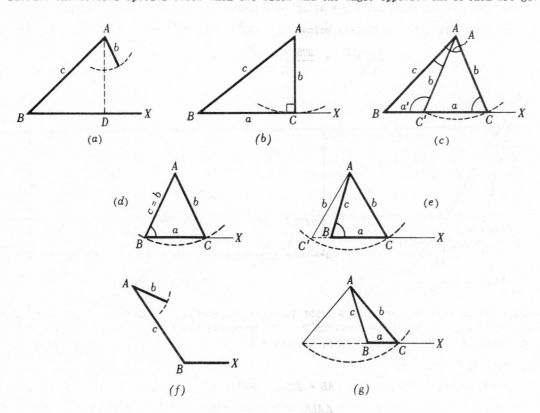

(a)          (b)          (c)

(d)          (e)

(f)          (g)

Let $b$, $c$, and $B$ be the given parts. Construct the given angle $B$ and lay off the side $BA = c$. With $A$ as center and radius equal to $b$ (the side opposite the given angle) describe an arc. Figures $(a)$-$(e)$ illustrate the special cases which may occur when the given angle $B$ is acute while Figures $(f)$-$(g)$ illustrate the cases when $B$ is obtuse.

*The given angle $B$ is acute.*

**Fig.** $(a)$. When $b < AD = c \sin B$, the arc does not meet $BX$ and no triangle is determined.

Fig. (b). When $b = AD$, the arc is tangent to $BX$ and one triangle — a right triangle with the right angle at $C$ — is determined.

Fig. (c). When $b > AD$ and $b < c$, the arc meets $BX$ in two points $C$ and $C'$ on the same side of $B$. Two triangles $ABC$, in which $C$ is acute, and $ABC'$ in which $C' = 180° - C$ is obtuse, are determined.

Fig. (d). When $b > AD$ and $b = c$, the arc meets $BX$ in $C$ and $B$. One triangle (isosceles) is determined.

Fig. (e). When $b > c$, the arc meets $BX$ in $C$ and $BX$ extended in $C'$. Since the triangle $ABC'$ does not contain the given angle $B$, only one triangle $ABC$ is determined.

*The given angle is obtuse.*

Fig. (f). When $b < c$ or $b = c$, no triangle is formed.

Fig. (g). When $b > c$, only one triangle is formed as in Fig. (e).

## CASE II.

**10.** Solve the triangle $ABC$, given $c = 628$, $b = 480$, and $C = 55°10'$. Refer to Fig. (a) below.

Since $C$ is acute and $c > b$, there is only one solution.

For $B$:    $\sin B = \dfrac{b \sin C}{c} = \dfrac{480 \sin 55°10'}{628} = \dfrac{480(0.8208)}{628} = 0.6274$   and   $B = 38°50'$.

For $A$:    $A = 180° - (B + C) = 86°0'$.

For $a$:    $a = \dfrac{b \sin A}{\sin B} = \dfrac{480 \sin 86°0'}{\sin 38°50'} = \dfrac{480(0.9976)}{0.6271} = 764.$

Check:    $(a + b) \sin \tfrac{1}{2}C = c \cos \tfrac{1}{2}(A - B)$

$(a + b) \sin \tfrac{1}{2}C = 1244 \sin 27°35' = 1244(0.4630) = 576.0$
$c \cos \tfrac{1}{2}(A - B) = 628 \cos 23°35' = 628(0.9165) = 575.6$

If preferred, a projection formula may be used for the check.

The required parts are $B = 38°50'$, $A = 86°0'$, and $a = 764$.

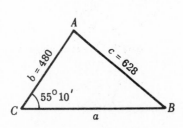

Fig. (a) Prob. 10

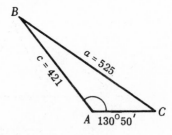

Fig. (b) Prob. 11

**11.** Solve the triangle $ABC$, given $a = 525$, $c = 421$, and $A = 130°50'$. Refer to Fig. (b) above.

Since $A$ is obtuse and $a > c$, there is one solution.

For $C$:    $\sin C = \dfrac{c \sin A}{a} = \dfrac{421 \sin 130°50'}{525} = \dfrac{421(0.7566)}{525} = 0.6067$   and   $C = 37°20'$.

For $B$:    $B = 180° - (C + A) = 11°50'$.

For $b$:    $b = \dfrac{a \sin B}{\sin A} = \dfrac{525 \sin 11°50'}{\sin 130°50'} = \dfrac{525(0.2051)}{0.7566} = 142.$

Check :    $(c + b) \sin \frac{1}{2}A = a \cos \frac{1}{2}(C - B)$

$(c + b) \sin \frac{1}{2}A = 563 \sin 65^\circ 25' = 563(0.9094) = 512.0$
$a \cos \frac{1}{2}(C - B) = 525 \cos 12^\circ 45' = 525(0.9754) = 512.1.$

The required parts are   $C = 37^\circ 20'$, $B = 11^\circ 50'$, and $b = 142.$

**12.** Solve the triangle $ABC$, given $a = 31.5$, $b = 51.8$, and $A = 33^\circ 40'$.  Refer to Fig. (c) below.

Since $A$ is acute and $a < b$, there is the possibility of two solutions.

For $B$:   $\sin B = \dfrac{b \sin A}{a} = \dfrac{51.8 \sin 33^\circ 40'}{31.5} = \dfrac{51.8(0.5544)}{31.5} = 0.9117.$

There are two solutions, $B = 65^\circ 40'$ and $B' = 180^\circ - 65^\circ 40' = 114^\circ 20'.$

For $C$:   $C = 180^\circ - (A + B) = 80^\circ 40'.$      For $C'$:   $C' = 180^\circ - (A + B') = 32^\circ 0'.$

For $c$:   $c = \dfrac{a \sin C}{\sin A} = \dfrac{31.5 \sin 80^\circ 40'}{\sin 33^\circ 40'}$    For $c'$:   $c' = \dfrac{a \sin C'}{\sin A} = \dfrac{31.5 \sin 32^\circ 0'}{\sin 33^\circ 40'}$

$= \dfrac{31.5(0.9868)}{0.5544} = 56.1.$          $= \dfrac{31.5(0.5299)}{0.5544} = 30.1.$

Check :   $(c + b) \sin \frac{1}{2}A = a \cos \frac{1}{2}(C - B)$     Check :   $(b + c') \sin \frac{1}{2}A = a \cos \frac{1}{2}(B' - C')$

$(c + b) \sin \frac{1}{2}A = 107.9 \sin 16^\circ 50'$             $(b + c') \sin \frac{1}{2}A = 81.9 \sin 16^\circ 50'$
$= 107.9(0.2896)$                         $= 81.9(0.2896)$
$= 31.25$                                 $= 23.72$

$a \cos \frac{1}{2}(C - B) = 31.5 \cos 7^\circ 30'$        $a \cos \frac{1}{2}(B' - C') = 31.5 \cos 41^\circ 10'$
$= 31.5(0.9914)$                          $= 31.5(0.7528)$
$= 31.23.$                              $= 23.71.$

The required parts are

for triangle $ABC$:   $B = 65^\circ 40'$,   $C = 80^\circ 40'$,   and   $c = 56.1.$

for triangle $ABC'$:   $B' = 114^\circ 20'$,   $C' = 32^\circ 0'$,   and   $c' = 30.1.$

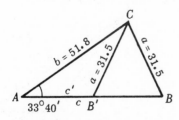
Fig. (c)  Prob. 12

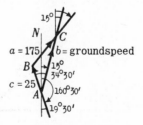
Fig. (d)  Prob. 13

**13.** A pilot wishes to track $15^\circ 0'$ against a wind of 25 mph from $160^\circ 30'$. Find his required heading and the groundspeed when the airspeed is 175 mph.  Refer to Fig. (d) above.

Since $A$ is acute and $a > c$, there is one solution.

$\sin C = \dfrac{c \sin A}{a} = \dfrac{25 \sin 34^\circ 30'}{175} = \dfrac{25(0.5664)}{175} = 0.0809$  and  $C = 4^\circ 40'.$

$B = 180^\circ - (A + C) = 140^\circ 50'.$      $b = \dfrac{a \sin B}{\sin A} = \dfrac{175 \sin 140^\circ 50'}{\sin 34^\circ 30'} = \dfrac{175(0.6316)}{0.5664} = 195.$

The groundspeed is 195 mph and the required heading is $19^\circ 40'.$

**14.** Derive the law of cosines.

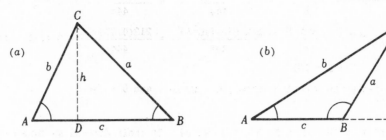

In the right triangle $ACD$ of either figure, $b^2 = h^2 + (AD)^2$.

In the right triangle $BCD$ of Fig. (a), $h = a \sin B$ and $DB = a \cos B$.

Then
$$AD = AB - DB = c - a \cos B$$

and
$$b^2 = h^2 + (AD)^2 = a^2 \sin^2 B + c^2 - 2ca \cos B + a^2 \cos^2 B$$
$$= a^2(\sin^2 B + \cos^2 B) + c^2 - 2ca \cos B = c^2 + a^2 - 2ca \cos B.$$

In the right triangle $BCD$ of Fig. (b), $h = a \sin \angle CBD = a \sin(180^\circ - B) = a \sin B$ and
$$BD = a \cos \angle CBD = a \cos(180^\circ - B) = -a \cos B.$$

Then $AD = AB + BD = c - a \cos B$ and $b^2 = c^2 + a^2 - 2ca \cos B.$

The remaining equations may be obtained by cyclic changes of the letters.

## CASE III.

**15.** Solve the triangle $ABC$, given $a = 132$, $b = 224$, and $C = 28^\circ 40'$. Refer to Fig. (c) below.

For $c$: $c^2 = a^2 + b^2 - 2ab \cos C$
$$= (132)^2 + (224)^2 - 2(132)(224) \cos 28^\circ 40'$$
$$= (132)^2 + (224)^2 - 2(132)(224)(0.8774) = 15714 \quad \text{and} \quad c = 125.$$

For $A$: $\sin A = \dfrac{a \sin C}{c} = \dfrac{132 \sin 28^\circ 40'}{125} = \dfrac{132(0.4797)}{125} = 0.5066$ and $A = 30^\circ 30'.$

For $B$: $\sin B = \dfrac{b \sin C}{c} = \dfrac{224 \sin 28^\circ 40'}{125} = \dfrac{224(0.4797)}{125} = 0.8596$ and $B = 120^\circ 40'.$

(Since $b > a$, $A$ is acute; since $A + C < 90^\circ$, $B > 90^\circ$.)

Check: $A + B + C = 179^\circ 50'$. The required parts are $A = 30^\circ 30'$, $B = 120^\circ 40'$, $c = 125.$

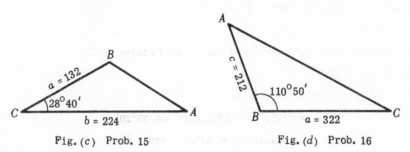

Fig. (c) Prob. 15    Fig. (d) Prob. 16

**16.** Solve the triangle $ABC$, given $a = 322$, $c = 212$, and $B = 110^\circ 50'$. Refer to Fig. (d) above.

For $b$: $b^2 = c^2 + a^2 - 2ca \cos B$  $[\cos 110^\circ 50' = -\cos(180^\circ - 110^\circ 50') = -\cos 69^\circ 10']$
$$= (212)^2 + (322)^2 - 2(212)(322)(-0.3557) = 197191 \quad \text{and} \quad b = 444.$$

For $A$ :   $\sin A = \dfrac{a \sin B}{b} = \dfrac{322 \sin 110^\circ 50'}{444} = \dfrac{322(0.9346)}{444} = 0.6778$   and   $A = 42^\circ 40'$.

For $C$ :   $\sin C = \dfrac{c \sin B}{b} = \dfrac{212 \sin 110^\circ 50'}{444} = \dfrac{212(0.9346)}{444} = 0.4463$   and   $C = 26^\circ 30'$.

Check :   $A + B + C = 180^\circ$.

The required parts are   $A = 42^\circ 40'$,   $C = 26^\circ 30'$,   and $b = 444$.

17. Two forces of 17.5 lb and 22.5 lb act on a body. If their directions make an angle of $50^\circ 10'$ with each other, find the magnitude of their resultant and the angle which it makes with the larger force.

Refer to Fig.($e$) below.

In the parallelogram $ABCD$,   $A + B = C + D = 180^\circ$   and   $B = 180^\circ - 50^\circ 10' = 129^\circ 50'$.

In the triangle $ABC$,

$$b^2 = c^2 + a^2 - 2ca \cos B \qquad [\cos 129^\circ 50' = -\cos(180^\circ - 129^\circ 50') = -\cos 50^\circ 10']$$

$$= (22.5)^2 + (17.5)^2 - 2(22.5)(17.5)(-0.6406) = 1317 \quad \text{and} \quad b = 36.3.$$

$$\text{Sin } A = \dfrac{a \sin B}{b} = \dfrac{17.5 \sin 129^\circ 50'}{36.3} = \dfrac{17.5(0.7679)}{36.3} = 0.3702 \quad \text{and} \quad A = 21^\circ 40'.$$

The resultant is a force of 36.3 lb; the required angle is $21^\circ 40'$.

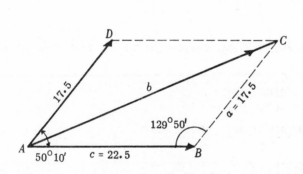

Fig.($e$)  Prob. 17

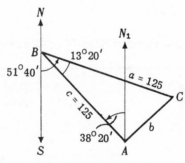

Fig.($f$)  Prob. 18

18. From $A$ a pilot flies 125 mi in the direction N $38^\circ 20'$ W  and turns back. Through an error, he then flies 125 mi in the direction S $51^\circ 40'$ E. How far and in what direction must he now fly to reach his intended destination $A$ ?

Refer to Fig.($f$) above.

Denote the turn back point as $B$ and his final position as $C$.

In the triangle $ABC$,

$$b^2 = c^2 + a^2 - 2ca \cos B$$

$$= (125)^2 + (125)^2 - 2(125)(125) \cos 13^\circ 20'$$

$$= 2(125)^2 (1 - 0.9730) = 843.7 \quad \text{and} \quad b = 29.0.$$

$$\text{Sin } A = \dfrac{a \sin B}{b} = \dfrac{125 \sin 13^\circ 20'}{29.0} = \dfrac{125(0.2306)}{29.0} = 0.9940 \quad \text{and} \quad A = 83^\circ 40'.$$

Since   $\angle CAN_1 = A - \angle N_1 AB = 45^\circ 20'$, the pilot must fly a course S $45^\circ 20'$ W for 29.0 miles in going from $C$ to $A$.

**CASE IV.**

**19.** Solve the triangle $ABC$, given $a = 25.2$, $b = 37.8$, and $c = 43.4$.  Refer to Fig. (g) below.

For $A$:   $\cos A = \dfrac{b^2 + c^2 - a^2}{2bc} = \dfrac{(37.8)^2 + (43.4)^2 - (25.2)^2}{2(37.8)(43.4)} = 0.8160$ and $A = 35°20'$.

For $B$:   $\cos B = \dfrac{c^2 + a^2 - b^2}{2ca} = \dfrac{(43.4)^2 + (25.2)^2 - (37.8)^2}{2(43.4)(25.2)} = 0.4982$ and $B = 60°10'$.

For $C$:   $\cos C = \dfrac{a^2 + b^2 - c^2}{2ab} = \dfrac{(25.2)^2 + (37.8)^2 - (43.4)^2}{2(25.2)(37.8)} = 0.0947$ and $C = 84°30'$.

Check :   $A + B + C = 180°$.

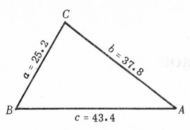

Fig. (g)  Prob. 19

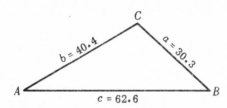

Fig. (h)  Prob. 20

**20.** Solve the triangle $ABC$, given $a = 30.3$, $b = 40.4$, and $c = 62.6$.  Refer to Fig. (h) above.

For $A$:   $\cos A = \dfrac{b^2 + c^2 - a^2}{2bc} = \dfrac{(40.4)^2 + (62.6)^2 - (30.3)^2}{2(40.4)(62.6)} = 0.9159$ and $A = 23°40'$.

For $B$:   $\cos B = \dfrac{c^2 + a^2 - b^2}{2ca} = \dfrac{(62.6)^2 + (30.3)^2 - (40.4)^2}{2(62.6)(30.3)} = 0.8448$ and $B = 32°20'$.

For $C$:   $\cos C = \dfrac{a^2 + b^2 - c^2}{2ab} = \dfrac{(30.3)^2 + (40.4)^2 - (62.6)^2}{2(30.3)(40.4)} = -0.5590$ and $C = 124°0'$.

Check :   $A + B + C = 180°$.

**21.** The distances of a point $C$ from two points $A$ and $B$, which cannot be  measured  directly, are required. The line $CA$ is continued through $A$ for a distance 175 ft to $D$, the line $CB$ is continued through $B$ for 225 ft to $E$, and the distances $AB = 300$ ft, $DB = 326$ ft, and $DE = 488$ ft are measured.  Find $AC$ and $BC$.

Triangle $ABC$ may be solved for the  required parts after the angles $\angle BAC$ and $\angle ABC$ have been found. The first angle is the supplement of $\angle BAD$ and the second is  the supplement of the sum of $\angle ABD$ and $\angle DBE$.

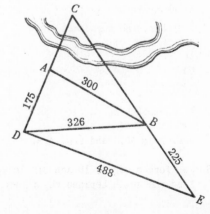

In the triangle $ABD$ whose sides are known,

$\cos \angle BAD = \dfrac{(175)^2 + (300)^2 - (326)^2}{2(175)(300)} = 0.1367$ and

$\angle BAD = 82°10'$,

$\cos \angle ABD = \dfrac{(300)^2 + (326)^2 - (175)^2}{2(300)(326)} = 0.8469$ and

$\angle ABD = 32°10'$.

In the triangle $BDE$ whose sides are known,

$$\cos \angle DBE = \frac{(225)^2 + (326)^2 - (488)^2}{2(225)(326)} = -0.5538 \quad \text{and} \quad \angle DBE = 123^\circ 40'.$$

In the triangle $ABC$:    $AB = 300$,   $\angle BAC = 180^\circ - \angle BAD = 97^\circ 50'$,

$$\angle ABC = 180^\circ - (\angle ABD + \angle DBE) = 24^\circ 10',$$
$$\angle ACB = 180^\circ - (\angle BAC + \angle ABC) = 58^\circ 0'.$$

Then
$$AC = \frac{AB \sin \angle ABC}{\sin \angle ACB} = \frac{300 \sin 24^\circ 10'}{\sin 58^\circ 0'} = \frac{300(0.4094)}{0.8480} = 145$$

and
$$BC = \frac{AB \sin \angle BAC}{\sin \angle ACB} = \frac{300 \sin 97^\circ 50'}{\sin 58^\circ 0'} = \frac{300(0.9907)}{0.8480} = 350.$$

The required distances are   $AC = 145$ ft   and   $BC = 350$ ft.

# SUPPLEMENTARY PROBLEMS

Solve each of the following oblique triangles $ABC$, given:

22. $a = 125$, $A = 54^\circ 40'$, $B = 65^\circ 10'$.       *Ans.* $b = 139$, $c = 133$, $C = 60^\circ 10'$

23. $b = 321$, $A = 75^\circ 20'$, $C = 38^\circ 30'$.       *Ans.* $a = 339$, $c = 218$, $B = 66^\circ 10'$

24. $b = 215$, $c = 150$, $B = 42^\circ 40'$.       *Ans.* $a = 300$, $A = 109^\circ 10'$, $C = 28^\circ 10'$

25. $a = 512$, $b = 426$, $A = 48^\circ 50'$.       *Ans.* $c = 680$, $B = 38^\circ 50'$, $C = 92^\circ 20'$

26. $b = 50.4$, $c = 33.3$, $B = 118^\circ 30'$.       *Ans.* $a = 25.1$, $A = 26^\circ 0'$, $C = 35^\circ 30'$

27. $b = 40.2$, $a = 31.5$, $B = 112^\circ 20'$.       *Ans.* $c = 15.7$, $A = 46^\circ 30'$, $C = 21^\circ 10'$

28. $b = 51.5$, $a = 62.5$, $B = 40^\circ 40'$.       *Ans.* $c = 78.9$, $A = 52^\circ 20'$, $C = 87^\circ 0'$
                                                $c' = 16.0$, $A' = 127^\circ 40'$, $C' = 11^\circ 40'$

29. $a = 320$, $c = 475$, $A = 35^\circ 20'$.       *Ans.* $b = 552$, $B = 85^\circ 30'$, $C = 59^\circ 10'$
                                                $b' = 224$, $B' = 23^\circ 50'$, $C' = 120^\circ 50'$

30. $b = 120$, $c = 270$, $A = 118^\circ 40'$.       *Ans.* $a = 344$, $B = 17^\circ 50'$, $C = 43^\circ 30'$

31. $a = 24.5$, $b = 18.6$, $c = 26.4$.       *Ans.* $A = 63^\circ 10'$, $B = 42^\circ 40'$, $C = 74^\circ 10'$

32. $a = 6.34$, $b = 7.30$, $c = 9.98$.       *Ans.* $A = 39^\circ 20'$, $B = 46^\circ 50'$, $C = 93^\circ 50'$

33. Two ships have radio equipment with a range of 200 miles. One is 155 miles N $42^\circ 40'$ E and the other is 165 miles N $45^\circ 10'$ W of a shore station. Can the two ships communicate directly? *Ans.* No; they are 222 miles apart.

34. A ship sails 15.0 miles on a course S $40^\circ 10'$ W and then 21.0 miles on a course N $28^\circ 20'$ W. Find the distance and direction of the last position from the first. *Ans.* 20.9 miles, N $70^\circ 40'$ W

35. A lighthouse is 10 miles northwest of a dock. A ship leaves the dock at 9 A.M. and steams west at 12 miles per hour. At what time will it be 8 miles from the lighthouse? *Ans.* 9:16 A.M. and 9:54 A.M.

36. Two forces of 115 lb and 215 lb acting on an object have a resultant of magnitude 275 lb. Find the angle between the directions in which the given forces act.     *Ans.* $70^\circ 50'$

**37.** A tower 150 ft high is situated at the top of a hill. At a point 650 ft down the hill the angle between the surface of the hill and the line of sight to the top of the tower is $12°30'$. Find the inclination of the hill to a horizontal plane.     *Ans.* $7°50'$

**38.** Three circles of radii 115, 150, and 225 ft respectively are tangent to each other externally. Find the angles of the triangle formed by joining the centers of the circles.
*Ans.* $43°10'$, $61°20'$, $75°30'$

**39.** A ship is supposed to leave $A$ and, by taking a straight course of $255°$, reach $D$ 525 miles distant in 25 hr. After completing 125 miles of the trip, the ship is ordered to stop at $C$ which is 225 miles from $A$ in the direction S $60°20'$ W, before continuing to $D$. If the ship changes course immediately, maintains the original speed for the remainder of the trip, and stops for 1 hr at $C$, how late will it be in reaching $D$? What is the course on each of the two last legs of the trip?     *Ans.* 2 hr; $223°30'$, $265°30'$
Hint: $BC = 109$, $\angle ABC = 148°30'$; $CD = 312$, $\angle BDC = 10°30'$, $B$ being the position at which the course is first changed.

# CHAPTER 14

# Logarithmic Solution of Oblique Triangles

**CASE I.** Given two angles and a side.

The triangle is solved by using the angle relation, $A + B + C = 180°$, and the law of sines twice. The solution is checked by using one of the Mollweide formulas.

**EXAMPLE 1.** Let $a$, $A$, and $B$ be given. Then

$$C = 180° - (A + B), \qquad b = \frac{a \sin B}{\sin A}, \qquad c = \frac{a \sin C}{\sin A}.$$

Check:    $(b + c) \sin \tfrac{1}{2}A = a \cos \tfrac{1}{2}(B - C)$, if $B > C$;

or    $(c + b) \sin \tfrac{1}{2}A = a \cos \tfrac{1}{2}(C - B)$, if $C > B$.

See Problems 1-3.

**CASE II.** Given two sides and the angle opposite one of them.

The triangle is solved by using the law of sines and the angle relation. The solution is checked by using one of the Mollweide formulas.

**EXAMPLE 2.** Let $a$, $b$, and $A$ be given with $a < b$. Then

$$\sin B = \frac{b \sin A}{a}, \qquad C = 180° - (A + B), \qquad c = \frac{a \sin C}{\sin A}.$$

When there are two solutions

$$B' = 180° - B, \qquad C' = 180° - (A + B'), \qquad c' = \frac{a \sin C'}{\sin A}.$$

Check:    $(b + a) \sin \tfrac{1}{2}C = c \cos \tfrac{1}{2}(B - A)$,

$(b + a) \sin \tfrac{1}{2}C' = c' \cos \tfrac{1}{2}(B' - A)$.

See Problems 4-5.

**LAW OF TANGENTS.** The law of cosines of the preceding chapter is not well adapted for logarithmic computation. In solving Case III, the law of tangents

$$\frac{a - b}{a + b} = \frac{\tan \tfrac{1}{2}(A - B)}{\tan \tfrac{1}{2}(A + B)}, \qquad \frac{b - c}{b + c} = \frac{\tan \tfrac{1}{2}(B - C)}{\tan \tfrac{1}{2}(B + C)}, \qquad \frac{c - a}{c + a} = \frac{\tan \tfrac{1}{2}(C - A)}{\tan \tfrac{1}{2}(C + A)}$$

will be used. For a proof of the law, see Problem 6.

Note. If, for example, $b > a$ it will be more convenient to write the first formula with the letters $a$ and $b$ (also $A$ and $B$) interchanged.

**CASE III.** Given two sides and the included angle.

    The triangle is solved by using the law of tangents to find the unknown angles and the law of sines to find the unknown side. The solution is checked by using one of the Mollweide formulas.

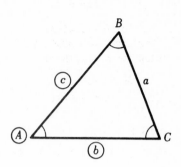

    **EXAMPLE 3.**   Let $c > b$ and $A$ be given.   Then

$$\tfrac{1}{2}(C+B) \;=\; 90^\circ - \tfrac{1}{2}A \,,$$

$$\tan \tfrac{1}{2}(C-B) \;=\; \frac{c-b}{c+b}\,\tan \tfrac{1}{2}(C+B)\,, \qquad a \;=\; \frac{c \sin A}{\sin C}\,.$$

    Check:    $(c+b)\sin\tfrac{1}{2}A \;=\; a\cos\tfrac{1}{2}(C-B).$

See Problems 7-9.

**HALF-ANGLE FORMULAS.**  In any triangle $ABC$

$$\tan \tfrac{1}{2}A \;=\; \frac{r}{s-a}\,, \qquad \tan \tfrac{1}{2}B \;=\; \frac{r}{s-b}\,, \qquad \tan \tfrac{1}{2}C \;=\; \frac{r}{s-c}$$

where $s = \tfrac{1}{2}(a+b+c)$ is the semi-perimeter of the triangle

and    $r = \sqrt{\dfrac{(s-a)(s-b)(s-c)}{s}}$   is the radius of the inscribed circle.

    For a proof of the formulas, see Problem 10. For the identification of $r$, see Problem 8, Chapter 15.

**CASE IV.** Given the three sides.

    The triangle is solved by using the half-angle formulas and is checked by using the angle relation.

See Problems 11-12.

## SOLVED PROBLEMS

**CASE I.**

**1.** Solve the triangle $ABC$, given $a = 38.124$, $A = 46^\circ 31.8'$, and $C = 79^\circ 17.4'$.

    $B = 180^\circ - (A+C) = 54^\circ 10.8'.$

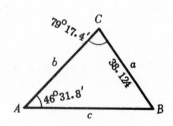

$$c = \frac{a \sin C}{\sin A} \qquad\qquad b = \frac{a \sin B}{\sin A}$$

| | |
|---|---|
| $\log a = 1.58120$ | $\log a = 1.58120$ |
| $\log \sin C = 9.99237-10$ | $\log \sin B = 9.90894-10$ |
| $\operatorname{colog} \sin A = 0.13922$ | $\operatorname{colog} \sin A = 0.13922$ |
| $\log c = 1.71279$ | $\log b = 1.62936$ |
| $c = 51.617$ | $b = 42.595$ |

Check:                   $(c+b)\sin\tfrac{1}{2}A \;=\; a\cos\tfrac{1}{2}(C-B)$

$c+b = 94.212,\;\; \tfrac{1}{2}A = 23^\circ 15.9'$          $a = 38.124,\;\; \tfrac{1}{2}(C-B) = 12^\circ 33.3'$

$\log(c+b) = 1.97411$                 $\log a = 1.58120$

$\log \sin \tfrac{1}{2}A = 9.59658-10$        $\log \cos \tfrac{1}{2}(C-B) = 9.98949-10$

                  $1.57069$                            $1.57069$

**2.** Solve the triangle $ABC$,  given $b = 282.66$, $A = 111°42.7'$, and $C = 24°25.8'$.

$B = 180° - (C+A) = 43°51.5'$.

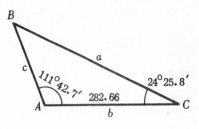

$$a = \frac{b \sin A}{\sin B} \qquad\qquad c = \frac{b \sin C}{\sin B}$$

| | |
|---|---|
| log $b$ = 2.45127 | log $b$ = 2.45127 |
| log sin $A$ = 9.96804 -10 | log sin $C$ = 9.61656 -10 |
| colog sin $B$ = 0.15934 | colog sin $B$ = 0.15934 |
| log $a$ = 2.57865 | log $c$ = 2.22717 |
| $a$ = 379.01 | $c$ = 168.72 |

Check :  $\qquad\qquad\qquad (a+c)\ \sin\tfrac{1}{2}B = b \cos \tfrac{1}{2}(A-C)$

$a+c = 547.73$,  $\tfrac{1}{2}B = 21°55.8'$ $\qquad\qquad b = 282.66$,  $\tfrac{1}{2}(A-C) = 43°38.4'$

| | |
|---|---|
| log $(a+c)$ = 2.73856 | log $b$ = 2.45127 |
| log sin $\tfrac{1}{2}B$ = 9.57226 -10 | log cos $\tfrac{1}{2}(A-C)$ = 9.85955 -10 |
| 2.31082 | 2.31082 |

**3.** In running a line $PQ$, S 38°42.4' E from the point $P$, a surveyor encounters a swamp. At a point $A$, on the line and at one edge of the swamp, he changes his direction to N 61°0.0' E for a distance of 1500.0 ft to a point $B$. He then sights to the other edge of the swamp, the direction being S 10°30.6' W. If this line meets $PQ$ in $C$, find the distance from $B$ to $C$, the angle through which he must turn from $BC$ to continue on the original line, and the distance $AC$ across the swamp.

In triangle $ABC$,
$A = 80°17.6'$, $B = 50°29.4'$, and $c = 1500.0$ ft.

Then  $C = 180° - (A+B) = 49°13.0'$.

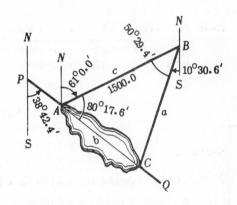

$$a = \frac{c \sin A}{\sin C} \qquad\qquad b = \frac{c \sin B}{\sin C}$$

| | |
|---|---|
| log $c$ = 3.17609 | log $c$ = 3.17609 |
| log sin $A$ = 9.99374 -10 | log sin $B$ = 9.88734 -10 |
| colog sin $C$ = 0.12080 | colog sin $C$ = 0.12080 |
| log $a$ = 3.29063 | log $b$ = 3.18423 |
| $a$ = 1952.7 | $b$ = 1528.4 |

The distance from $B$ to $C$ is 1952.7 ft.  The angle through which the surveyor must turn at $C$ is $\angle BCQ = 180° - \angle ACB = 130°47.0'$.  The distance across the swamp is 1528.4 ft.

## CASE II.

**4.** Solve the triangle $ABC$,  given $b = 67.246$, $c = 56.915$, and $B = 65°15.8'$.

Since $B$ is acute and $b > c$, there is one solution.

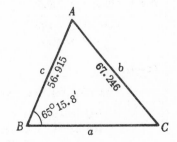

$$\sin C = \frac{c \sin B}{b} \qquad\qquad a = \frac{b \sin A}{\sin B}$$

| | |
|---|---|
| log $c$ = 1.75522 | log $b$ = 1.82767 |
| log sin $B$ = 9.95820 -10 | log sin $A$ = 9.95549 -10 |
| colog $b$ = 8.17233 -10 | colog sin $B$ = 0.04180 |
| log sin $C$ = 9.88575 -10 | log $a$ = 1.82496 |
| $C = 50°14.2'$ | $a$ = 66.828 |
| $A = 180° - (B+C) = 64°30.0'$ | |

Check : $(b+c) \sin \frac{1}{2}A = a \cos \frac{1}{2}(B-C)$

$b+c = 124.16,$    $\frac{1}{2}A = 32°15.0'$        $a = 66.828,$    $\frac{1}{2}(B-C) = 7°30.8'$

$$\log (b+c) = 2.09398$$
$$\log \sin \tfrac{1}{2}A = 9.72723-10$$
$$\overline{\phantom{\log \sin \tfrac{1}{2}A = } 1.82121}$$

$$\log a = 1.82496$$
$$\log \cos \tfrac{1}{2}(B-C) = 9.99625-10$$
$$\overline{\phantom{\log \cos \tfrac{1}{2}(B-C) = } 1.82121}$$

**5.** Solve the triangle $ABC$, given $a = 123.20$, $b = 155.37$, $A = 16°33.7'$.

Since $A$ is acute and $a < b$, there may be two solutions.

$$\sin B = \frac{b \sin A}{a}$$

$$\log b = 2.19137$$
$$\log \sin A = 9.45491-10$$
$$\text{colog } a = 7.90939-10$$
$$\overline{\log \sin B = 9.55567-10}$$
$$B = 21°4.1'$$
$$C = 180° - (A+B) = 142°22.2'$$

$B' = 180° - B = 158°55.9'$
$C' = 180° - (A+B') = 4°30.4'$

$$c = \frac{a \sin C}{\sin A}$$

$$\log a = 2.09061$$
$$\log \sin C = 9.78573-10$$
$$\text{colog } \sin A = 0.54509$$
$$\overline{\log c = 2.42143}$$
$$c = 263.89$$

$$c' = \frac{a \sin C'}{\sin A}$$

$$\log a = 2.09061$$
$$\log \sin C' = 8.89528-10$$
$$\text{colog } \sin A = 0.54509$$
$$\overline{\log c' = 1.53098}$$
$$c' = 33.961$$

Check :   $(b+a) \sin \frac{1}{2}C = c \cos \frac{1}{2}(B-A)$

$b+a = 278.57,$   $\frac{1}{2}C = 71°11.1'$
$c = 263.89,$   $\frac{1}{2}(B-A) = 2°15.2'$

$$\log (b+a) = 2.44494$$
$$\log \sin \tfrac{1}{2}C = 9.97615-10$$
$$\overline{\phantom{\log \sin \tfrac{1}{2}C = } 2.42109}$$

$$\log c = 2.42143$$
$$\log \cos \tfrac{1}{2}(B-A) = 9.99967-10$$
$$\overline{\phantom{\log \cos \tfrac{1}{2}(B-A) = } 2.42110}$$

Check :   $(b+a) \sin \frac{1}{2}C' = c' \cos \frac{1}{2}(B'-A)$

$b+a = 278.57,$   $\frac{1}{2}C' = 2°15.2'$
$c' = 33.961,$   $\frac{1}{2}(B'-A) = 71°11.1'$

$$\log (b+a) = 2.44494$$
$$\log \sin \tfrac{1}{2}C' = 8.59459-10$$
$$\overline{\phantom{\log \sin \tfrac{1}{2}C' = } 1.03953}$$

$$\log c' = 1.53098$$
$$\log \cos \tfrac{1}{2}(B'-A) = 9.50854-10$$
$$\overline{\phantom{\log \cos \tfrac{1}{2}(B'-A) = } 1.03952}$$

**6.** Derive the law of tangents.

In any triangle $ABC$, we have the Mollweide formulas

$$\frac{a-b}{c} = \frac{\sin \frac{1}{2}(A-B)}{\cos \frac{1}{2}C} \qquad \text{and} \qquad \frac{a+b}{c} = \frac{\cos \frac{1}{2}(A-B)}{\sin \frac{1}{2}C}.$$

Dividing the first by the second,

$$\frac{a-b}{c} \cdot \frac{c}{a+b} = \frac{\sin \frac{1}{2}(A-B)}{\cos \frac{1}{2}C} \cdot \frac{\sin \frac{1}{2}C}{\cos \frac{1}{2}(A-B)} = \tan \tfrac{1}{2}(A-B) \cdot \tan \tfrac{1}{2}C.$$

Since $C = 180^\circ - (A+B)$,    $\tfrac{1}{2}C = 90^\circ - \tfrac{1}{2}(A+B)$    and    $\tan \tfrac{1}{2}C = \cot \tfrac{1}{2}(A+B) = \dfrac{1}{\tan \tfrac{1}{2}(A+B)}$.

Thus,                $\dfrac{a-b}{a+b} = \tan \tfrac{1}{2}(A-B) \cdot \tan \tfrac{1}{2}C = \dfrac{\tan \tfrac{1}{2}(A-B)}{\tan \tfrac{1}{2}(A+B)}$.

The two other forms may be obtained in a similar manner or by cyclic changes of letters on the above form.

## CASE III.

**7.** Solve the triangle $ABC$,  given $a = 2526.4$, $c = 1387.6$,  $B = 54^\circ 24.2'$.

$A + C = 180^\circ - B = 125^\circ 35.8'$
$\tfrac{1}{2}(A+C) = 62^\circ 47.9'$

$\begin{aligned} a &= 2526.4 \\ c &= \underline{1387.6} \\ a - c &= 1138.8 \\ a + c &= 3914.0 \end{aligned}$

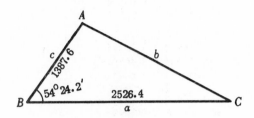

$\tan \tfrac{1}{2}(A-C) = \dfrac{a-c}{a+c} \tan \tfrac{1}{2}(A+C)$      $b = \dfrac{c \sin B}{\sin C}$

$\begin{aligned} \log (a-c) &= 3.05644 \\ \text{colog } (a+c) &= 6.40738-10 \\ \log \tan \tfrac{1}{2}(A+C) &= \underline{0.28907} \\ \log \tan \tfrac{1}{2}(A-C) &= 9.75289-10 \end{aligned}$

$\begin{aligned} \log c &= 3.14227 \\ \log \sin B &= 9.91016-10 \\ \text{colog } \sin C &= \underline{0.26058} \\ \log b &= 3.31301 \\ b &= 2056.0 \end{aligned}$

$\begin{aligned} \tfrac{1}{2}(A-C) &= 29^\circ 30.8' \\ \tfrac{1}{2}(A+C) &= \underline{62^\circ 47.9'} \\ A &= 92^\circ 18.7' \\ C &= 33^\circ 17.1' \end{aligned}$

Check : It is left to the student to check the solution by using the Mollweide formula
$(a+c) \sin \tfrac{1}{2}B = b \cos \tfrac{1}{2}(A-C)$.

**8.** Solve the triangle $ABC$,  given $b = 472.12$, $c = 607.44$, $A = 125^\circ 14.6'$.

$C + B = 180^\circ - A = 54^\circ 45.4'$
$\tfrac{1}{2}(C+B) = 27^\circ 22.7'$

$\begin{aligned} c &= 607.44 \\ b &= \underline{472.12} \\ c - b &= 135.32 \\ c + b &= 1079.56 = 1079.6 \end{aligned}$

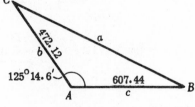

$\tan \tfrac{1}{2}(C-B) = \dfrac{c-b}{c+b} \tan \tfrac{1}{2}(C+B)$      $a = \dfrac{b \sin A}{\sin B}$

$\begin{aligned} \log (c-b) &= 2.13136 \\ \text{colog } (c+b) &= 6.96674-10 \\ \log \tan \tfrac{1}{2}(C+B) &= \underline{9.71422-10} \\ \log \tan \tfrac{1}{2}(C-B) &= 8.81232-10 \end{aligned}$

$\begin{aligned} \log b &= 2.67405 \\ \log \sin A &= 9.91207-10 \\ \text{colog } \sin B &= \underline{0.39644} \\ \log a &= 2.98256 \\ a &= 960.64 \end{aligned}$

$\begin{aligned} \tfrac{1}{2}(C-B) &= 3^\circ 42.8' \\ \tfrac{1}{2}(C+B) &= \underline{27^\circ 22.7'} \\ C &= 31^\circ 5.5' \\ B &= 23^\circ 39.9' \end{aligned}$

Check : To check the solution use the Mollweide formula  $(c+b) \sin \tfrac{1}{2}A = a \cos \tfrac{1}{2}(C-B)$.

**9.** Two adjacent sides of a parallelogram are 3472.7 and 4822.3 ft respectively and the angle between them is $72°14.8'$. Find the length of the longer diagonal.

In triangle $ABC$:    $B = 180° - 72°14.8' = 107°45.2'$
                   $C + A = 72°14.8'$,    $\frac{1}{2}(C+A) = 36°7.4'$

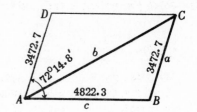

$$c = 4822.3$$
$$a = 3472.7$$
$$c - a = 1349.6$$
$$c + a = 8295.0.$$

$$\tan \tfrac{1}{2}(C-A) = \frac{c-a}{c+a} \tan \tfrac{1}{2}(C+A) \qquad\qquad b = \frac{c \sin B}{\sin C}$$

| | |
|---|---|
| $\log (c-a) = 3.13020$ | $\log c = 3.68326$ |
| $\mathrm{colog}\ (c+a) = 6.08118-10$ | $\log \sin B = 9.97881-10$ |
| $\log \tan \tfrac{1}{2}(C+A) = 9.86322-10$ | $\mathrm{colog} \sin C = 0.16707$ |
| $\log \tan \tfrac{1}{2}(C-A) = 9.07460-10$ | $\log b = 3.82914$ |

$$\tfrac{1}{2}(C-A) = 6°46.3'$$
$$\tfrac{1}{2}(C+A) = 36°\ 7.4'$$

$$b = 6747.4 \text{ ft}$$

$$C = 42°53.7'$$
$$A = 29°21.1'$$

Check : $\qquad\qquad\qquad (c+a)\sin \tfrac{1}{2}B = b \cos \tfrac{1}{2}(C-A)$

| | |
|---|---|
| $\log (c+a) = 3.91882$ | $\log b = 3.82914$ |
| $\log \sin \tfrac{1}{2}B = 9.90727-10$ | $\log \cos \tfrac{1}{2}(C-A) = 9.99696-10$ |
| $3.82609$ | $3.82610$ |

**10.** Derive the half-angle formulas.

Let $ABC$ be any triangle.   Then   $\tan \tfrac{1}{2}A = \sqrt{\dfrac{1 - \cos A}{1 + \cos A}}$   since $\tfrac{1}{2}A$ is always acute.

By the law of cosines,   $\cos A = \dfrac{b^2 + c^2 - a^2}{2bc}$   so that

$$1 - \cos A = 1 - \frac{b^2 + c^2 - a^2}{2bc} = \frac{2bc - b^2 - c^2 + a^2}{2bc} = \frac{a^2 - (b-c)^2}{2bc} = \frac{(a-b+c)(a+b-c)}{2bc}$$

and   $1 + \cos A = 1 + \dfrac{b^2 + c^2 - a^2}{2bc} = \dfrac{2bc + b^2 + c^2 - a^2}{2bc} = \dfrac{(b+c)^2 - a^2}{2bc} = \dfrac{(b+c+a)(b+c-a)}{2bc}.$

Let $a + b + c = 2s$;   then   $a - b + c = (a+b+c) - 2b = 2s - 2b = 2(s-b)$,   $a+b-c = 2(s-c)$, $b + c - a = 2(s-a)$,   and

$$\tan \tfrac{1}{2}A = \sqrt{\frac{1 - \cos A}{1 + \cos A}} = \sqrt{\frac{(a-b+c)(a+b-c)}{2bc} \cdot \frac{2bc}{(b+c+a)(b+c-a)}} = \sqrt{\frac{2(s-b)\cdot 2(s-c)}{2s \cdot 2(s-a)}}$$

$$= \sqrt{\frac{(s-b)(s-c)}{s(s-a)}} = \sqrt{\frac{(s-a)(s-b)(s-c)}{s(s-a)^2}} = \frac{1}{s-a}\sqrt{\frac{(s-a)(s-b)(s-c)}{s}}.$$

Finally, setting   $r = \sqrt{\dfrac{(s-a)(s-b)(s-c)}{s}}$,   $\tan \tfrac{1}{2}A = \dfrac{r}{s-a}$.   The remaining formulas may be obtained by cyclic changes of letters.

## CASE IV.

**11.** Solve the triangle $ABC$, given $a = 643.84$, $b = 778.72$, $c = 912.28$.

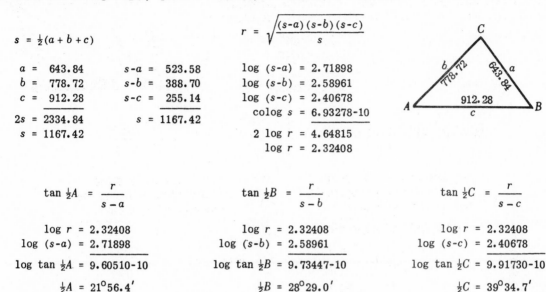

$$s = \tfrac{1}{2}(a + b + c)$$

| | | |
|---|---|---|
| $a =$ 643.84 | $s-a =$ 523.58 |
| $b =$ 778.72 | $s-b =$ 388.70 |
| $c =$ 912.28 | $s-c =$ 255.14 |
| $2s =$ 2334.84 | $s =$ 1167.42 |
| $s =$ 1167.42 | |

$$r = \sqrt{\frac{(s-a)(s-b)(s-c)}{s}}$$

$\log (s-a) = 2.71898$
$\log (s-b) = 2.58961$
$\log (s-c) = 2.40678$
$\text{colog } s = 6.93278-10$
$2 \log r = 4.64815$
$\log r = 2.32408$

| | | |
|---|---|---|
| $\tan \tfrac{1}{2}A = \dfrac{r}{s-a}$ | $\tan \tfrac{1}{2}B = \dfrac{r}{s-b}$ | $\tan \tfrac{1}{2}C = \dfrac{r}{s-c}$ |
| $\log r = 2.32408$ | $\log r = 2.32408$ | $\log r = 2.32408$ |
| $\log (s-a) = 2.71898$ | $\log (s-b) = 2.58961$ | $\log (s-c) = 2.40678$ |
| $\log \tan \tfrac{1}{2}A = 9.60510-10$ | $\log \tan \tfrac{1}{2}B = 9.73447-10$ | $\log \tan \tfrac{1}{2}C = 9.91730-10$ |
| $\tfrac{1}{2}A = 21°56.4'$ | $\tfrac{1}{2}B = 28°29.0'$ | $\tfrac{1}{2}C = 39°34.7'$ |
| $A = 43°52.8'$ | $B = 56°58.0'$ | $C = 79° 9.4'$ |

Check: $A + B + C = 180°0.2'$.

**12.** A triangular field has sides 2025.0, 2450.0, and 1575.0 ft respectively, as shown in the adjoining figure. If the bearing of $AB$ is S 35°30.4′ E, find the bearing of the other two sides.

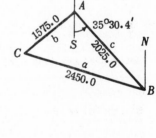

$$s = \tfrac{1}{2}(a + b + c)$$

| | |
|---|---|
| $a = 2450.0$ | $s-a = 575$ |
| $b = 1575.0$ | $s-b = 1450$ |
| $c = 2025.0$ | $s-c = 1000$ |
| $2s = 6050.0$ | $s = 3025$ |
| $s = 3025.0$ | |

$$r = \sqrt{\frac{(s-a)(s-b)(s-c)}{s}}$$

$\log (s-a) = 2.75967$
$\log (s-b) = 3.16137$
$\log (s-c) = 3.00000$
$\text{colog } s = 6.51927-10$
$2 \log r = 5.44031$
$\log r = 2.72016$

| | | |
|---|---|---|
| $\tan \tfrac{1}{2}A = \dfrac{r}{s-a}$ | $\tan \tfrac{1}{2}B = \dfrac{r}{s-b}$ | $\tan \tfrac{1}{2}C = \dfrac{r}{s-c}$ |
| $\log r = 2.72016$ | $\log r = 2.72016$ | $\log r = 2.72016$ |
| $\log (s-a) = 2.75967$ | $\log (s-b) = 3.16137$ | $\log (s-c) = 3.00000$ |
| $\log \tan \tfrac{1}{2}A = 9.96049-10$ | $\log \tan \tfrac{1}{2}B = 9.55879-10$ | $\log \tan \tfrac{1}{2}C = 9.72016-10$ |
| $\tfrac{1}{2}A = 42°23.8'$ | $\tfrac{1}{2}B = 19°54.2'$ | $\tfrac{1}{2}C = 27°42.0'$ |
| $A = 84°47.6'$ | $B = 39°48.4'$ | $C = 55°24.0'$ |

$\angle SAC = 84°47.6' - 35°30.4' = 49°17.2'$; the bearing of $AC$ is S 49°17.2′ W.

$\angle NBC = 35°30.4' + 39°48.4' = 75°18.8'$; the bearing of $BC$ is N 75°18.8′ W.

## SUPPLEMENTARY PROBLEMS

Solve and check each of the oblique triangles $ABC$, given:

13. $c = 78.753$, $A = 33°9.9'$, $C = 81°24.6'$.     *Ans.* $a = 43.571$, $b = 72.432$, $B = 65°25.5'$

14. $b = 730.80$, $B = 42°12.8'$, $C = 109°32.5'$.     *Ans.* $a = 514.73$, $c = 1025.0$, $A = 28°14.7'$

15. $a = 31.259$, $A = 57°59.9'$, $C = 23°36.6'$.     *Ans.* $b = 36.466$, $c = 14.763$, $B = 98°23.5'$

16. $b = 13.218$, $c = 10.004$, $B = 25°57.2'$.     *Ans.* $a = 21.467$, $A = 134°42.2'$, $C = 19°20.6'$

17. $b = 10.884$, $c = 35.730$, $C = 115°33.8'$.     *Ans.* $a = 29.658$, $A = 48°29.2'$, $B = 15°57.0'$

18. $b = 86.425$, $c = 73.463$, $C = 49°18.9'$.     *Ans.* $a = 89.534$, $B = 63°8.3'$, $A = 67°32.8'$
    $a' = 23.147$, $B' = 116°51.7'$, $A' = 13°49.4'$

19. $a = 12.695$, $c = 15.873$, $A = 24°7.4'$.     *Ans.* $b = 25.399$, $B = 125°8.7'$, $C = 30°43.9'$
    $b' = 3.5745$, $B' = 6°36.5'$, $C' = 149°16.1'$

20. $a = 482.33$, $c = 395.71$, $B = 137°31.2'$.     *Ans.* $b = 819.00$, $A = 23°26.2'$, $C = 19°2.6'$

21. $b = 561.23$, $c = 387.19$, $A = 56°43.8'$.     *Ans.* $a = 475.89$, $B = 80°24.4'$, $C = 42°51.8'$

22. $a = 123.79$, $b = 264.23$, $c = 256.04$.     *Ans.* $A = 27°28.2'$, $B = 79°57.0'$, $C = 72°34.8'$

23. $a = 1894.3$, $b = 2246.5$, $c = 3548.8$.     *Ans.* $A = 28°11.8'$, $B = 34°4.8'$, $C = 117°43.2'$

24. A pole, which leans $10°15'$ from the vertical toward the sun, casts a shadow 40.75 ft long when the angle of elevation of the sun is $40°35'$. Find the length of the pole.
    *Ans.* 41.97 ft

25. Two observers $A$ and $B$, on level ground 2875 ft apart, measure the angle of elevation of an airplane as it flies over the line joining them. The angle of elevation at $A$ is $62°45'$ and at $B$ is $50°54'$. Find the distance of the airplane from $A$, from $B$, and above the earth's surface.     *Ans.* 2436 ft, 2790 ft, 2165 ft

26. A tunnel is to be constructed through a mountain from $A$ to $B$. A point $C$, from which both $A$ and $B$ are visible, is 384.8 ft from $A$ and 555.6 ft from $B$. How long is the tunnel if $\angle ACB = 35°42'$?     *Ans.* 330.9 ft

27. Assuming the distance of the sun from the earth to be 92,897,000 mi and the distance of the sun from Mercury to be 35,960,000 mi, find the possible distances of Mercury from the earth when the angle made by Mercury and the sun with the earth as vertex is $8°24.6'$.
    *Ans.* 58,600,000 or 125,190,000 mi

28. A point $B$ is inaccessible and invisible from a point $A$. In order to find the distance $AB$, two points $C$ and $D$ on a line with $A$ and from which $B$ is visible are selected, and $\angle ADB = 55°18'$ and $\angle ACB = 41°36'$ are measured. If $AD = 432.3$ ft and $AC = 521.8$ ft, find $AB$.
    *Ans.* 529.1 ft

# CHAPTER 15

# Areas. Radii of Inscribed and Circumscribed Circles

THE AREA $K$ OF ANY TRIANGLE equals half the product of its base and altitude. Formulas applicable to the four cases of oblique triangles are listed below.

CASE III. Given two sides and the included angle.

$$K = \tfrac{1}{2}bc \sin A = \tfrac{1}{2}ca \sin B = \tfrac{1}{2}ab \sin C$$

For a derivation of these formulas, see Prob. 1. See also Prob. **4.**

CASE I.   Given two angles and a side.

$$K = \frac{a^2 \sin B \sin C}{2 \sin A} = \frac{b^2 \sin C \sin A}{2 \sin B} = \frac{c^2 \sin A \sin B}{2 \sin C}$$

For a derivation of these formulas, see Prob. 2. See also Prob. **5.**

CASE II.   Given two sides and the angle opposite one of them.

A second angle is obtained by using the law of sines and the appropriate formula under Case I is used.                See Prob. **6.**

CASE IV.   Given the three sides.

$$K = \sqrt{s(s-a)(s-b)(s-c)}, \quad \text{where} \quad s = \tfrac{1}{2}(a+b+c).$$

For a derivation, see Problem 3.   See also Problem 7.

FOR ANY TRIANGLE $ABC$,

a) the radius $R$ of the circumscribed circle is

$$R = \frac{a}{2 \sin A} = \frac{b}{2 \sin B} = \frac{c}{2 \sin C}.$$

b) the radius $r$ of the inscribed circle is

$$r = \sqrt{\frac{(s-a)(s-b)(s-c)}{s}}, \quad \text{where} \quad s = \tfrac{1}{2}(a+b+c).$$

For a proof of these statements, see Problem 8.   See also Problems 9-10.

## SOLVED PROBLEMS

$1.$ Derive the formula  $K = \frac{1}{2}bc \sin A$.

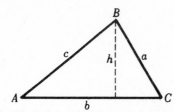 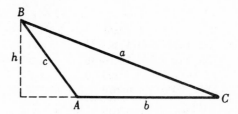

Denoting the altitude drawn to side $b$ of the triangle $ABC$ by $h$, we have from either figure $h = c \sin A$.  Thus,  $K = \frac{1}{2}bh = \frac{1}{2}bc \sin A$.

$2.$ Derive the formula  $K = \dfrac{c^2 \sin A \sin B}{2 \sin C}$.

From Problem 1,  $K = \frac{1}{2}bc \sin A$;  and by the law of sines  $b = \dfrac{c \sin B}{\sin C}$.

Then  $K = \frac{1}{2}bc \sin A = \frac{1}{2}\dfrac{c \sin B}{\sin C}\, c \sin A = \dfrac{c^2 \sin A \sin B}{2 \sin C}$.

$3.$ Derive the formula  $K = \sqrt{s(s-a)(s-b)(s-c)}$.

From the derivations in Problem 10, Chapter 14,

$$\sin^2 \tfrac{1}{2}A = \tfrac{1}{2}(1 - \cos A) = \frac{(a-b+c)(a+b-c)}{4bc} = \frac{2(s-b)\cdot 2(s-c)}{4bc} = \frac{(s-b)(s-c)}{bc}$$

and  $\cos^2 \tfrac{1}{2}A = \tfrac{1}{2}(1 + \cos A) = \dfrac{(b+c+a)(b+c-a)}{4bc} = \dfrac{2s\cdot 2(s-a)}{4bc} = \dfrac{s(s-a)}{bc}$.

Since $\tfrac{1}{2}A < 90°$,  $\sin \tfrac{1}{2}A = \sqrt{\dfrac{(s-b)(s-c)}{bc}}$  and  $\cos \tfrac{1}{2}A = \sqrt{\dfrac{s(s-a)}{bc}}$.  Then

$$K = \tfrac{1}{2}bc \sin A = bc \sin \tfrac{1}{2}A \cos \tfrac{1}{2}A = bc \sqrt{\dfrac{(s-b)(s-c)}{bc}} \sqrt{\dfrac{s(s-a)}{bc}} = \sqrt{s(s-a)(s-b)(s-c)}.$$

$4.$ Find the area of the triangle $ABC$,  given $a = 16.384$, $b = 55.726$, and $C = 27°15.3'$.

This is a Case III triangle  and $K = \frac{1}{2}ab \sin C$.

$$
\begin{aligned}
\log a &= 1.21442\\
\log b &= 1.74606\\
\log \sin C &= 9.66082\text{-}10\\
\text{colog } 2 &= 9.69897\text{-}10\\
\hline
\log K &= 2.32027\\
K &= 209.06
\end{aligned}
$$

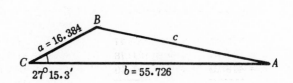

The area is 209.06 square units.

**5.** Find the area of the triangle $ABC$, given $A = 37°14.3'$, $C = 62°29.7'$, and $b = 34.876$.

$B = 180° - (A + C) = 80°16.0'$.

This is a Case I triangle and $K = \dfrac{b^2 \sin C \sin A}{2 \sin B}$.

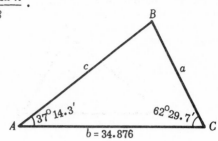

$$
\begin{aligned}
2 \log b &= 3.08506 \\
\log \sin C &= 9.94791\text{-}10 \\
\log \sin A &= 9.78185\text{-}10 \\
\text{colog } 2 &= 9.69897\text{-}10 \\
\text{colog } \sin B &= 0.00630 \\
\hline
\log K &= 2.52009 \\
K &= 331.20 \text{ square units}
\end{aligned}
$$

**6.** Find the area of the triangle $ABC$, given $b = 28.642$, $c = 44.280$, and $B = 23°18.6'$.

This is a Case II triangle in which there may be two solutions.

$$\sin C = \frac{c \sin B}{b}$$

$$
\begin{aligned}
\log c &= 1.64621 \\
\log \sin B &= 9.59737\text{-}10 \\
\text{colog } b &= 8.54300\text{-}10 \\
\hline
\log \sin C &= 9.78658\text{-}10
\end{aligned}
$$

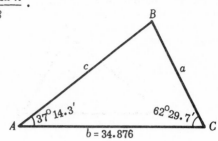

$C = 37°43.0'$  and  $C' = 180° - C = 142°17.0'$
$A = 180° - (B + C) = 118°58.4'$  and  $A' = 180° - (B + C') = 14°24.4'$

Area of $ABC$ is $K = \dfrac{c^2 \sin A \sin B}{2 \sin C}$.       Area of $ABC'$ is $K = \dfrac{c^2 \sin A' \sin B}{2 \sin C'}$.

$$
\begin{aligned}
2 \log c &= 3.29242 \\
\log \sin A &= 9.94193\text{-}10 \\
\log \sin B &= 9.59737\text{-}10 \\
\text{colog } 2 &= 9.69897\text{-}10 \\
\text{colog } \sin C &= 0.21342 \\
\hline
\log K &= 2.74411 \\
K &= 554.76
\end{aligned}
\qquad
\begin{aligned}
2 \log c &= 3.29242 \\
\log \sin A' &= 9.39586\text{-}10 \\
\log \sin B &= 9.59737\text{-}10 \\
\text{colog } 2 &= 9.69897\text{-}10 \\
\text{colog } \sin C' &= 0.21342 \\
\hline
\log K &= 2.19804 \\
K &= 157.78
\end{aligned}
$$

Two triangles are determined, their areas being 554.76 and 157.78 square units  respectively.

**7.** Find the area of the triangle $ABC$, given $a = 255.18$, $b = 290.87$, and $c = 419.25$.

This is a Case IV triangle and $K = \sqrt{s(s-a)(s-b)(s-c)}$.

$s_\bullet = \frac{1}{2}(a + b + c)$       $K = \sqrt{s(s-a)(s-b)(s-c)}$

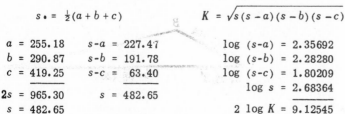
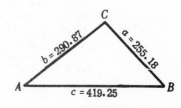

$$
\begin{aligned}
a &= 255.18 & s-a &= 227.47 \\
b &= 290.87 & s-b &= 191.78 \\
c &= 419.25 & s-c &= 63.40 \\
\hline
2s &= 965.30 & s &= 482.65 \\
s &= 482.65 &&
\end{aligned}
$$

$$
\begin{aligned}
\log (s-a) &= 2.35692 \\
\log (s-b) &= 2.28280 \\
\log (s-c) &= 1.80209 \\
\log s &= 2.68364 \\
\hline
2 \log K &= 9.12545 \\
\log K &= 4.56272 \\
K &= 36,536
\end{aligned}
$$

The area is 36,536 square units.

**8.** Show that, in any triangle $ABC$,

a) the radius $R$ of the circumscribed circle is given by $R = \dfrac{a}{2 \sin A} = \dfrac{b}{2 \sin B} = \dfrac{c}{2 \sin C}$ and

b) the radius $r$ of the inscribed circle is given by $r = \sqrt{(s-a)(s-b)(s-c)/s}$ .

a) Let $O$ be the center of the circumscribed circle. Join $O$ to $B$ and $C$, and draw through $O$ the perpendicular to $BC$ meeting it at $D$. Since triangle $OBC$ is isosceles, $OD$ bisects $BC$.

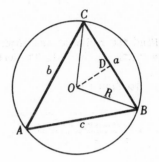

Since $A = \angle BAC$ and $\angle BOC$ intercept the same arc, $A = \frac{1}{2}\angle BOC = \angle BOD$. Then, in right triangle $BOD$,

$$R = OB = \frac{BD}{\sin \angle BOD} = \frac{a/2}{\sin A} = \frac{a}{2 \sin A} .$$

By the law of sines, $\dfrac{a}{\sin A} = \dfrac{b}{\sin B} = \dfrac{c}{\sin C}$ , it follows that

$$R = \frac{a}{2 \sin A} = \frac{b}{2 \sin B} = \frac{c}{2 \sin C} .$$

b) Denote the center of the inscribed circle (intersection of the bisectors of the interior angles of triangle $ABC$) by $I$ and the points of tangency of the circle and triangle by $D,E,F$. Since $I$ is equidistant from the sides of the triangle,

$$ID = IE = IF = r.$$

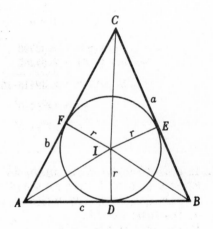

Now $K$ = area $ABC$ = area $AIB$ + area $BIC$ + area $CIA$

$$= \tfrac{1}{2}cr + \tfrac{1}{2}ar + \tfrac{1}{2}br$$

$$= \tfrac{1}{2}r(a+b+c) = \tfrac{1}{2}r(2s) = rs.$$

But by Prob. 3, $K = \sqrt{s(s-a)(s-b)(s-c)} = rs$. Hence,

$$r = \sqrt{\frac{s(s-a)(s-b)(s-c)}{s^2}} = \sqrt{\frac{(s-a)(s-b)(s-c)}{s}} .$$

**9.** Find the radii of the circumscribed and inscribed circles of triangle $ABC$, given $A = 41°15.8'$, $C = 69°52.4'$, and $b = 387.48$.

$$B = 180° - (C + A) = 68°51.8'$$

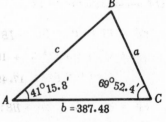

To find $R$:   $R = \dfrac{b}{2 \sin B}$

log $b$ = 2.58825
colog 2 = 9.69897-10
colog sin $B$ = 0.03025
―――――――
log $R$ = 2.31747
$R$ = 207.71

To find $r$ :

$$a = \frac{b \sin A}{\sin B} \qquad\qquad c = \frac{b \sin C}{\sin B}$$

log $b$ = 2.58825          log $b$ = 2.58825
log sin $A$ = 9.81923-10    log sin $C$ = 9.97264-10
colog sin $B$ = 0.03025     colog sin $B$ = 0.03025
――――――          ――――――
log $a$ = 2.43773          log $c$ = 2.59114
$a$ = 273.99              $c$ = 390.07

$$r = \sqrt{\frac{(s-a)(s-b)(s-c)}{s}}$$

| | | | |
|---|---|---|---|
| $a =$ | 273.99 | $s-a =$ 251.78 | $\log (s-a) =$ 2.40102 |
| $b =$ | 387.48 | $s-b =$ 138.29 | $\log (s-b) =$ 2.14079 |
| $c =$ | 390.07 | $s-c =$ 135.70 | $\log (s-c) =$ 2.13258 |
| $2s =$ | 1051.54 | $s =$ 525.77 | colog $s =$ 7.27920-10 |
| $s =$ | 525.77 | | $2 \log r =$ 3.95359 |
| | | | $\log r =$ 1.97680 |
| | | | $r =$ 94.798 |

**10.** Find the radii of the inscribed and circumscribed circles of the triangle $ABC$ whose sides are $a = 375.68$, $b = 512.37$, and $c = 742.05$.

To find $r$:

$$r = \sqrt{\frac{(s-a)(s-b)(s-c)}{s}}$$

| | | | |
|---|---|---|---|
| $a =$ | 375.68 | $s-a =$ 439.37 | $\log (s-a) =$ 2.64283 |
| $b =$ | 512.37 | $s-b =$ 302.68 | $\log (s-b) =$ 2.48098 |
| $c =$ | 742.05 | $s-c =$ 73.00 | $\log (s-c) =$ 1.86332 |
| $2s =$ | 1630.10 | $s =$ 815.05 | colog $s =$ 7.08882-10 |
| $s =$ | 815.05 | | $2 \log r =$ 4.07595 |
| | | | $\log r =$ 2.03798 |
| | | | $r =$ 109.14 |

To find $R$:

$$\tan \tfrac{1}{2}A = \frac{r}{s-a} \qquad\qquad R = \frac{a}{2 \sin A}$$

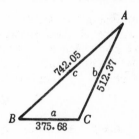

| | |
|---|---|
| $\log r = 2.03798$ | $\log a = 2.57482$ |
| $(-) \log (s-a) = 2.64283$ | colog $2 = 9.69897-10$ |
| $\log \tan \tfrac{1}{2}A = 9.39515-10$ | colog $\sin A = 0.32982$ |
| $\tfrac{1}{2}A = 13°57.0'$ | $\log R = 2.60361$ |
| $A = 27°54.0'$ | $R = 401.43$ |

**11.** In a quadrangular field $ABCD$, $AB$ runs N 62°10′ E 11.4 rd, $BC$ runs N 22°20′ W 19.8 rd, and $CD$ runs S 40°40′ W 15.3 rd. $DA$ runs S 32°10′ E but cannot be measured. Find:
a) the length of $DA$,
b) the area of the field.

In the figure $SN$ is the north-south line through $D$, the points $E, F, G$ are the feet of the perpendiculars to this line through $A, B, C$ respectively, and the lines $AH$ and $CI$ are perpendicular to $BF$.

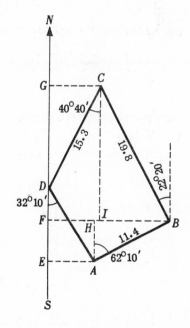

a) $FB = FI + IB = GC + IB$

$= 15.3 \sin 40°40′ + 19.8 \sin 22°20′$

$= 9.97 + 7.52 = 17.49.$

$FB = FH + HB = EA + HB$; hence

$EA = FB - HB$

$= 17.49 - 11.4 \sin 62°10′ = 17.49 - 10.08 = 7.41.$

Since $EA = DA \sin 32°10′$, $DA = \dfrac{7.41}{\sin 32°10′} = 13.9$ rd.

*b*) Area $ABCD$ = area $EABF$ + area $FBCG$ − area $EAD$ − area $GCD$

$$= \tfrac{1}{2}(EA + FB)AH + \tfrac{1}{2}(FB + GC)CI - \tfrac{1}{2}EA \cdot ED - \tfrac{1}{2}GC \cdot GD.$$

Now $EA$ = 7.41, $FB$ = 17.49, $AH$ = 11.4 cos $62°10'$ = 5.32, $GC$ = 9.97,
$CI$ = 19.8 cos $22°20'$ = 18.32, $ED$ = 13.9 cos $32°10'$ = 11.77,
$GD$ = 15.3 cos $40°40'$ = 11.61. Then:

Area $ABCD$ = $\tfrac{1}{2}(7.41 + 17.49)(5.32) + \tfrac{1}{2}(17.49 + 9.97)(18.32) - \tfrac{1}{2}(7.41)(11.77) - \tfrac{1}{2}(9.97)(11.61)$

= 66.23 + 251.53 − 43.61 − 57.88 = 216.27 or 216 sq. rd.

---

**12.** Prove that the area of a quadrilateral is equal to half the product of its diagonals and the sine of the included angle. See Fig. (*a*) below.

Let the diagonals of the quadrilateral $ABCD$ intersect in $O$, let $\theta$ be an angle of intersection of the diagonals, and let $O$ separate the diagonals into segments of length $p,q$; $r,s$ as in the figure.

Area $ABCD$ = area $AOB$ + area $AOD$ + area $BOC$ + area $DOC$

$$= \tfrac{1}{2}rp \sin \theta + \tfrac{1}{2}qr \sin (180° - \theta) + \tfrac{1}{2}ps \sin (180° - \theta) + \tfrac{1}{2}qs \sin \theta$$

$$= \tfrac{1}{2}(pr + qr + ps + qs) \sin \theta = \tfrac{1}{2}(p + q)(r + s) \sin \theta.$$

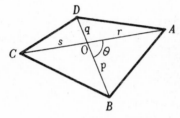
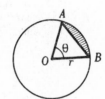

Fig. (*a*) Prob. 12            Fig. (*b*) Prob. 13

---

**13.** Prove that the area $K$ of the smaller segment (shaded) of a circle of radius $r$ and center $O$ cut off by the chord $AB$ of the Fig. (*b*) above is given by $K = \tfrac{1}{2}r^2(\theta - \sin \theta)$, where $\theta$ radians is the central angle intercepted by the chord.

The required area is the difference between the area of sector $AOB$ and triangle $AOB$.

The area $S$ of the sector $AOB$ is to the area of the circle as the arc $AB$ is to the circumference of the circle, that is, $\dfrac{S}{\pi r^2} = \dfrac{r\theta}{2\pi r}$ and $S = \tfrac{1}{2}r^2\theta$.

The area of triangle $AOB$ = $\tfrac{1}{2}r \cdot r \sin \theta = \tfrac{1}{2}r^2 \sin \theta.$

Thus,            $K = \tfrac{1}{2}r^2\theta - \tfrac{1}{2}r^2 \sin \theta = \tfrac{1}{2}r^2(\theta - \sin \theta).$

---

**14.** Three circles with centers $A,B,C$ have respective radii 50, 30, 20 in. and are tangent to each other externally. Find the area of the *curvilinear* triangle formed by the three circles.

Let the points of tangency of the circles be $R,S,T$ as in the figure. The required area is the difference between the area of triangle $ABC$ and the sum of the areas of the three sectors $ART$, $BRS$, and $SCT$.

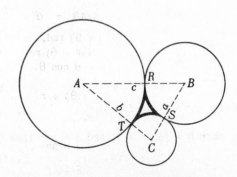

Since the join of the centers of any two circles passes through their point of tangency, $a = BC = 50$, $b = CA = 70$, and $c = AB = 80$ in. Then

$$s = \tfrac{1}{2}(a + b + c) = 100, \quad s - a = 50, \quad s - b = 30, \quad s - c = 20, \quad \text{and}$$

$$K = \text{area } ABC = \sqrt{s(s-a)(s-b)(s-c)} = \sqrt{100(50)(30)(20)} = 1000\sqrt{3} = 1732.$$

Since $r = K/s = 17.32$,

$$\tan \tfrac{1}{2}A = \frac{r}{s-a} = \frac{17.32}{50} = 0.3464, \quad \tfrac{1}{2}A = 19°6', \quad A = 38°12' = 0.667 \text{ rad},$$

$$\tan \tfrac{1}{2}B = \frac{r}{s-b} = \frac{17.32}{30} = 0.5773, \quad \tfrac{1}{2}B = 30°0', \quad B = 60°0' = 1.047 \text{ rad},$$

$$\tan \tfrac{1}{2}C = \frac{r}{s-c} = \frac{17.32}{20} = 0.8660, \quad \tfrac{1}{2}C = 40°54', \quad C = 81°48' = 1.428 \text{ rad}.$$

Area $ART = \tfrac{1}{2}r^2\theta = \tfrac{1}{2}(50)^2(0.667) = 833.75$, area $BRS = \tfrac{1}{2}(30)^2(1.047) = 471.15$, area $CST = \tfrac{1}{2}(20)^2(1.428) = 285.60$, and their sum is 1590.50.

The required area is $1732 - 1590.50 = 141.50$ or 142 square inches.

**15.** *a*) Derive a formula for the length $L$ of an open driving belt.

   *b*) Find, to the nearest tenth of an inch, the length of a driving belt running around two pulleys of radii 15 in. and 5 in. respectively if the distance between the centers of the wheels is 30 in.

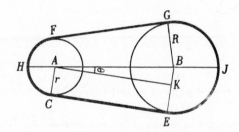

*a*) Let the two wheels of radii $r$ and $R$ be centered at $A$ and $B$ respectively, and let $d$ denote the distance between the centers. The required length of belting is

$$L = \text{arc } GJE + CE + \text{arc } CHF + FG = 2(\text{arc } JE + \text{arc } CH + CE).$$

Since $CE$ is tangent to the two circles, $AC$ and $BE$ are parallel. Let the line through $A$ parallel to $CE$ meet $BE$ at $K$. Denote the angle $BAK$, measured in radians, by $\theta$. Then

$$\sin \theta = BK/AB = (R-r)/d \quad \text{and} \quad \theta = \text{Arc sin } (R-r)/d.$$

$\angle JBE = \angle BAC = (\tfrac{1}{2}\pi + \theta)$ rad, and arc $JE = R(\tfrac{1}{2}\pi + \theta)$ length units.
$\angle HAC = \pi - \angle BAC = (\tfrac{1}{2}\pi - \theta)$ rad, and arc $CH = r(\tfrac{1}{2}\pi - \theta)$ length units.
$CE = AK = AB \cos \theta = d \cos \theta.$

Then
$$L = 2[R(\tfrac{1}{2}\pi + \theta) + r(\tfrac{1}{2}\pi - \theta) + d \cos \theta] = (R+r)\pi + 2(R-r)\theta + 2d \cos \theta,$$
where $\theta = \text{Arc sin } (R-r)/d$.

*b*) Here $R = 15$, $r = 5$, and $d = 30$; then $\theta = \text{Arc sin } (R-r)/d = \text{Arc sin } 1/3 = 0.340$ rad and

$$L = (15+5)(3.142) + 2(15-5)(0.340) + 2(30)(2\sqrt{2}/3)$$
$$= 62.84 + 6.80 + 56.56 = 126.2 \text{ in}.$$

**16.** *a*) Derive a formula for the length $L$ of a crossed driving belt.

   *b*) Find, to the nearest tenth of an inch, the length of a driving belt criss-crossed about two pulleys of radii 10 in. and 5 in. respectively if the distance between the centers of the wheels is 30 in.

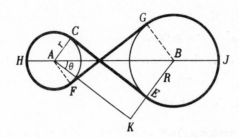

*a*) Let the two wheels of radii $r$ and $R$ be centered at $A$ and $B$ respectively, and let $d$ denote the distance between the centers. The required length of belting is

$$L = 2(\text{arc } JE + \text{arc } CH + CE).$$

Let the line through $A$ parallel to $CE$ meet $BE$ extended at $K$. Denote the angle $BAK$, measured in radians, by $\theta$. Then

$$\sin \theta = BK/AB = (R+r)/d \quad \text{and} \quad \theta = \text{Arc sin } (R+r)/d.$$

$\angle JBE = \pi - \angle ABE = (\frac{1}{2}\pi + \theta)$ rad, and arc $JE = R(\frac{1}{2}\pi + \theta)$ length units.
$\angle HAC = \angle JBE = (\frac{1}{2}\pi + \theta)$ rad, and arc $CH = r(\frac{1}{2}\pi + \theta)$ length units.
$CE = AK = d \cos \theta$.

Then

$$L = 2[R(\tfrac{1}{2}\pi + \theta) + r(\tfrac{1}{2}\pi + \theta) + d \cos \theta] = (R + r)(\pi + 2\theta) + 2d \cos \theta,$$

where $\theta = \text{Arc sin } (R+r)/d$.

*b*) Here $R = 10$, $r = 5$, and $d = 30$; then $\theta = \text{Arc sin } (R+r)/d = \text{Arc sin } \frac{1}{2} = 0.524$ rad    and

$$L = (10+5)(\pi + 1.048) + 2(30)(\tfrac{1}{2}\sqrt{3}) = 62.85 + 51.96 = 114.8 \text{ in.}$$

## SUPPLEMENTARY PROBLEMS

Find the area of the triangle $ABC$, given:

17. $b = 23.84$, $c = 35.26$, $A = 50°32'$.             *Ans.* 324.5 square units

18. $a = 456.32$, $b = 586.84$, $C = 28°16.6'$.     *Ans.* 63430 square units

19. $a = 512.32$, $B = 52°14.6'$, $C = 63°45.6'$.   *Ans.* 103,550 square units

20. $b = 444.85$, $A = 110°15.8'$, $B = 30°10.4'$.  *Ans.* 117,620 square units

21. $a = 384.22$, $b = 492.86$, $c = 677.98$.      *Ans.* 93,094 square units

22. $a = 28.165$, $b = 60.152$, $c = 51.177$.      *Ans.* 718.85 square units

23. Find the radius of the circumscribed circle of the triangle $ABC$, given that $b = 28.944$ and $B = 37°14.4'$.    *Ans.* 23.914

24. Find the radius of the inscribed circle of the triangle $ABC$, given that $a = 5.478$, $b = 4.823$ and $c = 6.019$.    *Ans.* 1.532

25. The sides of a triangular plot are 48.50, 64.70 and 88.80 ft, respectively. Find $a$) the minimum radius of action of an automatic lawn sprinkler which will water all parts of the plot and $b$) the radius of the largest circular flower bed which can be constructed on the plot. *Ans.* $a$) 45.46 ft, $b$) 15.17 ft

26. If $K$ is the area, $R$ is the radius of the circumscribed circle, and $r$ is the radius of the inscribed circle of the triangle $ABC$, prove:

$a$) $K = 2R^2 \sin A \sin B \sin C$,

$b$) $K = abc/4R$,

$c$) $K = rR(\sin A + \sin B + \sin C)$.

# CHAPTER 16

# Inverse Trigonometric Functions

**INVERSE TRIGONOMETRIC FUNCTIONS.** The equation

1)                                       $x = \sin y$

defines a unique value of $x$ for each given angle $y$. But when $x$ is given, the equation may have no solution or many solutions. For example: if $x = 2$, there is no solution, since the sine of an angle never exceeds 1; if $x = \frac{1}{2}$, there are many solutions $y = 30°$, $150°$, $390°$, $510°$, $-210°$, $-330°$, ....

To express $y$ as a function of $x$, we will write

2)                                       $y = \text{arc sin } x.$

In spite of the use of the word *arc*, 2) is to be interpreted as stating that '$y$ is an angle whose sine is $x$'. Similarly we shall write $y = \text{arc cos } x$ if $x = \cos y$, $y = \text{arc tan } x$ if $x = \tan y$, etc.

The notation $y = \sin^{-1} x$, $y = \cos^{-1} x$, etc., (to be read 'inverse sine of $x$, inverse cosine of $x$', etc.) are less frequently used since $\sin^{-1} x$ may be confused with $\dfrac{1}{\sin x} = (\sin x)^{-1}$.

**GRAPHS OF THE INVERSE TRIGONOMETRIC FUNCTIONS.** The graph of $y = \text{arc sin } x$ is the graph of $x = \sin y$ and differs from the graph of $y = \sin x$ of Chapter 9 in that the roles of $x$ and $y$ are interchanged. Thus, the graph of $y = \text{arc sin } x$ is a sine curve drawn on the $y$-axis instead of the $x$-axis.

Similarly the graphs of the remaining inverse trigonometric functions are those of the corresponding trigonometric functions except that the roles of $x$ and $y$ are interchanged.

**PRINCIPAL VALUES.** It is at times necessary to consider the inverse trigonometric functions as single valued (i.e., one value of $y$ corresponding to each admissible value of $x$). To do this, we agree to select one out of the many angles corresponding to the given value of $x$. For example, when $x = \frac{1}{2}$, we shall agree to select the value $y = 30°$ and when $x = -\frac{1}{2}$, we shall agree to select the value $y = -30°$. This selected value is called the *principal value* of arc sin $x$. When only the principal value is called for, we shall write Arc sin $x$, Arc cos $x$, etc. The portions of the graphs on which the principal values of each of the inverse trigonometric functions lie are shown in the figures below by a heavier line.

When $x$ is positive or zero and the inverse function exists, the principal value is defined as that value of $y$ which lies between 0 and $\frac{1}{2}\pi$ inclusive. For example:

$$\text{Arc sin } \sqrt{3}/2 = \pi/3 \quad \text{since} \quad \sin \pi/3 = \sqrt{3}/2 \quad \text{and} \quad 0 < \pi/3 < \pi/2 \,,$$

$$\text{Arc cos } \sqrt{3}/2 = \pi/6 \quad \text{since} \quad \cos \pi/6 = \sqrt{3}/2 \quad \text{and} \quad 0 < \pi/6 < \pi/2 \,,$$

$$\text{Arc tan } 1 = \pi/4 \qquad \text{since} \quad \tan \pi/4 = 1 \qquad \text{and} \quad 0 < \pi/4 < \pi/2 \,.$$

When $x$ is negative and the inverse function exists, the principal value is defined as follows:

$$-\tfrac{1}{2}\pi \leqq \text{Arc sin } x < 0 \qquad\qquad \tfrac{1}{2}\pi < \text{Arc cot } x < \pi$$

$$\tfrac{1}{2}\pi < \text{Arc cos } x \leqq \pi \qquad\qquad -\pi \leqq \text{Arc sec } x < -\tfrac{1}{2}\pi$$

$$-\tfrac{1}{2}\pi < \text{Arc tan } x < 0 \qquad\qquad -\pi < \text{Arc csc } x \leqq -\tfrac{1}{2}\pi$$

For example:

$$\text{Arc sin } (-\sqrt{3}/2) = -\pi/3 \qquad\qquad \text{Arc cot } (-1) = 3\pi/4$$

$$\text{Arc cos } (-1/2) = 2\pi/3 \qquad\qquad \text{Arc sec } (-2/\sqrt{3}) = -5\pi/6$$

$$\text{Arc tan } (-1/\sqrt{3}) = -\pi/6 \qquad\qquad \text{Arc csc } (-\sqrt{2}) = -3\pi/4$$

Note. Authors vary in defining the principal values of the inverse functions when $x$ is negative. The definitions given above are the most convenient for the calculus.

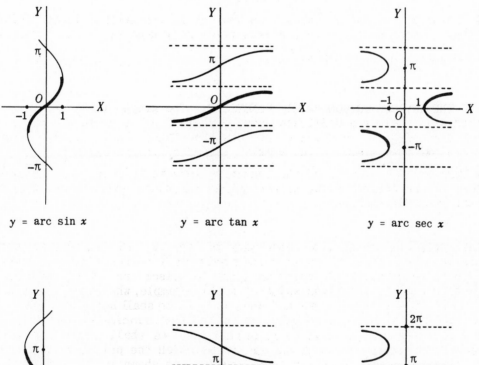

y = arc sin x          y = arc tan x          y = arc sec x

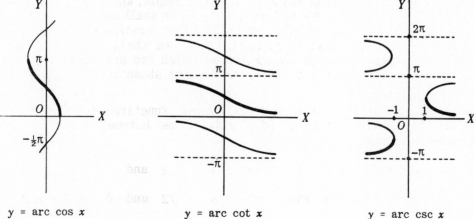

y = arc cos x          y = arc cot x          y = arc csc x

GENERAL VALUES OF THE INVERSE TRIGONOMETRIC FUNCTIONS. Let $y$ be an inverse trigonometric function of $x$. Since the value of a trigonometric function of $y$ is known, there are determined in general two positions for the terminal side of the angle $y$ (see Chapter 2). Let $y_1$ and $y_2$ respectively be angles determined by the two positions of the terminal side. Then the totality of values of $y$ consist of the angles $y_1$ and $y_2$, together with all angles coterminal with them, that is,

$$y_1 + 2n\pi \qquad \text{and} \qquad y_2 + 2n\pi$$

where $n$ is any positive or negative integer, or is zero.

One of the values $y_1$ or $y_2$ may always be taken as the principal value of the inverse trigonometric function.

EXAMPLE.  Write expressions for the general value of $a$) arc sin 1/2, $b$) arc cos (–1), $c$) arc tan (–1).

$a$) The principal value of arc sin 1/2 is $\pi/6$, and a second value (not coterminal with the principal value) is $5\pi/6$. The general value of arc sin 1/2 is given by

$$\pi/6 + 2n\pi, \quad 5\pi/6 + 2n\pi$$

where $n$ is any positive or negative integer, or is zero.

$b$) The principal value is $\pi$ and there is no other value not coterminal with it. Thus, the general value is given by $\pi + 2n\pi$, where $n$ is a positive or negative integer, or is zero.

$c$) The principal value is $-\pi/4$, and a second value (not coterminal with the principal value) is $3\pi/4$. Thus, the general value is given by

$$-\pi/4 + 2n\pi, \quad 3\pi/4 + 2n\pi$$

where $n$ is a positive or negative integer, or is zero.

## SOLVED PROBLEMS

1. Find the principal value of each of the following.

$a$) Arc sin $0 = 0$
$b$) Arc cos $(-1) = \pi$
$c$) Arc tan $\sqrt{3} = \pi/3$
$d$) Arc cot $\sqrt{3} = \pi/6$

$e$) Arc sec $2 = \pi/3$
$f$) Arc csc $(-\sqrt{2}) = -3\pi/4$
$g$) Arc cos $0 = \pi/2$
$h$) Arc sin $(-1) = -\pi/2$

$i$) Arc tan $(-1) = -\pi/4$
$j$) Arc cot $0 = \pi/2$
$k$) Arc sec $(-\sqrt{2}) = -3\pi/4$
$l$) Arc csc $(-2) = -5\pi/6$

2. Express the principal value of each of the following to the nearest minute.

$a$) Arc sin $0.3333 = 19°28'$
$b$) Arc cos $0.4000 = 66°25'$
$c$) Arc tan $1.5000 = 56°19'$
$d$) Arc cot $1.1875 = 40° 6'$
$e$) Arc sec $1.0324 = 14°24'$
$f$) Arc csc $1.5082 = 41°32'$

$g$) Arc sin $(-0.6439) = -40°5'$
$h$) Arc cos $(-0.4519) = 116°52'$
$i$) Arc tan $(-1.4400) = -55°13'$
$j$) Arc cot $(-0.7340) = 126°17'$
$k$) Arc sec $(-1.2067) = -145°58'$
$l$) Arc csc $(-4.1923) = -166°12'$

3. Verify each of the following.

$a$) sin ( Arc sin 1/2) = sin $\pi/6$ = 1/2
$b$) cos [Arc cos $(-1/2)$] = cos $2\pi/3$ = $-1/2$
$c$) cos [Arc sin$(-\sqrt{2}/2)$] = cos $(-\pi/4)$ = $\sqrt{2}/2$
$d$) Arc sin (sin $\pi/3$) = Arc sin $\sqrt{3}/2$ = $\pi/3$

$e$) Arc cos [cos $(-\pi/4)$] = Arc cos $\sqrt{2}/2$ = $\pi/4$
$f$) Arc sin (tan $3\pi/4$) = Arc sin $(-1)$ = $-\pi/2$
$g$) Arc cos [tan $(-5\pi/4)$] = Arc cos $(-1)$ = $\pi$

4. Verify each of the following.

   $a$) Arc sin $\sqrt{2}/2$ – Arc sin $1/2$ = $\pi/4$ – $\pi/6$ = $\pi/12$

   $b$) Arc cos $0$ + Arc tan $(-1)$ = $\pi/2$ + $(-\pi/4)$ = $\pi/4$ = Arc tan 1

5. Evaluate each of the following:
   $a$) cos (Arc sin 3/5),   $b$) sin [Arc cos (–2/3)],   $c$) tan [Arc sin (–3/4)].

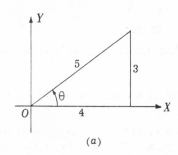
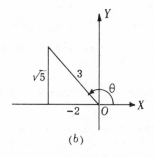
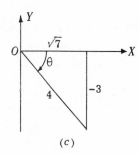

   (a)                    (b)                    (c)

   $a$) Let $\theta$ = Arc sin 3/5; then sin $\theta$ = 3/5, $\theta$ being a first quadrant angle.  From Fig.(a),
   
   cos (Arc sin 3/5) = cos $\theta$ = 4/5.

   $b$) Let $\theta$ = Arc cos (–2/3); then cos $\theta$ = –2/3, $\theta$ being a second quadrant angle.  From Fig.(b),
   
   sin [Arc cos (–2/3)] = sin $\theta$ = $\sqrt{5}/3$.

   $c$) Let $\theta$ = Arc sin (–3/4); then sin $\theta$ = –3/4, $\theta$ being a fourth quadrant angle.  From Fig.(c),
   
   tan [Arc sin (–3/4)] = tan $\theta$ = –3/$\sqrt{7}$ = –3$\sqrt{7}$/7.

6. Evaluate  sin ( Arc sin 12/13 + Arc sin 4/5).

   Let $\theta$ = Arc sin 12/13  and
       $\phi$ = Arc sin 4/5.

   Then sin $\theta$ = 12/13 and sin $\phi$ = 4/5, $\theta$ and $\phi$ being first quadrant angles.  From the adjoining figures,

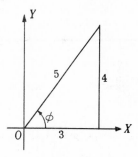

   sin (Arc sin 12/13 + Arc sin 4/5) = sin($\theta$ + $\phi$)

   = sin $\theta$ cos $\phi$ + cos $\theta$ sin $\phi$

   = $\dfrac{12}{13} \cdot \dfrac{3}{5}$ + $\dfrac{5}{13} \cdot \dfrac{4}{5}$ = $\dfrac{56}{65}$ .

7. Evaluate  cos ( Arc tan 15/8 – Arc sin 7/25).

   Let $\theta$ = Arc tan 15/8  and
       $\phi$ = Arc sin 7/25.

   Then tan $\theta$ = 15/8 and sin $\phi$ = 7/25, $\theta$ and $\phi$ being first quadrant angles.  From the adjoining figures,

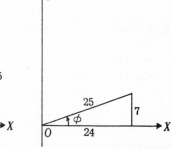

   cos (Arc tan 15/8 – Arc sin 7/25) = cos($\theta$ – $\phi$)

   = cos $\theta$ cos $\phi$ + sin $\theta$ sin $\phi$

   = $\dfrac{8}{17} \cdot \dfrac{24}{25}$ + $\dfrac{15}{17} \cdot \dfrac{7}{25}$ = $\dfrac{297}{425}$ .

**8.** Evaluate  sin (2 Arc tan 3).

Let $\theta$ = Arc tan 3; then tan $\theta$ = 3, $\theta$ being a first quadrant angle.

From the adjoining figure,    sin (2 Arc tan 3) = sin 2$\theta$

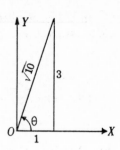

$$= 2 \sin \theta \cos \theta$$
$$= 2(3/\sqrt{10})(1/\sqrt{10})$$
$$= 3/5$$

**9.** Show that  Arc sin $1/\sqrt{5}$ + Arc sin $2/\sqrt{5}$ = $\pi/2$.

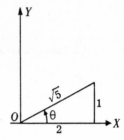

Let $\theta$ = Arc sin $1/\sqrt{5}$ and $\phi$ = Arc sin $2/\sqrt{5}$; then sin $\theta$ = $1/\sqrt{5}$ and sin $\phi$ = $2/\sqrt{5}$, each angle terminating in the first quadrant.  We are to show that $\theta+\phi$ = $\pi/2$ or, taking the sines of both members, that  sin ($\theta+\phi$) = sin $\pi/2$.

From the adjoining figures,

$$\sin (\theta + \phi) = \sin \theta \cos \phi + \cos \theta \sin \phi$$
$$= \frac{1}{\sqrt{5}} \cdot \frac{1}{\sqrt{5}} + \frac{2}{\sqrt{5}} \cdot \frac{2}{\sqrt{5}} = 1 = \sin \pi/2.$$

**10.** Show that  2 Arc tan 1/2 = Arc tan 4/3.

Let $\theta$ = Arc tan 1/2  and $\phi$ = Arc tan 4/3; then tan $\theta$ = 1/2  and tan $\phi$ = 4/3.

We are to show that 2$\theta$ = $\phi$ or,  taking the tangents of both members, that tan 2$\theta$ = tan $\phi$.

Now  $\tan 2\theta = \dfrac{2 \tan \theta}{1 - \tan^2 \theta} = \dfrac{2(1/2)}{1 - (1/2)^2} = 4/3 = \tan \phi.$

**11.** Show that  Arc sin 77/85 – Arc sin 3/5 = Arc cos 15/17.

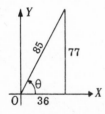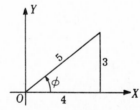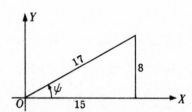

Let $\theta$ = Arc sin 77/85, $\phi$ = Arc sin 3/5, and $\psi$ = Arc cos 15/17; then sin $\theta$ = 77/85, sin $\phi$ = 3/5, and cos $\psi$ = 15/17, each angle terminating in the first quadrant. Taking the sine of both members of the given relation, we are to show that  sin ($\theta - \phi$) = sin $\psi$. From the figures,

$$\sin (\theta - \phi) = \sin \theta \cos \phi - \cos \theta \sin \phi = \frac{77}{85} \cdot \frac{4}{5} - \frac{36}{85} \cdot \frac{3}{5} = \frac{8}{17} = \sin \psi.$$

**12.** Show that  Arc cot 43/32 – Arc tan 1/4 = Arc cos 12/13.

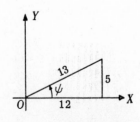

Let $\theta$ = Arc cot 43/32, $\phi$ = Arc tan 1/4, and $\psi$ = Arc cos 12/13; then cot $\theta$ = 43/32, tan $\phi$ = 1/4,  and cos $\psi$ = 12/13,  each angle terminating in the first quadrant. Taking the tangent of both members of the given relation, we are to show that  tan($\theta - \phi$) = tan $\psi$.

$$\tan (\theta - \phi) = \frac{\tan \theta - \tan \phi}{1 + \tan \theta \tan \phi} = \frac{32/43 - 1/4}{1 + (32/43)(1/4)} = \frac{5}{12} = \tan \psi.$$

**13.** Show that   Arc tan 1/2 + Arc tan 1/5 + Arc tan 1/8 = π/4.

We shall show that   Arc tan 1/2 + Arc tan 1/5  =  π/4 − Arc tan 1/8.

$$\tan(\text{Arc tan } 1/2 + \text{Arc tan } 1/5) \;=\; \frac{1/2 + 1/5}{1 - (1/2)(1/5)} \;=\; \frac{7}{9}$$

and

$$\tan(π/4 - \text{Arc tan } 1/8) \;=\; \frac{1 - 1/8}{1 + 1/8} \;=\; \frac{7}{9}.$$

**14.** Show that   2 Arc tan 1/3 + Arc tan 1/7 = Arc sec $\sqrt{34}/5$ + Arc csc $\sqrt{17}$.

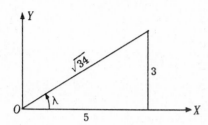   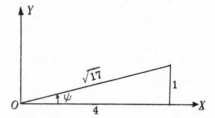

Let θ = Arc tan 1/3, φ = Arc tan 1/7, λ = Arc sec $\sqrt{34}/5$, and ψ = Arc csc $\sqrt{17}$; then tan θ = 1/3, tan φ = 1/7, sec λ = $\sqrt{34}/5$, and csc ψ = $\sqrt{17}$, each angle terminating in the first quadrant.

Taking the tangent of both members of the given relation, we are to show that
$$\tan(2θ + φ) \;=\; \tan(λ + ψ).$$

Now

$$\tan 2θ \;=\; \frac{2 \tan θ}{1 - \tan^2 θ} \;=\; \frac{2(1/3)}{1 - (1/3)^2} \;=\; 3/4,$$

$$\tan(2θ + φ) \;=\; \frac{\tan 2θ + \tan φ}{1 - \tan 2θ \tan φ} \;=\; \frac{3/4 + 1/7}{1 - (3/4)(1/7)} \;=\; 1$$

and, using the figures above,   $\tan(λ + ψ) \;=\; \dfrac{3/5 + 1/4}{1 - (3/5)(1/4)} \;=\; 1.$

**15.** Find the general value of each of the following.

 a) arc sin $\sqrt{2}/2$ = π/4 + 2nπ,   3π/4 + 2nπ       d) arc sin (−1)  = −π/2 + 2nπ
 b) arc cos 1/2 = π/3 + 2nπ,   5π/3 + 2nπ       e) arc cos 0  = π/2 + 2nπ,   3π/2 + 2nπ
 c) arc tan 0 = 2nπ,   (2n + 1)π       f) arc tan (−$\sqrt{3}$)  = −π/3 + 2nπ,   2π/3 + 2nπ

where n is a positive or negative integer, or is zero.

**16.** Show that the general value of   a) arc sin x = nπ + (−1)$^n$ Arc sin x ,
                                       b) arc cos x = 2nπ ± Arc cos x ,
                                       c) arc tan x = nπ + Arc tan x ,
where n is any positive or negative integer, or is zero.

a) Let θ = Arc sin x.   Then since  sin (π − θ) = sin θ, all values of arc sin x are given by
            1) θ + 2mπ       and       2) π − θ + 2mπ = (2m + 1)π − θ.

Now, when n = 2m, that is n is an even integer, 1) may be written as nπ + θ = nπ + (−1)$^n$θ; and when n = 2m + 1, that is n is an odd integer, 2) may be written as  nπ − θ = nπ + (−1)$^n$θ. Thus, arc sin x = nπ + (−1)$^n$ Arc sin x, where n is any positive or negative integer, or is zero.

b) Let θ = Arc cos x.   Then since  cos (−θ) = cos θ, all values of  arc cos x are given by θ + 2nπ and −θ + 2nπ  or  2nπ ± θ = 2nπ ± Arc cos x,  where n is any positive or negative integer, or is zero.

c) Let θ = Arc tan x.   Then since  tan(π + θ) = tan θ, all values of  arc tan x are given by θ + 2mπ and (π + θ) + 2mπ = θ + (2m + 1)π  or, as in a), by  nπ + Arc tan x,  where n is any positive or negative integer, or is zero.

17. Express the general value of each of the functions of Problem 15, using the form of Problem 16.

    *a)* arc sin $\sqrt{2}/2$ = $n\pi + (-1)^n \pi/4$           *d)* arc sin (−1) = $n\pi + (-1)^n (-\pi/2)$
    *b)* arc cos 1/2 = $2n\pi \pm \pi/3$                 *e)* arc cos 0 = $2n\pi \pm \pi/2$
    *c)* arc tan 0 = $n\pi$                          *f)* arc tan (−$\sqrt{3}$) = $n\pi - \pi/3$

where *n* is any positive or negative integer, or is zero.

# SUPPLEMENTARY PROBLEMS

18. Write the following in inverse function notation.
    *a)* $\sin \theta = 3/4$,   *b)* $\cos \alpha = -1$,   *c)* $\tan x = -2$,   *d)* $\cot \beta = 1/2$.
    *Ans. a)* $\theta$ = arc sin 3/4,   *b)* $\alpha$ = arc cos (−1),   *c)* $x$ = arc tan (−2),   *d)* $\beta$ = arc cot 1/2

19. Find the principal value of each of the following.
    *a)* Arc sin $\sqrt{3}/2$         *d)* Arc cot 1             *g)* Arc tan (− $\sqrt{3}$)        *j)* Arc csc (−1)
    *b)* Arc cos (− $\sqrt{2}/2$)        *e)* Arc sin (−1/2)        *h)* Arc cot 0
    *c)* Arc tan $1/\sqrt{3}$           *f)* Arc cos (−1/2)       *i)* Arc sec (− $\sqrt{2}$)

    *Ans. a)* $\pi/3$, *b)* $3\pi/4$, *c)* $\pi/6$, *d)* $\pi/4$, *e)* −$\pi/6$, *f)* $2\pi/3$, *g)* −$\pi/3$, *h)* $\pi/2$, *i)* −$3\pi/4$, *j)* −$\pi/2$

20. Evaluate each of the following.
    *a)* $\sin[$Arc sin (−1/2)$]$        *f)* $\sin($Arc cos 4/5$)$        *k)* Arc tan (cot 230°)
    *b)* $\cos($Arc cos $\sqrt{3}/2)$       *g)* $\cos[$Arc sin (−12/13)$]$     *l)* Arc cot (tan 100°)
    *c)* $\tan[$Arc tan (−1)$]$         *h)* $\sin($Arc tan 2$)$          *m)* $\sin($2 Arc sin 2/3$)$
    *d)* $\sin[$Arc cos (− $\sqrt{3}/2)]$      *i)* Arc cos (sin 220°)       *n)* $\cos($2 Arc sin 3/5$)$
    *e)* $\tan($Arc sin 0$)$            *j)* Arc sin $[\cos(-105°)]$     *o)* $\sin(\frac{1}{2}$ Arc cos 4/5$)$

    *Ans. a)* −1/2,   *b)* $\sqrt{3}/2$,   *c)* −1,   *d)* 1/2,   *e)* 0,   *f)* 3/5,   *g)* 5/13,   *h)* $2/\sqrt{5}$
          *i)* 130°,   *j)* −15°,   *k)* 40°,   *l)* 170°,   *m)* $4\sqrt{5}/9$,   *n)* 7/25,   *o)* $1/\sqrt{10}$

21. Show that
    *a)* $\sin($ Arc sin $\frac{5}{13}$ + Arc sin $\frac{4}{5}) = \frac{63}{65}$        *e)* $\cos($ Arc tan $\frac{-4}{3}$ + Arc sin $\frac{12}{13}) = \frac{63}{65}$

    *b)* $\cos($ Arc cos $\frac{15}{17}$ − Arc cos $\frac{7}{25}) = \frac{297}{425}$        *f)* $\tan($ Arc sin $\frac{-3}{5}$ − Arc cos $\frac{5}{13}) = \frac{63}{16}$

    *c)* $\sin($ Arc sin $\frac{1}{2}$ − Arc cos $\frac{1}{3}) = \frac{1 - 2\sqrt{6}}{6}$       *g)* $\tan($ 2 Arc sin $\frac{4}{5}$ + Arc cos $\frac{12}{13}) = -\frac{253}{204}$

    *d)* $\tan($ Arc tan $\frac{3}{4}$ + Arc cot $\frac{15}{8}) = \frac{77}{36}$         *h)* $\sin($ 2 Arc sin $\frac{4}{5}$ − Arc cos $\frac{12}{13}) = \cdot\frac{323}{325}$.

22. Show that
    *a)* Arc tan $\frac{1}{2}$ + Arc tan $\frac{1}{3} = \frac{\pi}{4}$           *e)* Arc cos $\frac{12}{13}$ + Arc tan $\frac{1}{4}$ = Arc cot $\frac{43}{32}$

    *b)* Arc sin $\frac{4}{5}$ + Arc tan $\frac{3}{4} = \frac{\pi}{2}$           *f)* Arc sin $\frac{3}{5}$ + Arc sin $\frac{15}{17}$ = Arc cos $\frac{-13}{85}$

    *c)* Arc tan $\frac{4}{3}$ − Arc tan $\frac{1}{7} = \frac{\pi}{4}$           *g)* Arc tan $a$ + Arc tan $\frac{1}{a} = \frac{\pi}{2}$   ($a > 0$).

    *d)* 2 Arc tan $\frac{1}{3}$ + Arc tan $\frac{1}{7} = \frac{\pi}{4}$

23. Prove: The area of the segment cut from a circle of radius *r* by a chord at a distance *d* from the center is given by $K = r^2$ Arc cos $\frac{d}{r} - d\sqrt{r^2 - d^2}$.

# CHAPTER 17

# Trigonometric Equations

TRIGONOMETRIC EQUATIONS, i.e., equations involving trigonometric functions of unknown angles, are called:

*a*) identical equations or *identities*, if they are satisfied by all values of the unknown angles for which the functions are defined;

*b*) conditional equations, or equations, if they are satisfied only by particular values of the unknown angles.

For example: *a*) $\sin x \csc x = 1$ is an identity, being satisfied by every value of $x$ for which $\csc x$ is defined;

*b*) $\sin x = 0$ is a conditional equation since it is not satisfied by $x = \frac{1}{4}\pi$ or $\frac{1}{2}\pi$.

Hereafter in this chapter we shall use the term "equation" instead of "conditional equation".

A SOLUTION OF A TRIGONOMETRIC EQUATION, as $\sin x = 0$, is a value of the angle $x$ which satisfies the equation. Two solutions of $\sin x = 0$ are $x = 0$ and $x = \pi$.

If a given equation has one solution, it has in general an unlimited number of solutions. Thus, the complete solution of $\sin x = 0$ is given by

$$x = 0 + 2n\pi, \quad x = \pi + 2n\pi$$

where $n$ is any positive or negative integer or is zero.

In this chapter we shall list only the particular solutions for which $0 \leqq x < 2\pi$.

PROCEDURES FOR SOLVING TRIGONOMETRIC EQUATIONS. There is no general method for solving trigonometric equations. Three standard procedures are illustrated below and other procedures are introduced in the solved problems.

*A*) The equation may be factorable.

EXAMPLE 1. Solve $\sin x - 2 \sin x \cos x = 0$.

Factoring, $\sin x - 2 \sin x \cos x = \sin x (1 - 2 \cos x) = 0$, and setting each factor equal to zero, we have

$$\sin x = 0 \quad \text{and} \quad x = 0, \pi;$$
$$1 - 2 \cos x = 0 \quad \text{or} \quad \cos x = \tfrac{1}{2} \quad \text{and} \quad x = \pi/3, 5\pi/3.$$

Check. For $x = 0$, $\quad \sin x - 2 \sin x \cos x = 0 - 2(0)(1) = 0$;

for $x = \pi/3$, $\quad \sin x - 2 \sin x \cos x = \tfrac{1}{2}\sqrt{3} - 2(\tfrac{1}{2}\sqrt{3})(\tfrac{1}{2}) = 0$;

for $x = \pi$, $\quad \sin x - 2 \sin x \cos x = 0 - 2(0)(-1) = 0$;

for $x = 5\pi/3$, $\sin x - 2 \sin x \cos x = -\tfrac{1}{2}\sqrt{3} - 2(-\tfrac{1}{2}\sqrt{3})(\tfrac{1}{2}) = 0$.

Thus, the required solutions ($0 \leqq x < 2\pi$) are $x = 0, \pi/3, \pi, 5\pi/3$.

*B*) The various functions occurring in the equation may be expressed in terms of a single function.

EXAMPLE 2. Solve $2 \tan^2 x + \sec^2 x = 2$. Replacing $\sec^2 x$ by $1 + \tan^2 x$, we have

125

$$2 \tan^2 x + (1 + \tan^2 x) = 2, \quad 3 \tan^2 x = 1, \quad \text{and} \quad \tan x = \pm 1/\sqrt{3}.$$

From $\tan x = 1/\sqrt{3}$, $x = \pi/6$ and $7\pi/6$; from $\tan x = -1/\sqrt{3}$, $x = 5\pi/6$ and $11\pi/6$. After checking each of these values in the original equation, we find that the required solutions ($0 \leqq x < 2\pi$) are $x = \pi/6$, $5\pi/6$, $7\pi/6$, $11\pi/6$.

The necessity of the check is illustrated in

EXAMPLE 3.  Solve $\sec x + \tan x = 0$.

Multiplying the equation $\sec x + \tan x = \dfrac{1}{\cos x} + \dfrac{\sin x}{\cos x} = 0$ by $\cos x$, we have

$1 + \sin x = 0$ or $\sin x = -1$; then $x = 3\pi/2$. However, neither $\sec x$ nor $\tan x$ is defined when $x = 3\pi/2$ and the equation has no solution.

$C$) Both members of the equation are squared.

EXAMPLE 4.  Solve $\sin x + \cos x = 1$.

If the procedure of $B$) were used, we would replace $\sin x$ by $\pm\sqrt{1 - \cos^2 x}$ or $\cos x$ by $\pm\sqrt{1 - \sin^2 x}$ and thereby introduce radicals. To avoid this, we write the equation in the form $\sin x = 1 - \cos x$ and square both members. We have

$$1) \qquad \sin^2 x = 1 - 2\cos x + \cos^2 x ,$$

$$1 - \cos^2 x = 1 - 2\cos x + \cos^2 x ,$$

$$2\cos^2 x - 2\cos x = 2\cos x (\cos x - 1) = 0.$$

From $\cos x = 0$, $x = \pi/2$, $3\pi/2$; from $\cos x = 1$, $x = 0$.

Check. For $x = 0$,     $\sin x + \cos x = 0 + 1 = 1$;
         for $x = \pi/2$,   $\sin x + \cos x = 1 + 0 = 1$;
         for $x = 3\pi/2$, $\sin x + \cos x = -1 + 0 \neq 1$.

Thus, the required solutions are $x = 0$, $\pi/2$.

The value $x = 3\pi/2$, called an *extraneous solution*, was introduced by squaring the two members. Note that 1) is also obtained when both members of $\sin x = \cos x - 1$ are squared and that $x = 3\pi/2$ satisfies this latter relation.


# SOLVED PROBLEMS

Solve each of the trigonometric equations 1-22 for all $x$ such that $0 \leq x < 2\pi$. (If all solutions are required, adjoin $+2n\pi$, where $n$ is zero or any positive or negative integer, to each result given.) In a number of the solutions, the details of the check have been omitted.

1. $2 \sin x - 1 = 0$.

Here $\sin x = 1/2$ and $x = \pi/6$, $5\pi/6$.

2. $\sin x \cos x = 0$.

From $\sin x = 0$, $x = 0$, $\pi$; from $\cos x = 0$, $x = \pi/2$, $3\pi/2$.
The required solutions are $x = 0$, $\pi/2$, $\pi$, $3\pi/2$.

3. $(\tan x - 1)(4 \sin^2 x - 3) = 0$.

From $\tan x - 1 = 0$, $\tan x = 1$ and $x = \pi/4$, $5\pi/4$; from $4 \sin^2 x - 3 = 0$, $\sin x = \pm\sqrt{3}/2$ and $x = \pi/3$, $2\pi/3$, $4\pi/3$, $5\pi/3$.
The required solutions are $x = \pi/4$, $\pi/3$, $2\pi/3$, $5\pi/4$, $4\pi/3$, $5\pi/3$.

4. $\sin^2 x + \sin x - 2 = 0$.

Factoring, $(\sin x + 2)(\sin x - 1) = 0$.

From $\sin x + 2 = 0$, $\sin x = -2$ and there is no solution; from $\sin x - 1 = 0$, $\sin x = 1$ and $x = \pi/2$. The required solution is $x = \pi/2$.

5. $3 \cos^2 x = \sin^2 x$.

First Solution. Replacing $\sin^2 x$ by $1 - \cos^2 x$, we have $3 \cos^2 x = 1 - \cos^2 x$ or $4 \cos^2 x = 1$. Then $\cos x = \pm 1/2$ and the required solutions are $x = \pi/3, 2\pi/3, 4\pi/3, 5\pi/3$.

Second Solution. Dividing the equation by $\cos^2 x$, we have $3 = \tan^2 x$. Then $\tan x = \pm\sqrt{3}$ and the solutions above are obtained.

6. $2 \sin x - \csc x = 1$.

Multiplying the equation by $\sin x$, $2 \sin^2 x - 1 = \sin x$, and rearranging, we have
$$2 \sin^2 x - \sin x - 1 = (2 \sin x + 1)(\sin x - 1) = 0.$$

From $2 \sin x + 1 = 0$, $\sin x = -1/2$ and $x = 7\pi/6, 11\pi/6$; from $\sin x = 1$, $x = \pi/2$.

Check. For $x = \pi/2$, $2 \sin x - \csc x = 2(1) - 1 = 1$;
for $x = 7\pi/6$ and $11\pi/6$, $2 \sin x - \csc x = 2(-1/2) - (-2) = 1$.

The solutions are $x = \pi/2, 7\pi/6, 11\pi/6$.

7. $2 \sec x = \tan x + \cot x$.

Transforming to sines and cosines, and clearing of fractions, we have
$$\frac{2}{\cos x} = \frac{\sin x}{\cos x} + \frac{\cos x}{\sin x} \quad \text{or} \quad 2 \sin x = \sin^2 x + \cos^2 x = 1.$$

Then $\sin x = 1/2$ and $x = \pi/6, 5\pi/6$.

8. $\tan x + 3 \cot x = 4$.

Multiplying by $\tan x$ and rearranging, $\tan^2 x - 4 \tan x + 3 = (\tan x - 1)(\tan x - 3) = 0$.

From $\tan x - 1 = 0$, $\tan x = 1$ and $x = \pi/4, 5\pi/4$; from $\tan x - 3 = 0$, $\tan x = 3$ and $x = 71°34', 251°34'$.

Check. For $x = \pi/4$ and $5\pi/4$, $\tan x + 3 \cot x = 1 + 3(1) = 4$;
for $x = 71°34'$ and $251°34'$, $\tan x + 3 \cot x = 3 + 3(1/3) = 4$.

The solutions are $45°, 71°34', 225°, 251°34'$.

9. $\csc x + \cot x = \sqrt{3}$.

First Solution. Writing the equation in the form $\csc x = \sqrt{3} - \cot x$ and squaring, we have
$$\csc^2 x = 3 - 2\sqrt{3} \cot x + \cot^2 x.$$

Replacing $\csc^2 x$ by $1 + \cot^2 x$ and combining, this becomes $2\sqrt{3} \cot x - 2 = 0$. Then $\cot x = 1/\sqrt{3}$ and $x = \pi/3, 4\pi/3$.

Check. For $x = \pi/3$, $\csc x + \cot x = 2/\sqrt{3} + 1/\sqrt{3} = \sqrt{3}$;
for $x = 4\pi/3$, $\csc x + \cot x = -2/\sqrt{3} + 1/\sqrt{3} \neq \sqrt{3}$. The required solution is $x = \pi/3$.

Second Solution. Upon making the indicated replacement, the equation becomes
$$\frac{1}{\sin x} + \frac{\cos x}{\sin x} = \sqrt{3} \quad \text{and, clearing of fractions,} \quad 1 + \cos x = \sqrt{3} \sin x.$$

Squaring both members, we have $1 + 2 \cos x + \cos^2 x = 3 \sin^2 x = 3(1 - \cos^2 x)$ or

$$4 \cos^2 x + 2 \cos x - 2 = 2(2 \cos x - 1)(\cos x + 1) = 0.$$

From $2 \cos x - 1 = 0$, $\cos x = 1/2$ and $x = \pi/3, 5\pi/3$; from $\cos x + 1 = 0$, $\cos x = -1$ and $x = \pi$.

Now $x = \pi/3$ is the solution. The values $x = \pi$ and $5\pi/3$ are to be excluded since $\csc \pi$ is not defined while $\csc 5\pi/3$ and $\cot 5\pi/3$ are both negative.

**10.** $\cos x - \sqrt{3} \sin x = 1$.

*First Solution.* Putting the equation in the form $\cos x - 1 = \sqrt{3} \sin x$ and squaring, we have
$$\cos^2 x - 2 \cos x + 1 = 3 \sin^2 x = 3(1 - \cos^2 x);$$

then, combining and factoring,
$$4 \cos^2 x - 2 \cos x - 2 = 2(2 \cos x + 1)(\cos x - 1) = 0.$$

From $2 \cos x + 1 = 0$, $\cos x = -1/2$ and $x = 2\pi/3, 4\pi/3$; from $\cos x - 1 = 0$, $\cos x = 1$ and $x = 0$.

Check. For $x = 0$,     $\cos x - \sqrt{3} \sin x = 1 - \sqrt{3}\,(0) = 1$;
      for $x = 2\pi/3$, $\cos x - \sqrt{3} \sin x = -1/2 - \sqrt{3}\,(\sqrt{3}/2) \neq 1$;
      for $x = 4\pi/3$, $\cos x - \sqrt{3} \sin x = -1/2 - \sqrt{3}\,(-\sqrt{3}/2) = 1$.

The required solutions are $x = 0, 4\pi/3$.

*Second Solution.* The left member of the given equation may be put in the form
$$\sin \theta \cos x + \cos \theta \sin x = \sin(\theta + x),$$

in which $\theta$ is a known angle, by dividing the given equation by $r > 0$, $\dfrac{1}{r} \cos x + \left(\dfrac{-\sqrt{3}}{r}\right) \sin x = \dfrac{1}{r}$, and setting $\sin \theta = \dfrac{1}{r}$ and $\cos \theta = \dfrac{-\sqrt{3}}{r}$. Since $\sin^2 \theta + \cos^2 \theta = 1$, $\left(\dfrac{1}{r}\right)^2 + \left(\dfrac{-\sqrt{3}}{r}\right)^2 = 1$

and $r = 2$. Now $\sin \theta = 1/2$, $\cos \theta = -\sqrt{3}/2$ so that the given equation may be written as $\sin(\theta + x) = 1/2$ with $\theta = 5\pi/6$. Then $\theta + x = 5\pi/6 + x = \arcsin 1/2 = \pi/6, 5\pi/6, 13\pi/6, 17\pi/6, \cdots$ and $x = -2\pi/3, 0, 4\pi/3, 2\pi, \cdots$. As before, the required solutions are $x = 0, 4\pi/3$.

Note that $r$ is the positive square root of the sum of the squares of the coefficients of $\cos x$ and $\sin x$ when the equation is written in the form $a \cos x + b \sin x = c$, that is,
$$r = \sqrt{a^2 + b^2}.$$
The equation will have no solution if $\dfrac{c}{\sqrt{a^2 + b^2}}$ is greater than 1 or less than $-1$.

**11.** $2 \cos x = 1 - \sin x$.

*First Solution.* As in Problem 10, we obtain
$$4 \cos^2 x = 1 - 2 \sin x + \sin^2 x,$$
$$4(1 - \sin^2 x) = 1 - 2 \sin x + \sin^2 x,$$
$$5 \sin^2 x - 2 \sin x - 3 = (5 \sin x + 3)(\sin x - 1) = 0.$$

From $5 \sin x + 3 = 0$, $\sin x = -3/5 = -0.6000$ and $x = 216°52', 323°8'$; from $\sin x - 1 = 0$, $\sin x = 1$ and $x = \pi/2$.

Check. For $x = \pi/2$,    $2(0) = 1 - 1$;
     for $x = 216°52'$,   $2(-4/5) \neq 1 - (-3/5)$;
     for $x = 323°8'$,    $2(4/5) = 1 - (-3/5)$.

The required solutions are $x = 90°, 323°8'$.

*Second Solution.* Writing the equation as $2 \cos x + \sin x = 1$ and dividing by $r = \sqrt{2^2 + 1^2} = \sqrt{5}$, we have

$$1) \qquad \frac{2}{\sqrt{5}} \cos x + \frac{1}{\sqrt{5}} \sin x = \frac{1}{\sqrt{5}} .$$

Let $\sin \theta = 2/\sqrt{5}$, $\cos \theta = 1/\sqrt{5}$; then 1) becomes

$$\sin \theta \cos x + \cos \theta \sin x = \sin(\theta + x) = \frac{1}{\sqrt{5}}$$

with $\theta = 63°26'$. Now $\theta + x = 63°26' + x = \arcsin (1/\sqrt{5}) = \arcsin (0.4472) = 26°34'$, $153°26'$, $386°34'$, $\cdots$ and $x = 90°$, $323°8'$ as before.

Equations Involving Multiple Angles.

12. $\sin 3x = -\frac{1}{2}\sqrt{2}$.

Since we require $x$ such that $0 \leqq x < 2\pi$, $3x$ must be such that $0 \leqq 3x < 6\pi$.

Then $3x = 5\pi/4,\ 7\pi/4,\ 13\pi/4,\ 15\pi/4,\ 21\pi/4,\ 23\pi/4$ and
$x = 5\pi/12,\ 7\pi/12,\ 13\pi/12,\ 5\pi/4,\ 7\pi/4,\ 23\pi/12$. Each of these values is a solution.

13. $\cos \frac{1}{2}x = \frac{1}{2}$.

Since we require $x$ such that $0 \leq x < 2\pi$, $\frac{1}{2}x$ must be such that $0 \leq \frac{1}{2}x < \pi$.

Then $\frac{1}{2}x = \pi/3$ and $x = 2\pi/3$.

14. $\sin 2x + \cos x = 0$.

Substituting for $\sin 2x$, we have $2 \sin x \cos x + \cos x = \cos x (2 \sin x + 1) = 0$.

From $\cos x = 0$, $x = \pi/2,\ 3\pi/2$; from $\sin x = -1/2$, $x = 7\pi/6,\ 11\pi/6$.

The required solutions are $x = \pi/2,\ 7\pi/6,\ 3\pi/2,\ 11\pi/6$.

15. $2 \cos^2 \frac{1}{2}x = \cos^2 x$.

Substituting $1 + \cos x$ for $2 \cos^2 \frac{1}{2}x$, the equation becomes $\cos^2 x - \cos x - 1 = 0$; then $\cos x = \dfrac{1 \pm \sqrt{5}}{2} = 1.6180,\ -0.6180$. Since $\cos x$ cannot exceed 1, we consider $\cos x = -0.6180$ and obtain the solutions $x = 128°10',\ 231°50'$.

Note. To solve $\sqrt{2} \cos \frac{1}{2}x = \cos x$ and $\sqrt{2} \cos \frac{1}{2}x = -\cos x$, we square and obtain the equation of this problem. The solution of the first of these equations is $231°50'$ and the solution of the second is $128°10'$.

16. $\cos 2x + \cos x + 1 = 0$.

Substituting $2 \cos^2 x - 1$ for $\cos 2x$, we have $2 \cos^2 x + \cos x = \cos x (2 \cos x + 1) = 0$.

From $\cos x = 0$, $x = \pi/2,\ 3\pi/2$; from $\cos x = -1/2$, $x = 2\pi/3,\ 4\pi/3$.

The required solutions are $x = \pi/2,\ 2\pi/3,\ 3\pi/2,\ 4\pi/3$.

17. $\tan 2x + 2 \sin x = 0$.

Using $\tan 2x = \dfrac{\sin 2x}{\cos 2x} = \dfrac{2 \sin x \cos x}{\cos 2x}$, we have

$$\frac{2 \sin x \cos x}{\cos 2x} + 2 \sin x = 2 \sin x \left( \frac{\cos x}{\cos 2x} + 1 \right) = 2 \sin x \left( \frac{\cos x + \cos 2x}{\cos 2x} \right) = 0.$$

From $\sin x = 0$, $x = 0, \pi$; from $\cos x + \cos 2x = \cos x + 2\cos^2 x - 1 = (2\cos x - 1)(\cos x + 1) = 0$, $x = \pi/3, 5\pi/3$, and $\pi$. The required solutions are $x = 0, \pi/3, \pi, 5\pi/3$.

**18.** $\sin 2x = \cos 2x$.

*First Solution.* Let $2x = \theta$; then we are to solve $\sin \theta = \cos \theta$ for $0 \le \theta < 4\pi$. Then $\theta = \pi/4, 5\pi/4, 9\pi/4, 13\pi/4$ and $x = \theta/2 = \pi/8, 5\pi/8, 9\pi/8, 13\pi/8$ are the solutions.

*Second Solution.* Dividing by $\cos 2x$, the equation becomes $\tan 2x = 1$ for which $2x = \pi/4$, $5\pi/4, 9\pi/4, 13\pi/4$ as in the first solution.

**19.** $\sin 2x = \cos 4x$.

Since $\cos 4x = \cos 2(2x) = 1 - 2\sin^2 2x$, the equation becomes
$$2\sin^2 2x + \sin 2x - 1 = (2\sin 2x - 1)(\sin 2x + 1) = 0.$$

From $2\sin 2x - 1 = 0$ or $\sin 2x = 1/2$, $2x = \pi/6, 5\pi/6, 13\pi/6, 17\pi/6$ and $x = \pi/12, 5\pi/12$, $13\pi/12, 17\pi/12$; from $\sin 2x + 1 = 0$ or $\sin 2x = -1$, $2x = 3\pi/2, 7\pi/2$ and $x = 3\pi/4, 7\pi/4$. All of these values are solutions.

**20.** $\sin 3x = \cos 2x$.

To avoid the substitution for $\sin 3x$, we use one of the procedures below.

*First Solution.* Since $\cos 2x = \sin(\frac{1}{2}\pi - 2x)$ and also $\cos 2x = \sin(\frac{1}{2}\pi + 2x)$, we consider
a) $\sin 3x = \sin(\frac{1}{2}\pi - 2x)$, obtaining $3x = \pi/2 - 2x, 5\pi/2 - 2x, 9\pi/2 - 2x, \cdots$,     and
b) $\sin 3x = \sin(\frac{1}{2}\pi + 2x)$, obtaining $3x = \pi/2 + 2x, 5\pi/2 + 2x, 9\pi/2 + 2x, \cdots$.

From a), $5x = \pi/2, 5\pi/2, 9\pi/2, 13\pi/2, 17\pi/2$ (since $5x < 10\pi$); and from b), $x = \pi/2$. The required solutions are $x = \pi/10, \pi/2, 9\pi/10, 13\pi/10, 17\pi/10$.

*Second Solution.* Since $\sin 3x = \cos(\frac{1}{2}\pi - 3x)$ and $\cos 2x = \cos(-2x)$, we consider
c) $\cos 2x = \cos(\frac{1}{2}\pi - 3x)$, obtaining $5x = \pi/2, 5\pi/2, 9\pi/2, 13\pi/2, 17\pi/2$,     and
d) $\cos(-2x) = \cos(\frac{1}{2}\pi - 3x)$, obtaining $x = \pi/2$, as before.

**21.** $\tan 4x = \cot 6x$.

Since $\cot 6x = \tan(\frac{1}{2}\pi - 6x)$, we consider the equation $\tan 4x = \tan(\frac{1}{2}\pi - 6x)$.

Then $4x = \pi/2 - 6x, 3\pi/2 - 6x, 5\pi/2 - 6x, \cdots$, the function $\tan \theta$ being of period $\pi$.

Thus, $10x = \pi/2, 3\pi/2, 5\pi/2, 7\pi/2, 9\pi/2, \cdots, 39\pi/2$ and the required solutions are
$$x = \pi/20, 3\pi/20, \pi/4, 7\pi/20, \ldots, 39\pi/20.$$

**22.** $\sin 5x - \sin 3x - \sin x = 0$.

Replacing $\sin 5x - \sin 3x$ by $2\cos 4x \sin x$ (Chapter 12), the given equation becomes
$$2\cos 4x \sin x - \sin x = \sin x (2\cos 4x - 1) = 0.$$

From $\sin x = 0$, $x = 0, \pi$; from $2\cos 4x - 1 = 0$ or $\cos 4x = 1/2$, $4x = \pi/3, 5\pi/3, 7\pi/3$, $11\pi/3, 13\pi/3, 17\pi/3, 19\pi/3, 23\pi/3$ and $x = \pi/12, 5\pi/12, 7\pi/12, 11\pi/12, 13\pi/12, 17\pi/12, 19\pi/12$, $23\pi/12$. Each of the values obtained is a solution.

**23.** Solve the system $\begin{array}{l}(1) \; r \sin \theta = 2 \\ (2) \; r \cos \theta = 3\end{array}$ for $r > 0$ and $0 \le \theta < 2\pi$.

Squaring the two equations and adding, $r^2 \sin \theta + r^2 \cos \theta = r^2 = 13$ and $r = \sqrt{13} = 3.606$.

When $r > 0$, $\sin \theta$ and $\cos \theta$ are both $> 0$ and $\theta$ is acute.

Dividing (1) by (2), $\tan \theta = 2/3 = 0.6667$ and $\theta = 33°41'$.

**24.** Solve the system $\begin{array}{l}(1)\ r\sin\theta = 3\\(2)\ r = 4(1+\sin\theta)\end{array}$ for $r > 0$ and $0 \leqq \theta < 2\pi$.

Dividing (2) by (1), $\dfrac{1}{\sin\theta} = \dfrac{4(1+\sin\theta)}{3}$ or $4\sin^2\theta + 4\sin\theta - 3 = 0$ and

$$(2\sin\theta + 3)(2\sin\theta - 1) = 0.$$

From $2\sin\theta - 1 = 0$, $\sin\theta = 1/2$, $\theta = \pi/6$ and $5\pi/6$; using (1), $r(1/2) = 3$ and $r = 6$. Note that $2\sin\theta + 3 = 0$ is excluded since when $r > 0$, $\sin\theta > 0$ by (1).

The required solutions are $\theta = \pi/6$, $r = 6$ and $\theta = 5\pi/6$, $r = 6$.

**25.** Solve the system $\begin{array}{l}(1)\ \sin x + \sin y = 1.2\\(2)\ \cos x + \cos y = 1.5\end{array}$ for $0 \leqq x, y < 2\pi$.

Since each sum on the left is greater than 1, each of the four functions is positive and both $x$ and $y$ are acute.

Using the appropriate formulas of Chapter 12, we obtain

$$(1')\quad 2\sin\tfrac{1}{2}(x+y)\cos\tfrac{1}{2}(x-y) = 1.2$$
$$(2')\quad 2\cos\tfrac{1}{2}(x+y)\cos\tfrac{1}{2}(x-y) = 1.5.$$

Dividing (1') by (2'), $\dfrac{\sin\tfrac{1}{2}(x+y)}{\cos\tfrac{1}{2}(x+y)} = \tan\tfrac{1}{2}(x+y) = \dfrac{1.2}{1.5} = 0.8000$ and $\tfrac{1}{2}(x+y) = 38^{\circ}40'$

since $\tfrac{1}{2}(x+y)$ is also acute.

Substituting for $\sin\tfrac{1}{2}(x+y) = 0.6248$ in (1'), we have $\cos\tfrac{1}{2}(x-y) = \dfrac{0.6}{0.6248} = 0.9603$ and $\tfrac{1}{2}(x-y) = 16^{\circ}12'$.

Then $x = \tfrac{1}{2}(x+y) + \tfrac{1}{2}(x-y) = 54^{\circ}52'$ and $y = \tfrac{1}{2}(x+y) - \tfrac{1}{2}(x-y) = 22^{\circ}28'$.

**26.** Solve $\text{Arc cos } 2x = \text{Arc sin } x$.

If $x$ is positive, $\alpha = \text{Arc cos } 2x$ and $\beta = \text{Arc sin } x$ terminate in quadrant I; if $x$ is negative, $\alpha$ terminates in quadrant II and $\beta$ terminates in quadrant IV. Thus, $x$ must be positive.

For $x$ positive, $\sin\beta = x$ and $\cos\beta = \sqrt{1-x^2}$. Taking the cosine of both members of the given equation, we have

$$\cos(\text{Arc cos } 2x) = \cos(\text{Arc sin } x) = \cos\beta \quad\text{or}\quad 2x = \sqrt{1-x^2}.$$

Squaring, $4x^2 = 1 - x^2$, $5x^2 = 1$, and $x = \sqrt{5}/5 = 0.4472$.

Check. $\text{Arc cos } 2x = \text{Arc cos } 0.8944 = 26^{\circ}30' = \text{Arc sin } 0.4472$, approximating the angle to the nearest $10'$.

**27.** Solve $\text{Arc cos}(2x^2 - 1) = 2\,\text{Arc cos }\tfrac{1}{2}$.

Let $\alpha = \text{Arc cos}(2x^2 - 1)$ and $\beta = \text{Arc cos }\tfrac{1}{2}$; then $\cos\alpha = 2x^2 - 1$ and $\cos\beta = \tfrac{1}{2}$.

Taking the cosine of both members of the given equation,

$$\cos\alpha = 2x^2 - 1 = \cos 2\beta = 2\cos^2\beta - 1 = 2(\tfrac{1}{2})^2 - 1 = -\tfrac{1}{2}.$$

Then $2x^2 = \tfrac{1}{2}$ and $x = \pm\tfrac{1}{2}$.

Check. For $x = \pm\tfrac{1}{2}$, $\text{Arc cos}(-\tfrac{1}{2}) = 2\,\text{Arc cos }\tfrac{1}{2}$ or $120^{\circ} = 2(60^{\circ})$.

**28.** Solve $\text{Arc cos } 2x - \text{Arc cos } x = \pi/3$.

If $x$ is positive, $0 < \text{Arc cos } 2x < \text{Arc cos } x$; if $x$ is negative, $\text{Arc cos } 2x > \text{Arc cos } x > 0$.

Thus, $x$ must be negative.

Let $\alpha$ = Arc cos $2x$ and $\beta$ = Arc cos $x$; then cos $\alpha$ = $2x$, sin $\alpha$ = $\sqrt{1-4x^2}$, cos $\beta$ = $x$ and sin $\beta$ = $\sqrt{1-x^2}$ since both $\alpha$ and $\beta$ terminate in quadrant II.

Taking the cosine of both members of the given equation,

$$\cos(\alpha-\beta) = \cos\alpha\cos\beta + \sin\alpha\sin\beta = 2x^2 + \sqrt{1-4x^2}\sqrt{1-x^2} = \cos\pi/3 = \tfrac{1}{2}$$

or
$$\sqrt{1-4x^2}\sqrt{1-x^2} = \tfrac{1}{2} - 2x^2.$$

Squaring, $1 - 5x^2 + 4x^4 = \dfrac{1}{4} - 2x^2 + 4x^4$, $3x^2 = \dfrac{3}{4}$, and $x = -\tfrac{1}{2}$.

Check. Arc cos $(-1)$ − Arc cos $(-\tfrac{1}{2})$ = $\pi - 2\pi/3$ = $\pi/3$.

29. Solve Arc sin $2x$ = $\tfrac{1}{4}\pi$ − Arc sin $x$.

Let $\alpha$ = Arc sin $2x$ and $\beta$ = arc sin $x$; then sin $\alpha$ = $2x$ and sin $\beta$ = $x$. If $x$ is negative, $\alpha$ and $\beta$ terminate in quadrant IV; thus, $x$ must be positive and $\beta$ acute.

Taking the sine of both members of the given equation,

$$\sin\alpha = \sin(\tfrac{1}{4}\pi - \beta) = \sin\tfrac{1}{4}\pi\cos\beta - \cos\tfrac{1}{4}\pi\sin\beta$$

or
$$2x = \tfrac{1}{2}\sqrt{2}\sqrt{1-x^2} - \tfrac{1}{2}\sqrt{2}\,x \quad \text{and} \quad (2\sqrt{2}+1)x = \sqrt{1-x^2}.$$

Squaring, $(8 + 4\sqrt{2} + 1)x^2 = 1 - x^2$, $x^2 = 1/(10 + 4\sqrt{2})$, and $x = 0.2527$.

Check. Arc sin $0.5054$ = $30°22'$; Arc sin $0.2527$ = $14°38'$ and $\tfrac{1}{4}\pi - 14°38'$ = $30°22'$.

30. Solve Arc tan $x$ + Arc tan $(1-x)$ = Arc tan $4/3$.

Let $\alpha$ = Arc tan $x$ and $\beta$ = Arc tan $(1-x)$; then tan $\alpha$ = $x$ and tan $\beta$ = $1-x$.

Taking the tangent of both members of the given equation,

$$\tan(\alpha+\beta) = \frac{\tan\alpha + \tan\beta}{1 - \tan\alpha\tan\beta} = \frac{x + (1-x)}{1 - x(1-x)} = \frac{1}{1 - x + x^2} = \tan(\text{Arc tan } 4/3) = 4/3.$$

Then $3 = 4 - 4x + 4x^2$, $4x^2 - 4x + 1 = (2x-1)^2 = 0$, and $x = \tfrac{1}{2}$.

Check. Arc tan $\tfrac{1}{2}$ + Arc tan $(1 - \tfrac{1}{2})$ = 2 Arc tan $0.5000$ = $53°8'$ and
Arc tan $4/3$ = Arc tan $1.3333$ = $53°8'$.

## SUPPLEMENTARY PROBLEMS

Solve each of the following equations for all $x$ such that $0 \leqq x < 2\pi$.

31. sin $x$ = $\sqrt{3}/2$.     *Ans.* $\pi/3$, $2\pi/3$

32. $\cos^2 x$ = $1/2$.     *Ans.* $\pi/4$, $3\pi/4$, $5\pi/4$, $7\pi/4$

33. sin $x$ cos $x$ = $0$.     *Ans.* $0$, $\pi/2$, $\pi$, $3\pi/2$

34. (tan $x$ − 1)(2 sin $x$ + 1) = $0$.     *Ans.* $\pi/4$, $7\pi/6$, $5\pi/4$, $11\pi/6$

35. $2\sin^2 x$ − sin $x$ − 1 = $0$.     *Ans.* $\pi/2$, $7\pi/6$, $11\pi/6$

36. sin $2x$ + sin $x$ = $0$.     *Ans.* $0$, $2\pi/3$, $\pi$, $4\pi/3$

37. cos $x$ + cos $2x$ = $0$.     *Ans.* $\pi/3$, $\pi$, $5\pi/3$

38. $2 \tan x \sin x - \tan x = 0$.      *Ans.* $0$, $\pi/6$, $5\pi/6$, $\pi$

39. $2 \cos x + \sec x = 3$.      *Ans.* $0$, $\pi/3$, $5\pi/3$

40. $2 \sin x + \csc x = 3$.      *Ans.* $\pi/6$, $\pi/2$, $5\pi/6$

41. $\sin x + 1 = \cos x$.      *Ans.* $0$, $3\pi/2$

42. $\sec x - 1 = \tan x$.      *Ans.* $0$

43. $2 \cos x + 3 \sin x = 2$.      *Ans.* $0°$, $112°37'$

44. $3 \sin x + 5 \cos x + 5 = 0$.      *Ans.* $180°$, $241°56'$

45. $1 + \sin x = 2 \cos x$.      *Ans.* $36°52'$, $270°$

46. $3 \sin x + 4 \cos x = 2$.      *Ans.* $103°18'$, $330°27'$

47. $\sin 2x = -\sqrt{3}/2$.      *Ans.* $2\pi/3$, $5\pi/6$, $5\pi/3$, $11\pi/6$

48. $\tan 3x = 1$.      *Ans.* $\pi/12$, $5\pi/12$, $3\pi/4$, $13\pi/12$, $17\pi/12$, $7\pi/4$

49. $\cos x/2 = \sqrt{3}/2$.      *Ans.* $\pi/3$

50. $\cot x/3 = -1/\sqrt{3}$.      *Ans.* No solution in given interval

51. $\sin x \cos x = 1/2$.      *Ans.* $\pi/4$, $5\pi/4$

52. $\sin x/2 + \cos x = 1$.      *Ans.* $0$, $\pi/3$, $5\pi/3$

53. $\sin 3x + \sin x = 0$.      *Ans.* $0$, $\pi/2$, $\pi$, $3\pi/2$

54. $\cos 2x + \cos 3x = 0$.      *Ans.* $\pi/5$, $3\pi/5$, $\pi$, $7\pi/5$, $9\pi/5$

55. $\sin 2x + \sin 4x = 2 \sin 3x$.      *Ans.* $0$, $\pi/3$, $2\pi/3$, $\pi$, $4\pi/3$, $5\pi/3$

56. $\cos 5x + \cos x = 2 \cos 2x$.      *Ans.* $0$, $\pi/4$, $2\pi/3$, $3\pi/4$, $5\pi/4$, $4\pi/3$, $7\pi/4$

57. $\sin x + \sin 3x = \cos x + \cos 3x$.      *Ans.* $\pi/8$, $\pi/2$, $5\pi/8$, $9\pi/8$, $3\pi/2$, $13\pi/8$

Solve each of the following systems for $r \geq 0$ and $0 \leq \theta < 2\pi$.

58. $r = a \sin \theta$
    $r = a \cos 2\theta$      *Ans.* $\theta = \pi/6$, $r = a/2$
         $\theta = 5\pi/6$, $r = a/2$ ; $\theta = 3\pi/2$, $r = -a$

59. $r = a \cos \theta$
    $r = a \sin 2\theta$      *Ans.* $\theta = \theta = \pi/2$, $r = 0$; $\theta = 3\pi/2$, $r = 0$
         $\theta = \pi/6$, $r = \sqrt{3}a/2$
         $\theta = 5\pi/6$, $r = -\sqrt{3}a/2$

60. $r = 4(1 + \cos \theta)$
    $r = 3 \sec \theta$      *Ans.* $\theta = \pi/3$, $r = 6$
         $\theta = 5\pi/3$, $r = 6$

Solve each of the following equations.

61. Arc tan $2x$ + Arc tan $x = \pi/4$.      *Ans.* $x = 0.281$

62. Arc sin $x$ + Arc tan $x = \pi/2$.      *Ans.* $x = 0.786$

63. Arc cos $x$ + Arc tan $x = \pi/2$.      *Ans.* $x = 0$

CHAPTER 18

# Complex Numbers

**PURE IMAGINARY NUMBERS.** The square root of a negative number (i.e., $\sqrt{-1}$, $\sqrt{-5}$, $\sqrt{-9}$) is called a *pure imaginary number*. Since by definition $\sqrt{-5}=\sqrt{5}\cdot\sqrt{-1}$ and $\sqrt{-9}=\sqrt{9}\cdot\sqrt{-1}=3\sqrt{-1}$, it is convenient to introduce the symbol $i=\sqrt{-1}$ and to adopt $\sqrt{-5}=i\sqrt{5}$ and $\sqrt{-9}=3i$ as the standard form for these numbers.

The symbol $i$ has the property $i^2=-1$; and for higher integral powers we have $i^3=i^2\cdot i=(-1)i=-i$, $i^4=(i^2)^2=(-1)^2=1$, $i^5=i^4\cdot i=i$, etc.

The use of the standard form simplifies the operations on pure imaginaries and eliminates the possibility of certain common errors. Thus, $\sqrt{-9}\cdot\sqrt{4}=\sqrt{-36}=6i$ since $\sqrt{-9}\cdot\sqrt{4}=3i(2)=6i$ but $\sqrt{-9}\cdot\sqrt{-4}\neq\sqrt{36}$ since $\sqrt{-9}\cdot\sqrt{-4}=(3i)(2i)=6i^2=-6$.

**COMPLEX NUMBERS.** A number $a+bi$, where $a$ and $b$ are real numbers, is called a *complex number*. The first term $a$ is called the *real part* of the complex number and the second term $bi$ is called the *pure imaginary part*.

Complex numbers may be thought of as including all real numbers and all pure imaginary numbers. For example, $5=5+0i$ and $3i=0+3i$.

Two complex numbers $a+bi$ and $c+di$ are said to be *equal* if and only if $a=c$ and $b=d$.

The *conjugate* of a complex number $a+bi$ is the complex number $a-bi$. Thus, $2+3i$ and $2-3i$, $-3+4i$ and $-3-4i$ are pairs of conjugate complex numbers.

**ALGEBRAIC OPERATIONS.**

1) *Addition.* To add two complex numbers, add the real parts and add the pure imaginary parts.

EXAMPLE 1. $(2+3i)+(4-5i)=(2+4)+(3-5)i=6-2i$.

2) *Subtraction.* To subtract two complex numbers, subtract the real parts and subtract the pure imaginary parts.

EXAMPLE 2. $(2+3i)-(4-5i)=(2-4)+[3-(-5)]i=-2+8i$.

3) *Multiplication.* To multiply two complex numbers, carry out the multiplication as if the numbers were ordinary binomials and replace $i^2$ by $-1$.

EXAMPLE 3. $(2+3i)(4-5i)=8+2i-15i^2=8+2i-15(-1)=23+2i$.

4) *Division.* To divide two complex numbers, multiply both numerator and denominator of the fraction by the conjugate of the denominator.

EXAMPLE 4. $\dfrac{2+3i}{4-5i}=\dfrac{(2+3i)(4+5i)}{(4-5i)(4+5i)}=\dfrac{(8-15)+(10+12)i}{16+25}=-\dfrac{7}{41}+\dfrac{22}{41}i$.

(Note the form of the result; it is neither $\dfrac{-7+22i}{41}$ nor $\dfrac{1}{41}(-7+22i)$.)

(See Problems 1-9.)

134

GRAPHIC REPRESENTATION OF COMPLEX NUMBERS. The complex number $x+yi$ may be represented graphically by the point $P$ (see Fig. 18-A) whose rectangular coordinates are $(x,y)$.

The point $O$, having coordinates $(0,0)$ represents the complex number $0+0i$ $=0$. All points on the $x$-axis have coordinates of the form $(x,0)$ and correspond to real numbers $x+0i=x$. For this reason, the $x$-axis is called the *axis of reals*. All points on the $y$-axis have coordinates of the form $(0,y)$ and correspond to pure imaginary numbers $0+yi=yi$. The $y$-axis is called the *axis of imaginaries*. The plane on which the complex numbers are represented is called the *complex plane*.

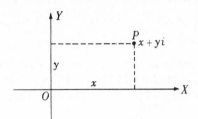

Fig. 18-A

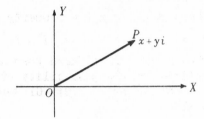

Fig. 18-B

In addition to representing a complex number by a point $P$ in the complex plane, the number may be represented (see Fig. 18-B) by the directed line segment or vector $OP$.

GRAPHIC REPRESENTATION OF ADDITION AND SUBTRACTION. Let $z_1 = x_1+iy_1$ and $z_2 = x_2+iy_2$ be two complex numbers. The vector representation of these numbers (Fig. 18-C) suggests the familiar parallelogram law for determining graphically the sum $z_1 + z_2 = (x_1+iy_1) + (x_2+iy_2)$.

Since $z_1-z_2 = (x_1+iy_1)-(x_2+iy_2) = (x_1+iy_1)+(-x_2-iy_2)$, the difference $z_1-z_2$ of the two complex numbers may be obtained graphically by applying the parallelogram law to $x_1+iy_1$ and $-x_2-iy_2$. (See Fig. 18-D.)

In Fig. 18-E both the sum $OR = z_1+z_2$ and the difference $OS = z_1-z_2$ are shown. Note that the segments $OS$ and $P_2P_1$ (the other diagonal of $OP_2RP_1$) are equal.

(See Problem 11.)

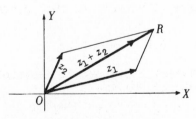

Fig. 18-C

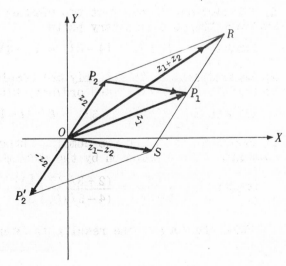

Fig. 18-E

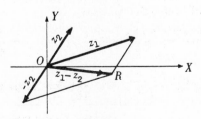

Fig. 18-D

POLAR OR TRIGONOMETRIC FORM OF COMPLEX NUMBERS.   Let the complex number $x + yi$ be represented (Fig. 18-F) by the vector $OP$.   This vector (and hence the complex number) may be described in terms of the length $r$ of the vector and *any* positive angle θ which the vector makes with the positive $x$-axis (axis of positive reals). The number $r = \sqrt{x^2 + y^2}$ is called the *modulus* or *absolute value* of the complex number. The angle θ, called the *amplitude* of the complex number, is usually chosen as the smallest positive angle for which $\tan θ = y/x$ but at times it will be found more convenient to choose some other angle coterminal with it.

From Fig. 18-F, $x = r \cos θ$ and $y = r \sin θ$; then $z = x + yi = r \cos θ + ir \sin θ = r(\cos θ + i \sin θ)$. We call $z = r(\cos θ + i \sin θ)$ the *polar or trigonometric form* and $z = x + yi$ the *rectangular form* of the complex number $z$.

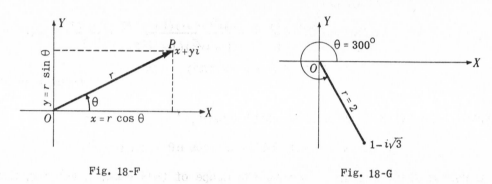

Fig. 18-F                                    Fig. 18-G

EXAMPLE 5.   Express $z = 1 - i\sqrt{3}$ in polar form.  (See Fig. 18-G above.)

The modulus is $r = \sqrt{(1)^2 + (-\sqrt{3})^2} = 2$. Since $\tan θ = y/x = -\sqrt{3}/1 = -\sqrt{3}$, the amplitude θ is either $120°$ or $300°$. Now we know that $P$ lies in quadrant IV; hence, $θ = 300°$ and the required polar form is $z = r(\cos θ + i \sin θ) = 2(\cos 300° + i \sin 300°)$. Note that $z$ may also be represented in polar form by $z = 2[\cos(300° + n\,360°) + i \sin(300° + n\,360°)]$, where $n$ is any integer.

EXAMPLE 6.   Express the complex number $z = 8(\cos 210° + i \sin 210°)$ in rectangular form.

Since  $\cos 210° = -\sqrt{3}/2$  and  $\sin 210° = -1/2$,

$z = 8(\cos 210° + i \sin 210°) = 8[-\sqrt{3}/2 + i(-1/2)] = -4\sqrt{3} - 4i$

is the required rectangular form.

(See Problems 12-13.)

MULTIPLICATION AND DIVISION IN POLAR FORM.

*Multiplication*. The modulus of the product of  two complex numbers is the product of their moduli, and the amplitude of the product is the sum of their amplitudes.

*Division*. The modulus of the quotient of two complex numbers is the modulus of the dividend divided by the modulus of the divisor, and the amplitude of the quotient is the amplitude of the dividend minus the amplitude of the divisor.   For a proof of these theorems, see Problem 14.

EXAMPLE 7.   Find $a)$ the product $z_1 z_2$, $b)$ the quotient $z_1/z_2$, and $c)$ the quotient $z_2/z_1$ where $z_1 = 2(\cos 300° + i \sin 300°)$ and $z_2 = 8(\cos 210° + i \sin 210°)$.

$a)$ The modulus of the product is $2(8) = 16$.  The amplitude is $300° + 210° = 510°$ but, following the convention, we shall use the smallest positive coterminal angle  $510° - 360° = 150°$.  Thus  $z_1 z_2 = 16(\cos 150° + i \sin 150°)$.

*b*) The modulus of the quotient $z_1/z_2$ is $2/8 = \frac{1}{4}$ and the amplitude is $300° - 210° = 90°$. Thus $z_1/z_2 = \frac{1}{4}(\cos 90° + i \sin 90°)$.

*c*) The modulus of the quotient $z_2/z_1$ is $8/2 = 4$.

The amplitude is $210° - 300° = -90°$ but we shall use the smallest positive coterminal angle $-90° + 360° = 270°$. Thus
$$z_2/z_1 = 4(\cos 270° + i \sin 270°).$$

Note. From Examples 5 and 6 the numbers are
$$z_1 = 1 - i\sqrt{3} \quad \text{and} \quad z_2 = -4\sqrt{3} - 4i$$

in rectangular form. Then
$$z_1 z_2 = (1 - i\sqrt{3})(-4\sqrt{3} - 4i) = -8\sqrt{3} + 8i = 16(\cos 150° + i \sin 150°)$$

as in *a*), and
$$z_2/z_1 = \frac{-4\sqrt{3} - 4i}{1 - i\sqrt{3}} = \frac{(-4\sqrt{3} - 4i)(1 + i\sqrt{3})}{(1 - i\sqrt{3})(1 + i\sqrt{3})} = \frac{-16i}{4} = -4i$$
$$= 4(\cos 270° + i \sin 270°) \qquad \text{as in } c).$$

(See Prob. 15-16.)

**DE MOIVRE'S THEOREM.** If *n* is any rational number,
$$\{r(\cos \theta + i \sin \theta)\}^n = r^n(\cos n\theta + i \sin n\theta).$$

A proof of this theorem is beyond the scope of this book; a verification for $n = 2$ and $n = 3$ is given in Problem 17.

**EXAMPLE 8.** $(\sqrt{3} - i)^{10} = \{2(\cos 330° + i \sin 330°)\}^{10}$
$$= 2^{10}(\cos 10 \cdot 330° + i \sin 10 \cdot 330°)$$
$$= 1024(\cos 60° + i \sin 60°) = 1024(1/2 + i\sqrt{3}/2)$$
$$= 512 + 512i\sqrt{3}.$$

(See Prob. 18.)

**ROOTS OF COMPLEX NUMBERS.** We state, without proof, the theorem: A complex number $a + bi = r(\cos \theta + i \sin \theta)$ has exactly *n* distinct *n*th roots.

The procedure for determining these roots is given in Example 9.

**EXAMPLE 9.** Find all fifth roots of $4 - 4i$.

The usual polar form of $4 - 4i$ is $4\sqrt{2}(\cos 315° + i \sin 315°)$ but we shall need the more general form
$$4\sqrt{2}[\cos(315° + k\,360°) + i \sin(315° + k\,360°)],$$

where *k* is any integer, including zero.

Using De Moivre's theorem, a fifth root of $4 - 4i$ is given by
$$\{4\sqrt{2}[\cos(315° + k\,360°) + i \sin(315° + k\,360°)]\}^{1/5}$$
$$= (4\sqrt{2})^{1/5}(\cos \frac{315° + k\,360°}{5} + i \sin \frac{315° + k\,360°}{5})$$
$$= \sqrt{2}[\cos(63° + k\,72°) + i \sin(63° + k\,72°)].$$

Assigning in turn the values $k = 0, 1, 2, \cdots$, we find

$k = 0$:   $\sqrt{2}\,(\cos 63^{\circ} \; + \; i \, \sin 63^{\circ}) \; = R_1$
$k = 1$:   $\sqrt{2}\,(\cos 135^{\circ} \; + \; i \, \sin 135^{\circ}) \; = R_2$
$k = 2$:   $\sqrt{2}\,(\cos 207^{\circ} \; + \; i \, \sin 207^{\circ}) \; = R_3$
$k = 3$:   $\sqrt{2}\,(\cos 279^{\circ} \; + \; i \, \sin 279^{\circ}) \; = R_4$
$k = 4$:   $\sqrt{2}\,(\cos 351^{\circ} \; + \; i \, \sin 351^{\circ}) \; = R_5$
$k = 5$:   $\sqrt{2}\,(\cos 423^{\circ} \; + \; i \, \sin 423^{\circ})$
$\qquad\qquad = \sqrt{2}\,(\cos 63^{\circ} \; + \; i \, \sin 63^{\circ}) \; = R_1$, etc.

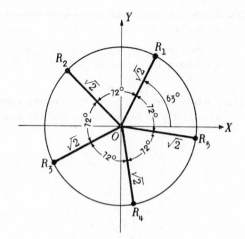

Thus, the five fifth roots are obtained by assigning the values $0,1,2,3,4$ (i.e., $0,1,2,3,\cdots,n-1$) to $k$.         (See also Problem 19.)

The modulus of each of the roots is $\sqrt{2}$; hence these roots lie on a circle of radius $\sqrt{2}$ with center at the origin. The difference in amplitude of two consecutive roots is $72^{\circ}$; hence the roots are equally spaced on this circle, as shown in the adjoining figure.

## SOLVED PROBLEMS

In Problems 1-6, perform the indicated operations, simplify, and write the result in the form $a + bi$.

1. $(3 - 4i) + (-5 + 7i) \; = \; (3 - 5) + (-4 + 7)i \; = \; -2 + 3i$

2. $(4 + 2i) - (-1 + 3i) \; = \; [4 - (-1)] + (2 - 3)i \; = \; 5 - i$

3. $(2 + i)(3 - 2i) \; = \; (6 + 2) + (-4 + 3)i \; = \; 8 - i$

4. $(3 + 4i)(3 - 4i) \; = \; 9 + 16 \; = \; 25$

5. $\dfrac{1 + 3i}{2 + i} \; = \; \dfrac{(1 + 3i)(2 - i)}{(2 + i)(2 - i)} \; = \; \dfrac{(2 + 3) + (-1 + 6)i}{4 + 1} \; = \; 1 + i$

6. $\dfrac{3 - 2i}{2 - 3i} \; = \; \dfrac{(3 - 2i)(2 + 3i)}{(2 - 3i)(2 + 3i)} \; = \; \dfrac{(6 + 6) + (9 - 4)i}{4 + 9} \; = \; \dfrac{12}{13} + \dfrac{5}{13}\,i$

7. Find $x$ and $y$ such that $2x - yi = 4 + 3i$.

Here $2x = 4$ and $-y = 3$; then $x = 2$ and $y = -3$.

8. Show that the conjugate complex numbers $2 + i$ and $2 - i$ are roots of the quadratic equation $x^2 - 4x + 5 = 0$.

For $x = 2 + i$:   $(2 + i)^2 - 4(2 + i) + 5 \; = \; 4 + 4i + i^2 - 8 - 4i + 5 \; = \; 0$.

For $x = 2 - i$:   $(2 - i)^2 - 4(2 - i) + 5 \; = \; 4 - 4i + i^2 - 8 + 4i + 5 \; = \; 0$.

Since each number satisfies the equation, it is a root of the equation.

9. Show that the conjugate of the sum of two complex numbers is equal to the sum of their conjugates.

Let the complex numbers be $a + bi$ and $c + di$. Their sum is $(a + c) + (b + d)i$ and the conjugate of the sum is $(a + c) - (b + d)i$.

The conjugates of the two given numbers are $a - bi$ and $c - di$, and their sum is
$$(a + c) + (-b - d)i = (a + c) - (b + d)i.$$

**10.** Represent graphically (as a vector) the following complex numbers:

   a) $3 + 2i$,   b) $2 - i$,   c) $-2 + i$,   d) $-1 - 3i$.

   We locate, in turn, the points whose coordinates are   (3,2),  (2,-1),  (-2,1),  (-1,-3)   and join each to the origin $O$.

**11.** Perform graphically the following operations:

   a) $(3 + 4i) + (2 + 5i)$,   b) $(3 + 4i) + (2 - 3i)$,   c) $(4 + 3i) - (2 + i)$,   d) $(4 + 3i) - (2 - i)$.

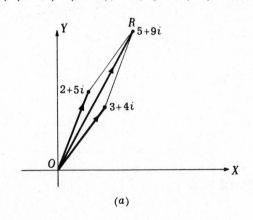

(a)

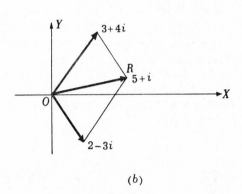

(b)

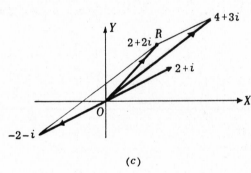

(c)

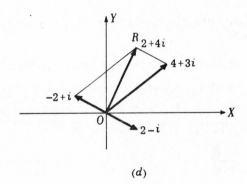

(d)

   For a) and b), draw as in Fig.(a) and (b) the two vectors and apply the parallelogram law.

   For c) draw the vectors representing $4 + 3i$ and $-2 - i$ and apply the parallelogram law as in Fig.(c).

   For d) draw the vectors representing $4 + 3i$ and $-2 + i$ and apply the parallelogram law as in Fig.(d).

**12.** Express each of the following complex numbers $z$ in polar form:

   a) $-1 + i\sqrt{3}$,   b) $6\sqrt{3} + 6i$,   c) $2 - 2i$,   d) $-3 = -3 + 0i$,   e) $4i = 0 + 4i$,   f) $-3 - 4i$.

   a) $P$ lies in the second quadrant; $r = \sqrt{(-1)^2 + (\sqrt{3})^2} = 2$; $\tan \theta = \sqrt{3}/-1 = -\sqrt{3}$ and $\theta = 120°$.
   Thus, $z = 2(\cos 120° + i \sin 120°)$.

   b) $P$ lies in the first quadrant; $r = \sqrt{(6\sqrt{3})^2 + 6^2} = 12$; $\tan \theta = 6/6\sqrt{3} = 1/\sqrt{3}$ and $\theta = 30°$.
   Thus, $z = 12(\cos 30° + i \sin 30°)$.

   c) $P$ lies in the fourth quadrant; $r = \sqrt{2^2 + (-2)^2} = 2\sqrt{2}$; $\tan \theta = -2/2 = -1$ and $\theta = 315°$.
   Thus, $z = 2\sqrt{2}(\cos 315° + i \sin 315°)$.

   d) $P$ lies on the negative $x$-axis and $\theta = 180°$; $r = \sqrt{(-3)^2 + 0^2} = 3$.
   Thus, $z = 3(\cos 180° + i \sin 180°)$.

*e*) *P* lies on the positive *y*-axis and $\theta = 90^\circ$; $r = \sqrt{0^2 + 4^2} = 4$.
Thus, $z = 4(\cos 90^\circ + i \sin 90^\circ)$.

*f*) *P* lies in the third quadrant; $r = \sqrt{(-3)^2 + (-4)^2} = 5$; $\tan \theta = -4/-3 = 1.3333$, $\theta = 233^\circ 8'$.
Thus, $z = 5(\cos 233^\circ 8' + i \sin 233^\circ 8')$.

**13.** Express each of the following complex numbers *z* in rectangular form:
  *a*) $4(\cos 240^\circ + i \sin 240^\circ)$      *c*) $3(\cos 90^\circ + i \sin 90^\circ)$
  *b*) $2(\cos 315^\circ + i \sin 315^\circ)$      *d*) $5(\cos 128^\circ + i \sin 128^\circ)$.

  *a*) $4(\cos 240^\circ + i \sin 240^\circ) = 4[-1/2 + i(-\sqrt{3}/2)] = -2 - 2i\sqrt{3}$

  *b*) $2(\cos 315^\circ + i \sin 315^\circ) = 2[1/\sqrt{2} + i(-1/\sqrt{2})] = \sqrt{2} - i\sqrt{2}$

  *c*) $3(\cos 90^\circ + i \sin 90^\circ) = 3[0 + i(1)] = 3i$

  *d*) $5(\cos 128^\circ + i \sin 128^\circ) = 5[-0.6157 + i(0.7880)] = -3.0785 + 3.9400i$

**14.** Prove: *a*) The modulus of the product of two complex numbers is the product of their moduli, and the amplitude of the product is the sum of their amplitudes.
  *b*) The modulus of the quotient of two complex numbers is the modulus of the dividend divided by the modulus of the divisor, and the amplitude of the quotient is the amplitude of the dividend minus the amplitude of the divisor.

Let $z_1 = r_1(\cos \theta_1 + i \sin \theta_1)$ and $z_2 = r_2(\cos \theta_2 + i \sin \theta_2)$.

*a*) $z_1 z_2 = r_1(\cos \theta_1 + i \sin \theta_1) \cdot r_2(\cos \theta_2 + i \sin \theta_2)$

$= r_1 r_2[(\cos \theta_1 \cos \theta_2 - \sin \theta_1 \sin \theta_2) + i(\sin \theta_1 \cos \theta_2 + \cos \theta_1 \sin \theta_2)]$

$= r_1 r_2[\cos(\theta_1 + \theta_2) + i \sin(\theta_1 + \theta_2)]$.

*b*) $\dfrac{r_1(\cos \theta_1 + i \sin \theta_1)}{r_2(\cos \theta_2 + i \sin \theta_2)} = \dfrac{r_1(\cos \theta_1 + i \sin \theta_1)(\cos \theta_2 - i \sin \theta_2)}{r_2(\cos \theta_2 + i \sin \theta_2)(\cos \theta_2 - i \sin \theta_2)}$

$= \dfrac{r_1}{r_2} \cdot \dfrac{(\cos \theta_1 \cos \theta_2 + \sin \theta_1 \sin \theta_2) + i(\sin \theta_1 \cos \theta_2 - \cos \theta_1 \sin \theta_2)}{\cos^2 \theta_2 + \sin^2 \theta_2}$

$= \dfrac{r_1}{r_2}[\cos(\theta_1 - \theta_2) + i \sin(\theta_1 - \theta_2)]$.

**15.** Perform the indicated operations, giving the result in both polar and rectangular form.
  *a*) $5(\cos 170^\circ + i \sin 170^\circ) \cdot (\cos 55^\circ + i \sin 55^\circ)$
  *b*) $2(\cos 50^\circ + i \sin 50^\circ) \cdot 3(\cos 40^\circ + i \sin 40^\circ)$
  *c*) $6(\cos 110^\circ + i \sin 110^\circ) \cdot \frac{1}{2}(\cos 212^\circ + i \sin 212^\circ)$
  *d*) $10(\cos 305^\circ + i \sin 305^\circ) \div 2(\cos 65^\circ + i \sin 65^\circ)$
  *e*) $4(\cos 220^\circ + i \sin 220^\circ) \div 2(\cos 40^\circ + i \sin 40^\circ)$
  *f*) $6(\cos 230^\circ + i \sin 230^\circ) \div 3(\cos 75^\circ + i \sin 75^\circ)$

  *a*) The modulus of the product is $5(1) = 5$ and the amplitude is $170^\circ + 55^\circ = 225^\circ$.
In polar form the product is $5(\cos 225^\circ + i \sin 225^\circ)$ and in rectangular form the product is $5(-\sqrt{2}/2 - i\sqrt{2}/2) = -5\sqrt{2}/2 - 5i\sqrt{2}/2$.

  *b*) The modulus of the product is $2(3) = 6$ and the amplitude is $50^\circ + 40^\circ = 90^\circ$.
In polar form the product is $6(\cos 90^\circ + i \sin 90^\circ)$ and in rectangular form it is $6(0 + i) = 6i$.

  *c*) The modulus of the product is $6(\frac{1}{2}) = 3$ and the amplitude is $110^\circ + 212^\circ = 322^\circ$.
In polar form the product is $3(\cos 322^\circ + i \sin 322^\circ)$ and in rectangular form it is $3(0.7880 - 0.6157i) = 2.3640 - 1.8471i$.

*d*) The modulus of the quotient is $10/2 = 5$ and the amplitude is $305° - 65° = 240°$.
In polar form the product is $5(\cos 240° + i \sin 240°)$ and in rectangular form it is
$5(-1/2 - i\sqrt{3}/2) = -5/2 - 5i\sqrt{3}/2$.

*e*) The modulus of the quotient is $4/2 = 2$ and the amplitude is $220° - 40° = 180°$.
In polar form the quotient is $2(\cos 180° + i \sin 180°)$ and in rectangular form it is
$2(-1 + 0i) = -2$.

*f*) The modulus of the quotient is $6/3 = 2$ and the amplitude is $230° - 75° = 155°$.
In polar form the quotient is $2(\cos 155° + i \sin 155°)$ and in rectangular form it is
$2(-0.9063 + 0.4226i) = -1.8126 + 0.8452i$.

**16.** Express each of the numbers in polar form, perform the indicated operation, and give the result in rectangular form.

*a*) $(-1 + i\sqrt{3})(\sqrt{3} + i)$      *d*) $-2 \div (-\sqrt{3} + i)$      *g*) $(3 + 2i)(2 + i)$

*b*) $(3 - 3i\sqrt{3})(-2 - 2i\sqrt{3})$      *e*) $6i \div (-3 - 3i)$

*c*) $(4 - 4i\sqrt{3}) \div (-2\sqrt{3} + 2i)$      *f*) $(1 + i\sqrt{3})(1 + i\sqrt{3})$      *h*) $(2 + 3i) \div (2 - 3i)$

*a*) $(-1 + i\sqrt{3})(\sqrt{3} + i) = 2(\cos 120° + i \sin 120°) \cdot 2(\cos 30° + i \sin 30°)$
$= 4(\cos 150° + i \sin 150°) = 4(-\sqrt{3}/2 + \tfrac{1}{2}i) = -2\sqrt{3} + 2i$

*b*) $(3 - 3i\sqrt{3})(-2 - 2i\sqrt{3}) = 6(\cos 300° + i \sin 300°) \cdot 4(\cos 240° + i \sin 240°)$
$= 24(\cos 540° + i \sin 540°) = 24(-1 + 0i) = -24$

*c*) $(4 - 4i\sqrt{3}) \div (-2\sqrt{3} + 2i) = 8(\cos 300° + i \sin 300°) \div 4(\cos 150° + i \sin 150°)$
$= 2(\cos 150° + i \sin 150°) = 2(-\sqrt{3}/2 + \tfrac{1}{2}i) = -\sqrt{3} + i$

*d*) $-2 \div (-\sqrt{3} + i) = 2(\cos 180° + i \sin 180°) \div 2(\cos 150° + i \sin 150°)$
$= \cos 30° + i \sin 30° = \tfrac{1}{2}\sqrt{3} + \tfrac{1}{2}i$

*e*) $6i \div (-3 - 3i) = 6(\cos 90° + i \sin 90°) \div 3\sqrt{2}(\cos 225° + i \sin 225°)$
$= \sqrt{2}(\cos 225° + i \sin 225°) = -1 - i$

*f*) $(1 + i\sqrt{3})(1 + i\sqrt{3}) = 2(\cos 60° + i \sin 60°) \cdot 2(\cos 60° + i \sin 60°)$
$= 4(\cos 120° + i \sin 120°) = 4(-\tfrac{1}{2} + \tfrac{1}{2}i\sqrt{3}) = -2 + 2i\sqrt{3}$

*g*) $(3 + 2i)(2 + i) = \sqrt{13}(\cos 33°41' + i \sin 33°41') \cdot \sqrt{5}(\cos 26°34' + i \sin 26°34')$
$= \sqrt{65}(\cos 60°15' + i \sin 60°15')$
$= \sqrt{65}(0.4962 + 0.8682i) = 4.001 + 7.000i = 4 + 7i$

*h*) $\dfrac{2 + 3i}{2 - 3i} = \dfrac{\sqrt{13}(\cos 56°19' + i \sin 56°19')}{\sqrt{13}(\cos 303°41' + i \sin 303°41')} = \dfrac{\cos 416°19' + i \sin 416°19'}{\cos 303°41' + i \sin 303°41'}$

$= \cos 112°38' + i \sin 112°38' = -0.3849 + 0.9230i$

**17.** Verify De Moivre's theorem for $n = 2$ and $n = 3$.

Let $z = r(\cos \theta + i \sin \theta)$.

For $n = 2$:   $z^2 = [r(\cos \theta + i \sin \theta)][r(\cos \theta + i \sin \theta)]$
$= r^2[(\cos \theta - \sin \theta) + i(2 \sin \theta \cos \theta)] = r^2(\cos 2\theta + i \sin 2\theta)$

For $n = 3$:   $z^3 = z^2 \cdot z = [r^2(\cos 2\theta + i \sin 2\theta)][r(\cos \theta + i \sin \theta)]$
$= r^3[(\cos 2\theta \cos \theta - \sin 2\theta \sin \theta) + i(\sin 2\theta \cos \theta + \cos 2\theta \sin \theta)]$
$= r^3(\cos 3\theta + i \sin 3\theta)$.

The theorem may be established for $n$ a positive integer by mathematical induction.

**18.** Evaluate each of the following using De Moivre's theorem and express each result in rectangular form: $a$) $(1 + i\sqrt{3})^4$,    $b$) $(\sqrt{3} - i)^5$,    $c$) $(-1 + i)^{10}$,    $d$) $(2 + 3i)^4$.

$a$) $(1 + i\sqrt{3})^4 = [2(\cos 60^\circ + i \sin 60^\circ)]^4 = 2^4(\cos 4\cdot 60^\circ + i \sin 4\cdot 60^\circ)$
$\qquad = 2^4(\cos 240^\circ + i \sin 240^\circ) = -8 - 8i\sqrt{3}$

$b$) $(\sqrt{3} - i)^5 = [2(\cos 330^\circ + i \sin 330^\circ)]^5 = 32(\cos 1650^\circ + i \sin 1650^\circ)$
$\qquad = 32(\cos 210^\circ + i \sin 210^\circ) = -16\sqrt{3} - 16i$

$c$) $(-1 + i)^{10} = [\sqrt{2}(\cos 135^\circ + i \sin 135^\circ)]^{10} = 32(\cos 270^\circ + i \sin 270^\circ) = -32i$

$d$) $(2 + 3i)^4 = [\sqrt{13}(\cos 56^\circ 19' + i \sin 56^\circ 19')]^4 = 13^2(\cos 225^\circ 16' + i \sin 225^\circ 16')$
$\qquad = 169(-0.7038 - 0.7104i) = -118.9 - 120.1i$

**19.** Find the indicated roots in rectangular form, except when this would necessitate the use of tables.

$a$) Square roots of $2 - 2i\sqrt{3}$          $e$) Fourth roots of $i$
$b$) Fourth roots of $-8 - 8i\sqrt{3}$       $f$) Sixth roots of $-1$
$c$) Cube roots of $-4\sqrt{2} + 4i\sqrt{2}$     $g$) Fourth roots of $-16i$
$d$) Cube roots of $1$                 $h$) Fifth roots of $1 + 3i$

$a$) $\qquad 2 - 2i\sqrt{3} = 4[\cos(300^\circ + k\,360^\circ) + i \sin(300^\circ + k\,360^\circ)]$

and $\qquad (2 - 2i\sqrt{3})^{1/2} = 2[\cos(150^\circ + k\,180^\circ) + i \sin(150^\circ + k\,180^\circ)]$.

Putting $k = 0$ and $1$, the required roots are
$$R_1 = 2(\cos 150^\circ + i \sin 150^\circ) = 2(-\tfrac{1}{2}\sqrt{3} + \tfrac{1}{2}i) = -\sqrt{3} + i$$
$$R_2 = 2(\cos 330^\circ + i \sin 330^\circ) = 2(\tfrac{1}{2}\sqrt{3} - \tfrac{1}{2}i) = \sqrt{3} - i.$$

$b$) $\qquad -8 - 8i\sqrt{3} = 16[\cos(240^\circ + k\,360^\circ) + i \sin(240^\circ + k\,360^\circ)]$

and $\qquad (-8 - 8i\sqrt{3})^{1/4} = 2[\cos(60^\circ + k\,90^\circ) + i \sin(60^\circ + k\,90^\circ)]$.

Putting $k = 0, 1, 2, 3$, the required roots are
$$R_1 = 2(\cos 60^\circ + i \sin 60^\circ) = 2(\tfrac{1}{2} + i\tfrac{1}{2}\sqrt{3}) = 1 + i\sqrt{3}$$
$$R_2 = 2(\cos 150^\circ + i \sin 150^\circ) = 2(-\tfrac{1}{2}\sqrt{3} + \tfrac{1}{2}i) = -\sqrt{3} + i$$
$$R_3 = 2(\cos 240^\circ + i \sin 240^\circ) = 2(-\tfrac{1}{2} - i\tfrac{1}{2}\sqrt{3}) = -1 - i\sqrt{3}$$
$$R_4 = 2(\cos 330^\circ + i \sin 330^\circ) = 2(\tfrac{1}{2}\sqrt{3} - \tfrac{1}{2}i) = \sqrt{3} - i.$$

$c$) $\qquad -4\sqrt{2} + 4i\sqrt{2} = 8[\cos(135^\circ + k\,360^\circ) + i \sin(135^\circ + k\,360^\circ)]$

and $\qquad (-4\sqrt{2} + 4i\sqrt{2})^{1/3} = 2[\cos(45^\circ + k\,120^\circ) + i \sin(45^\circ + k\,120^\circ)]$.

Putting $k = 0, 1, 2$, the required roots are
$$R_1 = 2(\cos 45^\circ + i \sin 45^\circ) = 2(1/\sqrt{2} + i/\sqrt{2}) = \sqrt{2} + i\sqrt{2}$$
$$R_2 = 2(\cos 165^\circ + i \sin 165^\circ)$$
$$R_3 = 2(\cos 285^\circ + i \sin 285^\circ).$$

$d$) $1 = \cos(0^\circ + k\,360^\circ) + i \sin(0^\circ + k\,360^\circ)$ and $1^{1/3} = \cos(k\,120^\circ) + i \sin(k\,120^\circ)$.

Putting $k = 0, 1, 2$, the required roots are
$$R_1 = \cos 0^\circ + i \sin 0^\circ = 1$$
$$R_2 = \cos 120^\circ + i \sin 120^\circ = -\tfrac{1}{2} + i\tfrac{1}{2}\sqrt{3}$$
$$R_3 = \cos 240^\circ + i \sin 240^\circ = -\tfrac{1}{2} - i\tfrac{1}{2}\sqrt{3}.$$

Note that $R_2^2 = \cos 2(120^\circ) + i \sin 2(120^\circ) = R_3$ ,

$\qquad R_3^2 = \cos 2(240^\circ) + i \sin 2(240^\circ) = R_2$ , and

$\qquad R_2 R_3 = (\cos 120^\circ + i \sin 120^\circ)(\cos 240^\circ + i \sin 240^\circ) = \cos 0^\circ + i \sin 0^\circ = R_1.$

e) $i = \cos(90^\circ + k\,360^\circ) + i \sin(90^\circ + k\,360^\circ)$ and $i^{1/4} = \cos(22\frac{1}{2}^\circ + k\,90^\circ) + i \sin(22\frac{1}{2}^\circ + k\,90^\circ).$

Thus, the required roots are

$\qquad R_1 = \cos 22\frac{1}{2}^\circ + i \sin 22\frac{1}{2}^\circ \qquad\qquad R_3 = \cos 202\frac{1}{2}^\circ + i \sin 202\frac{1}{2}^\circ$

$\qquad R_2 = \cos 112\frac{1}{2}^\circ + i \sin 112\frac{1}{2}^\circ \qquad\quad R_4 = \cos 292\frac{1}{2}^\circ + i \sin 292\frac{1}{2}^\circ.$

f) $-1 = \cos(180^\circ + k\,360^\circ) + i \sin(180^\circ + k\,360^\circ)$ and $(-1)^{1/6} = \cos(30^\circ + k\,60^\circ) + i \sin(30^\circ + k\,60^\circ).$

Thus, the required roots are

$\qquad R_1 = \cos 30^\circ + i \sin 30^\circ \quad = \frac{1}{2}\sqrt{3} + \frac{1}{2}i$

$\qquad R_2 = \cos 90^\circ + i \sin 90^\circ \quad = i$

$\qquad R_3 = \cos 150^\circ + i \sin 150^\circ = -\frac{1}{2}\sqrt{3} + \frac{1}{2}i$

$\qquad R_4 = \cos 210^\circ + i \sin 210^\circ = -\frac{1}{2}\sqrt{3} - \frac{1}{2}i$

$\qquad R_5 = \cos 270^\circ + i \sin 270^\circ = -i$

$\qquad R_6 = \cos 330^\circ + i \sin 330^\circ = \frac{1}{2}\sqrt{3} - \frac{1}{2}i.$

Note that $R_2^2 = R_5^2 = \cos 180^\circ + i \sin 180^\circ$ and thus $R_2$ and $R_5$ are the square roots of $-1$; that $R_1^3 = R_3^3 = R_5^3 = \cos 90^\circ + i \sin 90^\circ = i$ and thus $R_1, R_3, R_5$ are the cube roots of $i$; and that $R_2^3 = R_4^3 = R_6^3 = \cos 270^\circ + i \sin 270^\circ = -i$ and thus $R_2, R_4, R_6$ are the cube roots of $-i$.

g) $-16i = 16[\cos(270^\circ + k\,360^\circ) + i \sin(270^\circ + k\,360^\circ)]$ and
$(-16i)^{1/4} = 2[\cos(67\frac{1}{2}^\circ + k\,90^\circ) + i \sin(67\frac{1}{2}^\circ + k\,90^\circ)].$ Thus, the required roots are

$\qquad R_1 = 2(\cos 67\frac{1}{2}^\circ + i \sin 67\frac{1}{2}^\circ) \qquad R_3 = 2(\cos 247\frac{1}{2}^\circ + i \sin 247\frac{1}{2}^\circ)$

$\qquad R_2 = 2(\cos 157\frac{1}{2}^\circ + i \sin 157\frac{1}{2}^\circ) \qquad R_4 = 2(\cos 337\frac{1}{2}^\circ + i \sin 337\frac{1}{2}^\circ).$

h) $1 + 3i = \sqrt{10}\,[\cos(71^\circ 34' + k\,360^\circ) + i \sin(71^\circ 34' + k\,360^\circ)]$ and
$(1 + 3i)^{1/5} = \sqrt[10]{10}\,[\cos(14^\circ 19' + k\,72^\circ) + i \sin(14^\circ 19' + k\,72^\circ)].$ The required roots are

$\qquad R_1 = \sqrt[10]{10}(\cos 14^\circ 19' + i \sin 14^\circ 19')$

$\qquad R_2 = \sqrt[10]{10}(\cos 86^\circ 19' + i \sin 86^\circ 19')$

$\qquad R_3 = \sqrt[10]{10}(\cos 158^\circ 19' + i \sin 158^\circ 19')$

$\qquad R_4 = \sqrt[10]{10}(\cos 230^\circ 19' + i \sin 230^\circ 19')$

$\qquad R_5 = \sqrt[10]{10}(\cos 302^\circ 19' + i \sin 302^\circ 19').$

## SUPPLEMENTARY PROBLEMS

20. Perform the indicated operations, writing the results in the form $a + bi$.

a) $(6 - 2i) + (2 + 3i) = 8 + i$

b) $(6 - 2i) - (2 + 3i) = 4 - 5i$

c) $(3 + 2i) + (-4 - 3i) = -1 - i$

d) $(3 - 2i) - (4 - 3i) = -1 + i$

e) $3(2 - i) = 6 - 3i$

f) $2i(3 + 4i) = -8 + 6i$

g) $(2 + 3i)(1 + 2i) = -4 + 7i$

h) $(2 - 3i)(5 + 2i) = 16 - 11i$

i) $(3 - 2i)(-4 + i) = -10 + 11i$

j) $(2 + 3i)(3 + 2i) = 13i$

k) $(2 + \sqrt{-5})(3 - 2\sqrt{-4}) = (6 + 4\sqrt{5}) + (3\sqrt{5} - 8)i$

l) $(1 + 2\sqrt{-3})(2 - \sqrt{-3}) = 8 + 3\sqrt{3}\,i$

m) $(2 - i)^2 = 3 - 4i$

n) $(4 + 2i)^2 = 12 + 16i$

o) $(1 + i)^2(2 + 3i) = -6 + 4i$

p) $\dfrac{2 + 3i}{1 + i} = \dfrac{5}{2} + \dfrac{1}{2}i$

q) $\dfrac{3 - 2i}{3 - 4i} = \dfrac{17}{25} + \dfrac{6}{25}i$     r) $\dfrac{3 - 2i}{2 + 3i} = -i$

21. Show that $3 + 2i$ and $3 - 2i$ are roots of $x^2 - 6x + 13 = 0$.

22. Perform graphically the following operations.

a) $(2 + 3i) + (1 + 4i)$

b) $(4 - 2i) + (2 + 3i)$

c) $(2 + 3i) - (1 + 4i)$

d) $(4 - 2i) - (2 + 3i)$

23. Express each of the following complex numbers in polar form.

a) $3 + 3i = 3\sqrt{2}(\cos 45^\circ + i \sin 45^\circ)$

b) $1 + \sqrt{3}i = 2(\cos 60^\circ + i \sin 60^\circ)$

c) $-2\sqrt{3} - 2i = 4(\cos 210^\circ + i \sin 210^\circ)$

d) $\sqrt{2} - i\sqrt{2} = 2(\cos 315^\circ + i \sin 315^\circ)$

e) $-8 = 8(\cos 180^\circ + i \sin 180^\circ)$

f) $-2i = 2(\cos 270^\circ + i \sin 270^\circ)$

g) $-12 + 5i = 13(\cos 157^\circ 23' + i \sin 157^\circ 23')$

h) $-4 - 3i = 5(\cos 216^\circ 52' + i \sin 216^\circ 52')$

24. Perform the indicated operation and express the results in the form $a + bi$.

a) $3(\cos 25^\circ + i \sin 25^\circ)\, 8(\cos 200^\circ + i \sin 200^\circ) = -12\sqrt{2} - 12\sqrt{2}\,i$

b) $4(\cos 50^\circ + i \sin 50^\circ)\, 2(\cos 100^\circ + i \sin 100^\circ) = -4\sqrt{3} + 4i$

c) $\dfrac{4(\cos 190^\circ + i \sin 190^\circ)}{2(\cos 70^\circ + i \sin 70^\circ)} = -1 + i\sqrt{3}$

d) $\dfrac{12(\cos 200^\circ + i \sin 200^\circ)}{3(\cos 350^\circ + i \sin 350^\circ)} = -2\sqrt{3} - 2i$

25. Use the polar form in finding each of the following products and quotients, and express each result in the form $a + bi$.

a) $(1 + i)(\sqrt{2} - i\sqrt{2}) = 2\sqrt{2}$

b) $(-1 - i\sqrt{3})(-4\sqrt{3} + 4i) = 8\sqrt{3} + 8i$

c) $\dfrac{1 - i}{1 + i} = -i$

d) $\dfrac{4 + 4\sqrt{3}\,i}{\sqrt{3} + i} = 2\sqrt{3} + 2i$

e) $\dfrac{-1 + i\sqrt{3}}{\sqrt{2} + i\sqrt{2}} = 0.2588 + 0.9659i$

f) $\dfrac{3 + i}{2 + i} = 1.4 - 0.2i$

26. Use De Moivre's Theorem to evaluate each of the following and express each result in the form $a + bi$.

a) $[2(\cos 6^\circ + i \sin 6^\circ)]^5 = 16\sqrt{3} + 16i$

b) $[\sqrt{2}(\cos 75^\circ + i \sin 75^\circ)]^4 = 2 - 2\sqrt{3}i$

c) $(1 + i)^8 = 16$

d) $(1 - i)^6 = 8i$

e) $(1/2 - i\sqrt{3}/2)^{20} = -1/2 - i\sqrt{3}/2$

f) $(\sqrt{3}/2 + i/2)^9 = -i$

g) $(3 + 4i)^4 = -526.9 - 336.1i$

h) $\dfrac{(1 - i\sqrt{3})^3}{(-2 + 2i)^4} = \dfrac{1}{8}$     i) $\dfrac{(1 + i)(\sqrt{3} + i)^3}{(1 - i\sqrt{3})^3} = 1 - i$

27. Find all the indicated roots, expressing the results in the form $a + bi$ unless tables would be needed to do so.

   a) The square roots of $i$.              Ans. $\sqrt{2}/2 + i\sqrt{2}/2, \quad -\sqrt{2}/2 - i\sqrt{2}/2$

   b) The square roots of $1 + i\sqrt{3}$.       Ans. $\sqrt{6}/2 + i\sqrt{2}/2, \quad -\sqrt{6}/2 - i\sqrt{2}/2$

   c) The cube roots of $-8$.               Ans. $1 + i\sqrt{3}, \quad -2, \quad 1 - i\sqrt{3}$

   d) The cube roots of $27i$.              Ans. $3\sqrt{3}/2 + 3i/2, \quad -3\sqrt{3}/2 + 3i/2, \quad -3i$

   e) The cube roots of $-4\sqrt{3} + 4i$.
      Ans. $2(\cos 50^\circ + i \sin 50^\circ), \quad 2(\cos 170^\circ + i \sin 170^\circ), \quad 2(\cos 290^\circ + i \sin 290^\circ)$

   f) The fifth roots of $1 + i$.   Ans. $\sqrt[10]{2}(\cos 9^\circ + i \sin 9^\circ), \quad \sqrt[10]{2}(\cos 81^\circ + i \sin 81^\circ)$, etc.

   g) The sixth roots of $-\sqrt{3} + i$. Ans. $\sqrt[6]{2}(\cos 25^\circ + i \sin 25^\circ), \quad \sqrt[6]{2}(\cos 85^\circ + i \sin 85^\circ)$, etc.

28. Find the tenth roots of $i$ and show that the product of any two of them is again one of the tenth roots of 1.

29. Show that the reciprocal of any one of the tenth roots of 1 is again a tenth root of 1.

30. Denote either of the complex cube roots of 1 (Problem 19 $d$) by $\omega_1$ and the other by $\omega_2$. Show that $\omega_1^2 \omega_2 = \omega_1$ and $\omega_1 \omega_2^2 = \omega_2$.

31. Show that $(\cos \theta + i \sin \theta)^{-n} = \cos n\theta - i \sin n\theta$.

32. Use the fact that the segments $OS$ and $P_2 P_1$ in Fig. 18-E are equal to devise a second procedure for constructing the difference $OS = z_1 - z_2$ of two complex numbers $z_1$ and $z_2$.

# CHAPTER 19

# Topics from Solid Geometry

THE POINT OF INTERSECTION of a line with a plane is called the *foot* of the line.

A given line is perpendicular to a given plane, which it intersects, if every line in the plane through the foot of the given line is perpendicular to that line.

If a line is perpendicular to each of two intersecting lines at their point of intersection, it is perpendicular to the plane of the two lines. See Fig. 19-A.

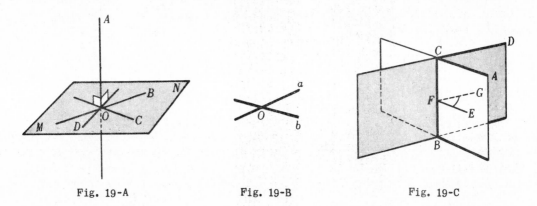

| Fig. 19-A | Fig. 19-B | Fig. 19-C |

DIHEDRAL ANGLES. When two lines have one and only one point in common (Fig. 19-B), they define four plane angles. When two planes have one and only one line in common (Fig. 19-C), they define four *dihedral angles*. We shall restrict our attention to the dihedral angle *A-BC-D* indicated by the heavy lines of Figure 19-C. The planes *ABC* and *DBC* are called the *faces* and the line of intersection *BC* is called the *edge* of this dihedral angle.

The plane angle formed by two lines, one in each face of a dihedral angle, perpendicular to the edge at a common point is called the *plane angle* of the dihedral angle. The plane angle, as ∠*EFG* of Fig. 19-C, is taken as the measure of the dihedral angle *A-BC-D*.

Dihedral angles are called acute, right, or obtuse according as their plane angles are acute, right, or obtuse.

TRIHEDRAL ANGLES. When three planes have one and only one point in common, they define eight *trihedral angles*. We shall restrict our attention to the trihedral angle *O-XYZ* indicated by the heavy lines of Fig. 19-D. The common point *O* is called the *vertex* and the planes *OXY*, *OYZ*, and *OZX* are called the *faces* of this trihedral angle. The faces, taken in pairs, form three dihedral angles whose edges *OX*, *OY*, *OZ* are called the *edges* of the trihedral angle. The plane angles *XOY*,

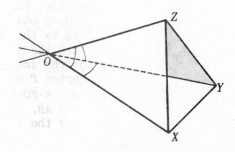

Fig. 19-D

*YOZ, ZOX* in the faces of the trihedral angle are called its *face angles*.

The sum of any two face angles of a trihedral angle is greater than the third face angle. For a proof, see Problem 1.

The sum of the face angles of a trihedral angle is less than 360°. For a proof, see Problem 2.

SPHERICAL ANGLES. The plane section of a sphere is a circle. This circle (Fig. 19-E) is called a *great circle* if the intersecting plane passes through the center of the sphere; otherwise, a *small circle*. The poles of such a circle (great or small) are the two points of intersection with the sphere of that diameter of the sphere which is perpendicular to the plane of the circle. In Fig. 19-E, *P* and *P'* are poles of both the great and small circles illustrated. Note that while *P* is the pole of many small circles (all small circles defined by planes parallel to *MM'*) it is the pole of only one great circle.

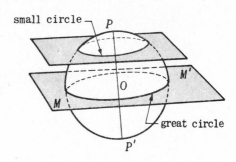

Fig. 19-E

Two distinct points on a sphere (as *A* and *B* of Fig. 19-F) which are not the extremities of a diameter lie on one and only one great circle. The shorter arc *AB* of this great circle is the shortest curve on the sphere joining the two points.

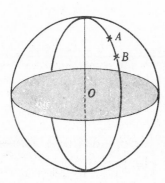

Fig. 19-F

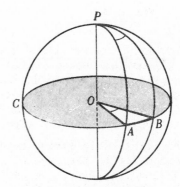

Fig. 19-G

The angle formed by two intersecting arcs of great circles on a sphere is called a *spherical angle*. The great circle arcs are called the *sides* and their point of intersection is called the *vertex* of the spherical angle. A spherical angle is measured by the dihedral angle formed by the planes of the great circles whose arcs are the *sides* of the spherical angle. In Fig. 19-G, *APB* is a spherical angle on the sphere of center *O* and the circle *ABC* is the great circle having the vertex *P* of the spherical angle as pole. Since the corresponding dihedral angle *A-PO-B* is measured by the plane angle *AOB* which in turn is measured by the arc *AB*, it follows that a spherical angle is measured by the arc intercepted by the sides on the great circle whose pole is the vertex of the angle.

SPHERICAL TRIANGLES. The portion of the surface of a sphere bounded by the arcs of three great circles on it is called a *spherical triangle*. The bounding arcs are called the *sides* and the vertices of the three spherical angles are called the *vertices* of the spherical triangle. We shall usually designate the vertices by $A, B, C$, and the corresponding opposite sides by $a, b, c$ respectively.

When the vertices $A, B, C$ of a spherical triangle (Fig. 19-H) are joined to the center of the sphere, a trihedral angle $O\text{-}ABC$ is formed. The sides $a, b, c$ of the spherical triangle are measured by the face angles $BOC, COA, AOB$ of this trihedral angle. The angles $A, B, C$ of the spherical triangle are measured by the dihedral angles of the trihedral angle — angle $A$ is measured by the dihedral angle $B\text{-}OA\text{-}C$, etc.

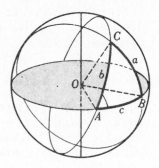

Fig. 19-H

Unless otherwise specified, the spherical triangles to be considered will be restricted to those for which each side and angle is less than 180°. For such triangles:

1) The sum of any two sides is greater than the third side.
2) The sum of the three sides is less than 360°.
   (These theorems follow from corresponding theorems regarding the face angles of a trihedral angle.)

3) If two sides are equal, the angles opposite are equal and conversely.
4) If two sides are unequal, the angles opposite are unequal and the greater angle is opposite the greater side, and conversely.
   (These theorems are intuitively evident and no formal proof is given.)

5) The sum of the three angles is greater than 180° and less than 540°.
   (A proof of this theorem in Problem 8 requires the use of the polar triangle discussed in the next section.)

The *spherical excess E* of a spherical triangle is the amount by which the sum of its angles exceeds 180°. For example, for the spherical triangle whose angles are $A = 65°$, $B = 75°$, $C = 112°$,

$$E = 65° + 75° + 112° - 180° = 72°.$$

**POLAR TRIANGLES.** Let $A, B, C$ be the vertices of a spherical triangle and construct the three great circles having these vertices as poles. Denote by $A'$ that intersection of the great circles having $B$ and $C$ as poles which lies on the same side of $BC$ as does $A$; by $B'$ that intersection of the great circles having $C$ and $A$ as poles which lies on the same side of $CA$ as does $B$; and by $C'$ that intersection of the great circles having $A$ and $B$ as poles which lies on the same side of $AB$ as does $C$. The spherical triangle $A'B'C'$ is called the *polar triangle* of $ABC$. We shall denote its sides by $a', b', c'$ as in Fig. 19-I.

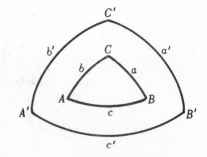

Fig. 19-I

The fundamental theorems concerning polar triangles are:

1) If $A'B'C'$ is the polar triangle of $ABC$, then $ABC$ is the polar triangle of $A'B'C'$. (For a proof, see Problem 5.)

2) In two polar triangles, each angle of one of the triangles is equal to the supplement of the corresponding side of the other triangle; thus,

$$A = 180° - a' \qquad B = 180° - b' \qquad C = 180° - c'$$
$$A' = 180° - a \qquad B' = 180° - b \qquad C' = 180° - c .$$

(For a proof, see Problem 6.)

APPLICATIONS. In order to simplify certain calculations, it is customary to consider the earth as a sphere. The axis of rotation of this sphere intersects its surface in the *north* and *south* poles, $P_n$ and $P_s$ of Fig. 19-J. The great circle having $P_n$ and $P_s$ as poles is called the *equator*. For any point $A$ on the earth's surface distinct from the poles, the half circle $P_nAP_s$ is called the *meridian* of $A$. The *first* or *prime* meridian passes through the astronomical observatory at Greenwich, England.

The *latitude* (lat.) of $A$ is the angular distance from the equator to $A$. It is measured either by the angle $A'OA$ or by the arc $A'A$ of the meridian of $A$. Latitude is designated north or south according as the point in question is in the northern or southern hemisphere. The difference in latitude between two points of latitudes $L_1$ and $L_2$, $(L_1 > L_2)$, respectively is $L_1-L_2$ if the points are in the same hemisphere and is $L_1+L_2$ if they are in different hemispheres.

Small circles cut by planes perpendicular to the axis are called *parallels of latitude* or *parallels*. All points on a parallel have the same latitude.

The *longitude* (long.) of $A$ is the angle (not greater than 180°) between the prime meridian and the meridian of $A$. It is measured either by the arc $G'A'$ intercepted on the equator by the two meridians or by the spherical angle $G'P_nA'$. Longitude is designated east or west according as the point in question is east or west of the prime meridian. The difference in longitude between two points of longitudes $\lambda_1$ and $\lambda_2$, $(\lambda_1 > \lambda_2)$, respectively is $\lambda_1-\lambda_2$ if the points are in the same direction from the prime meridian and is the smaller of $\lambda_1+\lambda_2$ and $360°-(\lambda_1+\lambda_2)$ if they are in different directions.

The equator and prime meridian act as a pair of coordinate axes on the earth's surface, the equator corresponding to the $x$-axis and the prime meridian corresponding to the $y$-axis of a system of rectangular coordinates in a plane. The latitude and longitude of a point $A$ are the coordinates of $A$ with respect to these axes, latitude corresponding to the $y$-coordinate and longitude to the $x$-coordinate. The designations north and south latitude and east and west longitude correspond to positive and negative coordinates of a point in a plane.

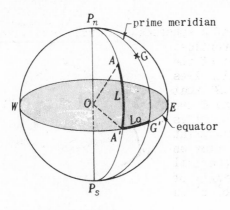

Fig. 19-J

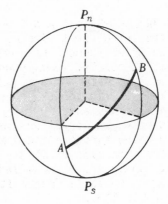

Fig. 19-K

The meridians through two points on the earth's surface and the smaller of the great circle arcs joining the points (Fig. 19-K) form two spherical triangles $AP_nB$ and $AP_sB$. In a later chapter, one of these triangles will be used to determine the great circle distance (length of arc $AB$) between the points. These distances along great circle arcs are usually given in nautical miles where, by definition,

$1'$ of great circle arc = 1 nautical mile = 6080 feet.

If a ship or airplane is following a great circle track between two points, its *course* is the angle which the track makes with the meridian of the ship or plane. In naval and air usage, the course is measured from the north around through the east.

EXAMPLE. *a)* In Fig. 19-L, a ship is to travel from $A$ to $B$.
The *initial course* (course at $A$) is angle $P_n AB$ and the *course on arrival* (course at $B$) is angle $P_n BC$ as marked.

*b)* In Fig. 19-M, a ship is to travel from $B$ to $A$.
The initial course (at $B$) is angle $P_n BA$ and the course on arrival (at $A$) is angle $P_n AC$ as marked.

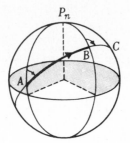

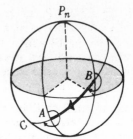

Fig. 19-L          Fig. 19-M

## SOLVED PROBLEMS

1. Prove: The sum of any two face angles of a trihedral angle is greater than the third face angle.

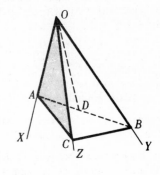

The theorem is true if the three face angles are equal. We shall consider then a trihedral angle $O$-$XYZ$ in which $\angle XOY$ is greater than either of the other two face angles. On $OX$ take any point $A$, on $OY$ take any point $B$, and on $AB$ take $D$ such that $\angle AOD = \angle XOZ$. On $OZ$ take $C$ such that $OC = OD$. Join $A$ and $B$ to $C$.

In the triangle $ABC$, $AC + CB > AB$, $AB = AD + DB$, and $AC + CB > AD + DB$. Since the triangles $AOC$ and $AOD$ are congruent, $AD = AC$; hence, $AC + CB > AC + DB$ and $CB > DB$.

Then, since sides $OD$ and $OB$ of triangle $ODB$ are equal respectively to sides $OC$ and $OB$ of triangle $OCB$, $\angle COB > \angle DOB$. By construction, $\angle AOC = \angle AOD$. Hence,

$$\angle AOC + \angle COB > \angle AOD + \angle DOB = \angle AOB$$

which was to be proved.

2. Prove: The sum of the face angles of a trihedral angle is less than $360°$.

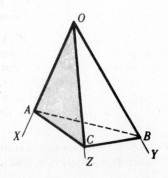

On the edges of the trihedral angle $O$-$XYZ$ take points $A, B, C$. We first note that there are three triangles with vertex $O$ and that the sum of the angles of these triangles is $3 \cdot 180° = 540°$; that is,

$$\angle AOB + \angle BOC + \angle COA + (\angle OAB + \angle OAC)$$
$$+ (\angle OBA + \angle OBC) + (\angle OCA + \angle OCB) = 540°.$$

By Problem 1,  $\angle OAB + \angle OAC > \angle BAC$,
$\angle OBA + \angle OBC > \angle ABC$,  and
$\angle OCA + \angle OCB > \angle ACB$.

Then                 $\angle AOB + \angle BOC + \angle COA + \angle BAC + \angle ABC + \angle ACB < 540^\circ$   or

$\angle AOB + \angle BOC + \angle COA < 540^\circ - (\angle BAC + \angle ABC + \angle ACB)$.

Since the sum in parentheses is the sum of the angles of the triangle $ABC$,

$\angle AOB + \angle BOC + \angle COA < 540^\circ - 180^\circ = 360^\circ$         which was to be proved.

3. Let $A$ and $B$ be two points of a great circle on a sphere of center $O$ and let $P$ be the pole of the great circle. Construct and solve the spherical triangle $ABP$ when (a) $AB = 75^\circ$ and (b) $AB = 90^\circ$.

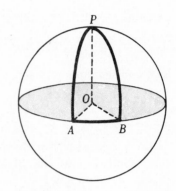

Join $A$ and $P$, also $B$ and $P$, by great circle arcs. Since every point on a great circle is at a distance $90^\circ$ from the pole of the great circle, $AP = BP = 90^\circ$. The spherical angles $PAB$ or $A$ and $PBA$ or $B$ are measured by the dihedral angles $P\text{-}AO\text{-}B$ and $P\text{-}BO\text{-}A$ whose respective faces are perpendicular planes. Thus, $A = B = 90^\circ$.

a) The spherical angle $APB$ or $P$ is measured by the plane angle $AOB$ which has the same measure as the arc $AB$. The sides of the triangle $APB$ are $AP = BP = 90^\circ$, $AB = 75^\circ$; and the angles are $A = B = 90^\circ$, $P = 75^\circ$.

b) For the spherical triangle $APB$, $AP = BP = AB = 90^\circ$ and $A = B = P = 90^\circ$.

4. In each of the following state whether a spherical triangle $ABC$ having the given parts is possible: (a) $AB = 50^\circ$, $BC = 70^\circ$, $CA = 100^\circ$; (b) $AB = 35^\circ$, $BC = 65^\circ$, $CA = 120^\circ$; (c) $AB = 150^\circ$, $BC = 100^\circ$, $CA = 120^\circ$.

a) Yes; $AB + BC + CA < 360^\circ$ and the sum of any two sides is greater than the third.
b) No; $AB + BC < CA$.
c) No; $AB + BC + CA > 360^\circ$.

5. Prove: If $A'B'C'$ is the polar triangle of $ABC$, then $ABC$ is the polar triangle of $A'B'C'$.

Since $A$ is the pole of $B'C'$ and $C$ is the pole of $A'B'$, $B'$ is a quadrant's distance ($90^\circ$) from $A$ and $C$. Thus, $B'$ is the pole of arc $AC$. In a similar manner it may be shown that $A'$ is the pole of arc $BC$ and $C'$ is the pole of arc $AB$. Then the triangle $ABC$ is one of the eight triangles formed by the great circles whose poles are $A', B', C'$. If $ABC$ is that one of the eight triangles called the polar triangle of $A'B'C'$, it is necessary that $A$ and $A'$ lie on the same side of $B'C'$, that $B$ and $B'$ lie on the same side of $C'A'$, and that $C$ and $C'$ lie on the same side of $A'B'$.

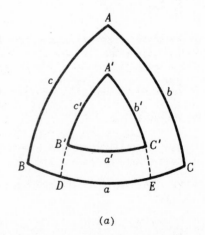

(a)

By definition, $B$ and $B'$ lie on the same side of $AC$ and $B$ is less than $180^\circ$ from any point on $AC$. Then, since $B'$ (the pole of $AC$) is $90^\circ$ from any point of $AC$, $B$ and $B'$ are less than $90^\circ$ apart. Finally, since $B$ (the pole of $A'C'$) is $90^\circ$ from any point on $A'C'$, $B$ and $B'$ lie on the same side of $A'C'$. Similarly it may be shown that $A$ and $A'$ lie on the same side of $B'C'$ and that $C$ and $C'$ lie on the same side of $A'B'$.

6. Prove: In two polar triangles, each angle of one triangle is equal to the supplement of the corresponding opposite side of the other triangle.

For the polar triangles $ABC$ and $A'B'C'$ of Fig. (a) above, we shall prove that $A' = 180^\circ - a$.

Extend the arcs $A'B'$ and $A'C'$ to meet $BC$ in $D$ and $E$ respectively. Then arc $DE$ is the measure of angle $A'$. Now $BE + DC = BC + DE = a + A'$ and, since $B$ is the pole of $A'E$ and $C$ is the pole of $A'D$, $BE = DC = 90°$. Thus, $a + A' = 180°$ and $A' = 180° - a$.

**7.** Find the parts of the polar triangle of the spherical triangle for which :
  a) $A = 156°56'$, $B = 83°11'$, $C = 90°$; $a = 157°55'$, $b = 72°22'$, $c = 106°18'$.
  b) $A = 44°59'$, $B = 112°47'$, $C = 85°7'$; $a = 43°17'$, $b = 116°36'$, $c = 105°15'$.

Using the theorem of Problem 6 :

  a) $A' = 180° - a = 22°5'$,   $B' = 180° - b = 107°38'$,   $C' = 180° - c = 73°42'$;
     $a' = 180° - A = 23°4'$,   $b' = 180° - B = 96°49'$,   $c' = 180° - C = 90°$.

  b) $A' = 180° - a = 136°43'$,   $B' = 180° - b = 63°24'$,   $C' = 180° - c = 74°45'$;
     $a' = 180° - A = 135°1'$,   $b' = 180° - B = 67°13'$,   $c' = 180° - C = 94°53'$.

**8.** Prove : The sum of the angles of a spherical triangle is greater than $180°$ and less than $540°$.

Let $ABC$ be the given spherical triangle (see Fig.$(a)$ of Prob. 6) and let $A'B'C'$ be its polar triangle.  From the theorem of Problem 6,

$$A + a' = B + b' = C + c' = 180°;  \text{ hence, } A + B + C + a' + b' + c' = 540°.$$

Now   $a' + b' + c' > 0°$    so that   $A + B + C < 540°$
and    $a' + b' + c' < 360°$    so that   $A + B + C > 180°$.

**9.** In each of the following state whether a spherical triangle $ABC$ having the given parts is possible: $(a)$ $A = 60°$, $B = 70°$, $C = 90°$; $(b)$ $A = 60°$, $B = 115°$, $C = 145°$; $(c)$ $A = 60°$, $B = 20°$, $C = 90°$.

a) Yes;  $A + B + C = 220°$ is between $180°$ and $540°$ while the sides $a' = 120°$, $b' = 110°$, $c' = 90°$ of the polar triangle satisfy the condition that the sum of any two sides is greater than the third side.

b) No;  the sides $a' = 120°$, $b' = 65°$, $c' = 35°$ of the polar triangle do not satisfy the condition $b' + c' > a'$.

c) No;  $A + B + C < 180°$.

**10.** Solve the spherical triangle,, given $a = b = 90°$, $c = 60°$.

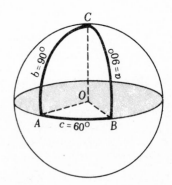

Assume the spherical triangle on a sphere of center $O$. Now $C$, being a quadrant's distance from both $A$ and $B$, is the pole of the great circle of which $AB = c$ is an arc. Then angle $C = 60°$ since it is measured by the arc $AB$.

The planes $AOC$ and $BOC$ are perpendicular to the plane $AOB$ since their intersection $OC$ is perpendicular to $AOB$; hence, angles $A$ and $B$ are right angles.

Thus, the required parts are : $A = B = 90°$, $C = 60°$.

**11.** Solve the spherical triangle, given $A = B = C = 90°$.

Since each vertex is the pole of the great circle through the other two, each vertex is a quadrant's distance from the other two vertices.  Thus, $a = b = c = 90°$.

**12.** Find the difference in longitude between :
  a) New York (long. $74°1.0'$ W) and Pearl Harbor (long. $157°58.3'$ W).
  b) New York and Moscow (long. $37°34.3'$ E).
  c) New York and Sydney (long. $151°13.0'$ E).
  d) Sydney and Moscow.

The required distances are:

*a*) $\lambda_1 - \lambda_2 = 157°58.3' - 74°1.0' = 83°57.3'$,   since both are west.

*b*) $\lambda_1 + \lambda_2 = 74°1.0' + 37°34.3' = 111°35.3'$,   since one is east and the other west.

*c*) $360° - (\lambda_1 + \lambda_2) = 360° - (151°13.0' + 74°1.0') = 134°46.0'$.

*d*) $\lambda_1 - \lambda_2 = 151°13.0' - 37°34.3' = 113°38.7'$.

**13.** Find the distance (in n.m.) between each pair of points on the earth's surface:
  *a*) $A$(lat. 30°25' N, long. 40° W)  and $B$(lat. 75°10' N, long. 40° W).
  *b*) $A$(lat. 30°25' S, long. 50° E)  and $B$(lat. 75°10' N, long. 50° E).

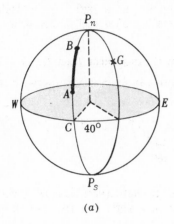

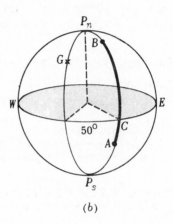

(*a*)                                    (*b*)

*a*) In Fig. (*a*),  $CA = 30°25'$, $CB = 75°10'$,  and $AB = CB - CA = 44°45' = 2685' = 2685$ n.m.

*b*) In Fig. (*b*),  $CA = 30°25'$, $CB = 75°10'$,  and $AB = CA + CB = 105°35' = 6335' = 6335$ n.m.

## SUPPLEMENTARY PROBLEMS

14. Show that any face angle of a trihedral angle is greater than the difference of the other two face angles. Hint: Use $A + B > C$.

15. What can be said of the third face angle of a trihedral angle, given:
    a) two of the face angles are $70^\circ$ and $50^\circ$ respectively?      Ans. $>20^\circ$; $<120^\circ$
    b) two of the face angles are $130^\circ$ and $150^\circ$ respectively?      Ans. $>20^\circ$; $<80^\circ$

16. Is it possible to have a spherical triangle $ABC$ whose sides are:
    a) $160^\circ$, $110^\circ$, $85^\circ$ ?    b) $170^\circ$, $150^\circ$, $10^\circ$ ?    c) $170^\circ$, $150^\circ$, $50^\circ$ ?    d) $30^\circ$, $50^\circ$, $70^\circ$ ?
    Ans. a) yes; b) no; c) no; d) yes

17. Find the parts of the polar triangle $A'B'C'$ of the spherical triangle $ABC$ for which:
    a) $A = 67^\circ19'$, $B = 48^\circ29'$, $C = 77^\circ17'$; $a = 43^\circ18'$, $b = 33^\circ49'$, $c = 46^\circ28'$.
    b) $A = 122^\circ7'$, $B = 32^\circ24'$, $C = 41^\circ36'$; $a = 73^\circ44'$, $b = 37^\circ25'$, $c = 48^\circ48'$.

    Ans. a) $A' = 136^\circ42'$, $B' = 146^\circ11'$, $C' = 133^\circ32'$; $a' = 112^\circ41'$, $b' = 131^\circ31'$, $c' = 102^\circ43'$
          b) $A' = 106^\circ16'$, $B' = 142^\circ35'$, $C' = 131^\circ12'$; $a' = 57^\circ53'$, $b' = 147^\circ36'$, $c' = 138^\circ24'$

18. Is it possible to have a spherical triangle $ABC$ whose angles are:
    a) $30^\circ$, $37^\circ$, $128^\circ$ ?    b) $30^\circ$, $37^\circ$, $111^\circ$ ?    c) $37^\circ$, $51^\circ$, $131^\circ$ ?    d) $40^\circ$, $85^\circ$, $140^\circ$ ?
    Ans. a) yes; b) no; c) yes; d) no

19. The area of the surface of a sphere of radius $R$ is equal to $4\pi R^2$. The area $K$ of a spherical triangle on this sphere is given by $K = \pi R^2 E/180$, where $E$ is the spherical excess in **degrees** of the triangle. What portion of the area of a sphere of radius 10 is bounded by each of the spherical triangles with angles:
    a) $A = B = C = 110^\circ$ ?             Ans. $K = 250\pi/3$; $5/24$
    b) $A = 150^\circ$, $B = 138^\circ$, $C = 132^\circ$ ?     Ans. $K = 400\pi/3$; $1/3$

20. Find the difference in longitude between
    a) San Francisco (long. $122^\circ15.7'$ W) and Dakar (long. $17^\circ25.0'$ W),
    b) San Francisco and Melbourne (long. $144^\circ58.5'$ E),
    c) Dakar and Cape Town (long. $18^\circ26.0'$ E),
    d) Melbourne and Cape Town.
    Ans. a) $104^\circ50.7'$,   b) $92^\circ45.8'$,   c) $35^\circ51.0'$,   d) $126^\circ32.5'$

21. Find the distance (in n.m.) between each pair of points on the earth's surface:
    a) $A$(lat. $40^\circ40'$ N; long. $120^\circ$ W) and $B$(lat. $75^\circ25'$ N; long. $120^\circ$ W).     Ans. 2085 n.m.
    b) $A$(lat. $50^\circ20'$ N; long. $80^\circ$ W) and $B$(lat. $30^\circ50'$ S; long. $80^\circ$ W).     Ans. 4870 n.m.
    c) $A$(lat. $10^\circ30'$ S; long. $40^\circ$ E) and $B$(lat. $50^\circ20'$ S; long. $40^\circ$ E).     Ans. 2390 n.m.

# CHAPTER 20

# Right Spherical Triangles

**FORMULAS.** A spherical triangle having one right angle is called a *right spherical triangle*. For any such triangle *ABC*, with the right angle *always* at *C*, the following ten fundamental relations hold:

Fig. 20-A

(1) $\sin a = \sin A \sin c$
(2) $\tan a = \tan A \sin b$
(3) $\tan a = \cos B \tan c$
(4) $\cos c = \cos b \cos a$
(5) $\cos A = \sin B \cos a$

(6) $\sin b = \sin B \sin c$
(7) $\tan b = \tan B \sin a$
(8) $\tan b = \cos A \tan c$
(9) $\cos c = \cot A \cot B$
(10) $\cos B = \sin A \cos b$.

For a derivation of these formulas, see Problem 1.

**LAWS OF QUADRANTS.** If in a right spherical triangle the parts *A* and *c* are known, the value of sin *a* is given by formula (1), $\sin a = \sin A \sin c$. Additional information is needed, however, to determine whether *a* is less than or greater than 90°. Such information is given by the laws of quadrants:

1. Side *a* and angle *A* (also side *b* and angle *B*) are in the same quadrant.

2. If *c* < 90°, then sides *a* and *b* (also angles *A* and *B*) are in the same quadrant; if *c* > 90°, then sides *a* and *b* (also angles *A* and *B*) are in different quadrants. (For a proof of these laws, see Problem 2.)

**EXAMPLE 1.** (*a*) If *A* < 90° and *c* < 90° then *a,b,B* are < 90° but if *c* > 90° then *a* < 90° and *b,B* are > 90°.

(*b*) If *A* > 90° and *c* < 90° then *a,b,B* are > 90° but if *c* > 90° then *a* > 90° and *b,B* are < 90°.

**NAPIER'S RULES.** Using either of the devices shown in Figures 20-B and 20-C, Napier gave rules for writing down the ten fundamental formulas. Figure 20-B shows a schematic triangle obtained from the spherical triangle of Figure 20-A by replacing c by co-c = 90° − c, A by co-A = 90° − A, and B by co-B = 90° − B. Note that the letter *C* is omitted. Fig. 20-C shows the five essential parts in Fig. 20-B arranged in a circle.

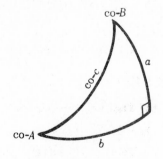

Fig. 20-B

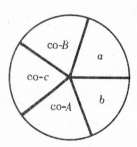

Fig. 20-C

Select any one of the five parts and call it the *middle part*, call the two parts next to it *adjacent parts*, and call the two parts remaining *opposite parts*. Then Napier's Rules are:

1. The sine of any middle part is equal to the product of the tangents of the adjacent parts.

2. The sine of any middle part is equal to the product of the cosines of the opposite parts.

EXAMPLE 2.    Take $b$ as the middle part; then co-$A$ and $a$ are the adjacent parts while co-$c$ and co-$B$ are the opposite parts.  Using Rule 1, we obtain

$$\sin b = \tan (\text{co-}A) \tan a = \cot A \tan a$$

or    (2)               $\tan a = \tan A \sin b.$

Using Rule 2, we have  (6) :  $\sin b = \cos (\text{co-}B) \cos (\text{co-}c) = \sin B \sin c.$

EXAMPLE 3.    Take co-$B$ as the middle part; then co-$c$ and $a$ are the adjacent parts and co-$A$ and $b$ are the opposite parts.  Using Rule 1, we obtain

$$\sin (\text{co-}B) = \tan (\text{co-}c) \tan a.$$

Then                       $\cos B = \cot c \tan a$

or    (3)               $\tan a = \cos B \tan c.$

Using Rule 2,  $\sin (\text{co-}B) = \cos (\text{co-}A) \cos b$  or  (10)  $\cos B = \sin A \cos b.$

Considering each of the five parts in turn as the middle part and proceeding as in Examples 2 and 3 above, we obtain the ten fundamental formulas. See Problems 3-4.

SOLUTIONS OF RIGHT SPHERICAL TRIANGLES.  In addition to the right angle, two other parts of a right spherical triangle must be given in order to determine it. When, however, two measures are chosen at random as these parts then no triangle, one triangle, or two triangles may be determined.  Two triangles (the ambiguous case) will be possible only when the given parts consist of a side ($a$ or $b$) and the opposite angle.  See Problems 5-6.

The following steps are suggested for solving right spherical triangles:

*A*) Draw a schematic triangle (Fig. 20-B) and encircle the given parts.
*B*) Write a formula relating the two given parts and an unknown part by applying the appropriate rule of Napier.
*C*) Write a check formula connecting the three unknown parts.
*D*) Apply the Laws of Quadrants especially when a part is to be found from its sine.

EXAMPLE 4.    Suppose $A = 65°$ and $B = 118°$ of a right spherical triangle $ABC$ are given.

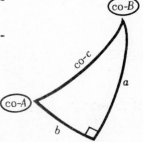

To find $a$: Consider $a$, co-$A$, and co-$B$; co-$A$ is the middle part and $a$ and co-$B$ are opposite parts.  Then

$$\sin (\text{co-}A) = \cos a \cos (\text{co-}B),$$

$$\cos A = \cos a \sin B,$$

and   (a)      $\cos a = \cos A \csc B.$

To find $b$: Consider $b$, co-$A$, and co-$B$; co-$B$ is the middle part and $b$ and co-$A$ are opposite parts.  Then

$$\sin(\text{co-}B) = \cos b \cos(\text{co-}A), \qquad \cos B = \cos b \sin A,$$

and    (b) $$\cos b = \cos B \csc A.$$

To find $c$: Consider co-$c$, co-$A$ and co-$B$; co-$c$ is the middle part and co-$A$ and co-$B$ are adjacent parts. Then

$$\sin(\text{co-}c) = \tan(\text{co-}A) \tan(\text{co-}B),$$

and    (c) $$\cos c = \cot A \cot B.$$

To check: Consider the unknown parts $a$, $b$ and co-$c$; co-$c$ is the middle part and $a$ and $b$ are opposite parts. Then    $\sin(\text{co-}c) = \cos a \cos b,$

and                        $\cos c = \cos a \cos b.$

The following computing form, found in official publications of the U.S. Navy Department, will be used here.

| | (a) | | (b) | | (c) | |
|---|---|---|---|---|---|---|
| $A = 65°$ | l cos | | l csc | | l cot | |
| $B = 118°$ | l csc | ———— | l cos | (n) | l cot | (n) |
| $a =$ | l cos | | | ———— | | |
| $b =$ | l cos | (n) | l cos | (n) | | |
| | | ———— | | | | ———— |
| $c =$ | l cos | (n) | | | l cos | (n) |

Note 1. The computing form is to be filled in by *rows*.

Note 2. The −10 after logarithms of numbers less than 1 will be omitted.

Note 3. If tables of log sec $\theta$ and log csc $\theta$ are not available, use

$$\log \sec \theta = \log \frac{1}{\cos \theta} = \text{colog} \cos \theta$$

and $$\log \csc \theta = \log \frac{1}{\sin \theta} = \text{colog} \sin \theta.$$

Note 4. The letter (n) following a logarithm indicates that the anti-logarithm (natural function) is negative. The absence of this letter indicates that the anti-logarithm is positive. In finding $a$, cos $A$ and csc $B$ are both positive; hence, their product cos $a$ is positive and $a < 90°$. In finding $b$, csc $A$ is positive and cos $B$ is negative; hence, their product cos $b$ is negative and $b > 90°$. In finding $c$, cot $A$ is positive and cot $B$ is negative; hence, their product cos $c$ is negative and $c > 90°$.

Note 5. The check, obtained by comparing the two entries in the last row, assures that the logarithms of the unknown parts are correct.

See Problems 7 - 10.

A QUADRANTAL TRIANGLE is one which has one side equal to 90°. It is solved by means of its polar triangle (a right spherical triangle). See Problem 11.

AN ISOSCELES SPHERICAL TRIANGLE is solved by dividing it into two right spherical triangles. See Problem 12.

## SOLVED PROBLEMS

1. Derive the ten fundamental formulas for right spherical triangles.

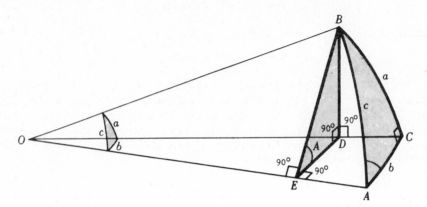

On a sphere of center $O$, let $ABC$ be a right spherical triangle with sides $a$ and $b$ less than $90^\circ$. Join $O$ to the vertices of the triangle to form the trihedral angle $O$-$ABC$. Through $B$ pass a plane perpendicular to $OA$ meeting $OC$ in $D$ and $OA$ in $E$.

Since $OE$ is perpendicular to the plane $BDE$, it is perpendicular to the lines $EB$ and $ED$. Thus the triangles $BEO$ and $DEO$ are right triangles with the right angles at $E$. Also $\angle BED$ is a plane angle of the dihedral angle $B$-$OA$-$C$ and thus measures angle $A$ of the spherical triangle.

Since the plane $BDE$ is perpendicular to $OE$, it is perpendicular to the plane $OAC$ through $OE$. Now $BD$, the intersection of the two planes $OBC$ and $BDE$ both perpendicular to the plane $OAC$, is itself perpendicular to $OAC$. Thus, triangles $BDO$ and $BDE$ are right triangles with the right angles at $D$.

In the right triangles $BDO$, $BDE$, and $BEO$:   $\sin a = \dfrac{DB}{OB} = \dfrac{DB}{EB} \cdot \dfrac{EB}{OB} = \sin A \sin c$   (1)

In the right triangles $BDO$, $BDE$, and $DEO$:   $\tan a = \dfrac{DB}{OD} = \dfrac{DB}{ED} \cdot \dfrac{ED}{OD} = \tan A \sin b$   (2)

In the right triangles $BEO$, $DEO$, and $BDO$:   $\cos c = \dfrac{OE}{OB} = \dfrac{OE}{OD} \cdot \dfrac{OD}{OB} = \cos b \cos a$   (4)

In the right triangles $DEO$, $BDE$, and $BEO$:   $\tan b = \dfrac{ED}{OE} = \dfrac{ED}{EB} \cdot \dfrac{EB}{OE} = \cos A \tan c$   (8)

Now by passing a plane through $A$ perpendicular to $OB$ and carrying through an argument similar to that above, we will obtain a set of four formulas which may be derived from the above four by interchanging $a$ and $b$, and $A$ and $B$. Formula (4) yields nothing new but from (1) we obtain (6), from (2) we obtain (7), and from (8) we obtain (3).

The product of (2) and (7) is:  $\tan a \tan b = \tan A \tan B \sin a \sin b$. Replacing $\tan a$ by $\dfrac{\sin a}{\cos a}$ and $\tan b$ by $\dfrac{\sin b}{\cos b}$, and dividing by $\sin a \sin b$, we have  $\dfrac{1}{\cos a \cos b} = \tan A \tan B$.

Substituting from (4), this becomes  $\dfrac{1}{\cos c} = \tan A \tan B$   or

$$\cos c = \cot A \cot B. \qquad (9)$$

The product of (6) and (8) is: $\sin b \cos A \tan c = \tan b \sin B \sin c$. Then

$$\cos A = \frac{\sin B \tan b \sin c}{\sin b \tan c} = \frac{\sin B \cos c}{\cos b} = \frac{\sin B (\cos a \cos b)}{\cos b}$$

or
$$\cos A = \sin B \cos a. \qquad (5)$$

In a similar manner, using the product of (1) and (3), we obtain

$$\cos B = \sin A \cos b. \qquad (10)$$

Consider next a right spherical triangle in which $a > 90°$ and $b < 90°$, as shown in Fig. $(a)$ below. Produce the arcs $BA$ and $BC$ to meet at $B'$ and consider the right spherical triangle $AB'C$ in which both $b$ and $180° - a$ are $< 90°$. Formula (1) for this triangle reads $\sin(180° - a) = \sin(180° - A)\sin(180° - c)$ or $\sin a = \sin A \sin c$; formula (4) reads $\cos(180° - c) = \cos b \cos(180° - a)$ which reduces to $-\cos c = (\cos b)(-\cos a)$ or $\cos c = \cos b \cos a$; and so on through the remaining formulas.

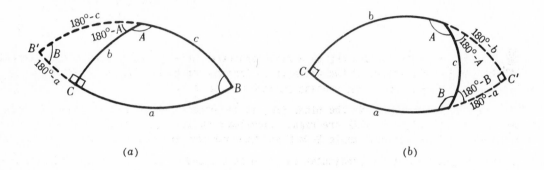

$(a)$ $(b)$

Consider finally a right spherical triangle in which both $a$ and $b$ are $> 90°$, as shown in Fig. $(b)$ above. Produce the arcs $CB$ and $CA$ to meet in $C'$ and consider the right spherical triangle $ABC'$ in which both $180° - a$ and $180° - b$ are $< 90°$. Formula (1) for this triangle reads $\sin(180° - a) = \sin(180° - A)\sin c$ or $\sin a = \sin A \sin c$; formula (4) reads $\cos c = \cos(180° - b)\cos(180° - a) = (-\cos b)(-\cos a) = \cos b \cos a$; and so on for the remaining formulas.

**2.** Derive the Laws of Quadrants.

From formula (5), $\sin B = \dfrac{\cos A}{\cos a}$. Since $B < 180°$, $\sin B$ is positive; hence, either $\cos a$ and $\cos A$ are both positive (i.e., both $a$ and $A$ are $< 90°$) or they are both negative (i.e., both $a$ and $A$ are $> 90°$). A similar argument using formula (10) establishes the second part of the first law.

From formula (4), $\cos c = \cos b \cos a$. If $c < 90°$, $\cos c$ is positive; hence, either $\cos b$ and $\cos a$ are both positive or both negative (i.e., $a$ and $b$ are in the same quadrant). If $c > 90°$, $\cos c$ is negative; hence, $\cos b$ and $\cos a$ have opposite signs (i.e., $a$ and $b$ are in different quadrants).

**3.** Write the formulas for finding $b$, $B$, and $c$ when $a$ and $A$ are given. Also the formula involving the three required parts.

For $b$: Applying Rule 1 to the parts co-$A$, $b$, and $a$,

$$\sin b = \tan (\text{co-}A) \tan a = \cot A \tan a.$$

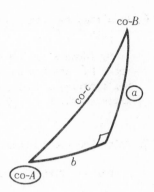

For $B$:  Applying Rule 2 to the parts co-$B$, co-$A$, and $a$,

$$\sin(\text{co-}A) = \cos(\text{co-}B)\cos a.$$

Then $\qquad\qquad \cos A = \sin B \cos a$

and $\qquad\qquad \sin B = \cos A \sec a.$

For $c$:  Applying Rule 2 to the parts co-$A$, $a$, and co-$c$,

$$\sin a = \cos(\text{co-}A)\cos(\text{co-}c) = \sin A \sin c$$

and $\qquad\qquad \sin c = \sin a \csc A.$

Applying Rule 2 to the required parts co-$B$, $b$, and co-$c$,

$$\sin b = \cos(\text{co-}B)\cos(\text{co-}c) = \sin B \sin c.$$

**4.** Write the formulas for finding $a$, $A$, and $b$ when $c$ and $B$ are given. Also the formula involving the three required parts.

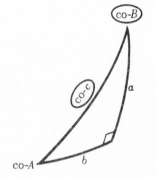

For $A$:  Applying Rule 1 to the parts co-$A$, co-$c$, and co-$B$,

$$\sin(\text{co-}c) = \tan(\text{co-}A)\tan(\text{co-}B).$$

Then  $\cos c = \cot A \cot B$  and  $\cot A = \cos c \tan B.$

For $a$:  Applying Rule 1 to the parts co-$c$, co-$B$, and $a$,

$$\sin(\text{co-}B) = \tan(\text{co-}c)\tan a.$$

Then  $\cos B = \cot c \tan a$  and  $\tan a = \cos B \tan c.$

For $b$:  Applying Rule 2 to the parts co-$B$, $b$, and co-$c$,

$$\sin b = \cos(\text{co-}B)\cos(\text{co-}c) = \sin B \sin c.$$

Applying Rule 1 to the unknown parts co-$A$, $b$, and $a$,

$$\sin b = \tan(\text{co-}A)\tan a = \cot A \tan a.$$

**5.** Show that no right spherical triangle exists satisfying any one of the following conditions:
a)  $A + B < 90^\circ$      c)  $A - B > 90^\circ$ or $B - A > 90^\circ$
b)  $A + B > 270^\circ$      d)  $\sin a > \sin c$ or $\sin b > \sin c$.

a) By formula (5), $\cos a = \dfrac{\cos A}{\sin B} = \dfrac{\sin(90^\circ - A)}{\sin B}$.  When  $A + B < 90^\circ$,  $90^\circ - A > B$.  Since $90^\circ - A$ is acute, $\sin(90^\circ - A) > \sin B$; then  $\cos a > 1$  and no triangle exists.

b) By formula (5), $\cos a = \dfrac{\cos A}{\sin B} = -\dfrac{\cos(180^\circ - A)}{\cos(B - 90^\circ)}$.  When  $A + B > 270^\circ$, both $A$ and $B$ are obtuse; then $180^\circ - A$ and $B - 90^\circ$ are acute. Now subtracting each side of the given inequality from $180^\circ + B$, we have  $180^\circ - A < B - 90^\circ$. Then  $\dfrac{\cos(180^\circ - A)}{\cos(B - 90^\circ)} > 1$  and no triangle exists.

c) By formula (10), $\cos b = \dfrac{\cos B}{\sin A} = \dfrac{\sin(90^\circ - B)}{\sin(180^\circ - A)}$.  When  $A > 90^\circ + B$,  $180^\circ - A < 90^\circ - B$; then $\cos b > 1$  and no triangle is determined. In a similar manner, using formula (5), it can be shown that no triangle exists when $B > 90^\circ + A$.

d) By formula (1), $\sin A = \sin a / \sin c$. When $\sin a > \sin c$, $\sin A > 1$ and no triangle exists. In a similar manner, using formula (6), it can be shown that no triangle exists when $\sin b > \sin c$.

**6.** Show that two right spherical triangles may be determined when $a$ and $A$ (or $b$ and $B$), both $< 90^\circ$ or both $> 90^\circ$, are given.

By formula (1), $\sin c = \sin a / \sin A$. When $\sin a < \sin A$, $\sin c < 1$ and two values of $c$, one $< 90^\circ$ and the other $> 90^\circ$, are determined. The laws of quadrants show that for the triangle having $c < 90^\circ$, $b$ and $B$ terminate in the same quadrant as $a$ and $A$, while for the triangle in which $c > 90^\circ$, $b$ and $a$ (also $B$ and $A$) terminate in different quadrants.

**7.** Solve the right spherical triangle $ABC$, given $a = 46^\circ 12.3'$, $c = 75^\circ 48.6'$.

For $A$: With $a$ as middle part and co-$A$ and co-$c$ as opposite parts,

$$\sin a = \cos (\text{co-}A) \cos (\text{co-}c) = \sin A \sin c$$

and

$$\sin A = \sin a \csc c.$$

For $B$: With co-$B$ as middle part and co-$c$ and $a$ as adjacent parts,

$$\sin (\text{co-}B) = \tan (\text{co-}c) \tan a$$

and

$$\cos B = \cot c \tan a.$$

For $b$: With co-$c$ as middle part and $b$ and $a$ as opposite parts,

$$\sin (\text{co-}c) = \cos b \cos a, \qquad \cos c = \cos b \cos a,$$

and

$$\cos b = \cos c \sec a.$$

For check: With co-$B$ as middle part and co-$A$ and $b$ as opposite parts,

$$\sin (\text{co-}B) = \cos (\text{co-}A) \cos b$$

and

$$\cos B = \sin A \cos b.$$

| | | $(A)$ | $(b)$ | $(B)$ |
|---|---|---|---|---|
| $a = 46^\circ 12.3'$ | | l sin 9.85843 | l sec 0.15984 | l tan 0.01828 |
| $c = 75^\circ 48.6'$ | | l csc 0.01346 | l cos 9.38941 | l cot 9.40287 |
| $A = 48^\circ 7.2'$ | | l sin 9.87189 | | |
| $b = 69^\circ 15.3'$ | | l cos 9.54925 | l cos 9.54925 | |
| $B = 74^\circ 42.5'$ | | l cos 9.42114 | | l cos 9.42115 |

Since $c < 90^\circ$, all parts terminate in the same quadrant as $a$.

Note. The order of the columns is determined by the check formula, that part of the triangle appearing on the left of the check formula being found in the last column of the computing form.

**8.** Solve the right spherical triangle $ABC$, given $a = 109^\circ 15.8'$, $B = 38^\circ 45.4'$.

For $A$: With co-$A$ as middle part and co-$B$ and $a$ as opposite parts,

$$\sin (\text{co-}A) = \cos (\text{co-}B) \cos a$$

and

$$\cos A = \sin B \cos a.$$

For $b$: With $a$ as middle part and co-$B$ and $b$ as adjacent parts,

$$\sin a = \tan (\text{co-}B) \tan b, \qquad \sin a = \cot B \tan b,$$

and

$$\tan b = \sin a \tan B.$$

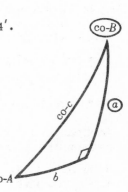

For $c$:  With co-$B$ as middle part and $a$ and co-$c$ as adjacent parts,

$$\sin(\text{co-}B) = \tan a \tan(\text{co-}c), \qquad \cos B = \tan a \cot c,$$

and $\qquad\qquad\qquad \cot c = \cos B \cot a.$

For check:  With co-$A$ as middle part and $b$ and co-$c$ as adjacent parts,

$$\sin(\text{co-}A) = \tan b \tan(\text{co-}c)$$

and $\qquad\qquad\qquad \cos A = \tan b \cot c.$

|  | (b) | (c) | (A) |
|---|---|---|---|
| $a = 109^\circ 15.8'$ | l sin 9.97498 | l cot 9.54342 (n) | l cos 9.51840 (n) |
| $B = 38^\circ 45.4'$ | l tan 9.90459 | l cos 9.89199 | l sin 9.79658 |
| $b = 37^\circ 9.3'$ | l tan 9.87957 | | |
| $c = 105^\circ 14.7'$ | l cot 9.43541 (n) | l cot 9.43541 (n) | |
| $A = 101^\circ 55.1'$ | l cos 9.31498 (n) | | l cos 9.31498 (n) |

The required parts agree as to quadrants:  $A > 90^\circ$ since $a > 90^\circ$;  $b < 90^\circ$ since $B < 90^\circ$; $c > 90^\circ$ since $a$ and $b$ terminate in different quadrants.

**9.** Solve the right spherical triangle $ABC$, given $c = 72^\circ 12.5'$, $A = 156^\circ 17.2'$.

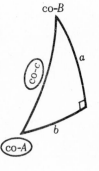

For $a$:  With $a$ as middle part and co-$A$ and co-$c$ as opposite parts,

$$\sin a = \cos(\text{co-}A)\cos(\text{co-}c) = \sin A \sin c.$$

For $b$:  With co-$A$ as middle part and $b$ and co-$c$ as adjacent parts,

$$\sin(\text{co-}A) = \tan b \tan(\text{co-}c), \qquad \cos A = \tan b \cot c,$$

and $\qquad\qquad\qquad \tan b = \cos A \tan c.$

For $B$:  With co-$c$ as middle part and co-$A$ and co-$B$ as adjacent parts,

$$\sin(\text{co-}c) = \tan(\text{co-}A)\tan(\text{co-}B), \qquad \cos c = \cot A \cot B,$$

and $\qquad\qquad\qquad \cot B = \cos c \tan A.$

For check:  With $a$ as middle part and co-$B$ and $b$ as adjacent parts,

$$\sin a = \tan(\text{co-}B)\tan b = \cot B \tan b.$$

|  | (B) | (b) | (a) |
|---|---|---|---|
| $A = 156^\circ 17.2'$ | l tan 9.64271 (n) | l cos 9.96169 (n) | l sin 9.60440 |
| $c = 72^\circ 12.5'$ | l cos 9.48510 | l tan 0.49362 | l sin 9.97872 |
| $B = 97^\circ 38.7'$ | l cot 9.12781 (n) | | |
| $b = 109^\circ 19.0'$ | l tan 0.45531 (n) | l tan 0.45531 (n) | |
| $a = 157^\circ 29.1'$ | l sin 9.58312 | | l sin 9.58312 |

**Note.**  $a > 90^\circ$ since $A > 90^\circ$.

**10.** Solve the right spherical triangle $ABC$, given $b = 138°46.4'$, $B = 125°10.6'$.

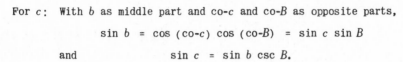

For $a$: With $a$ as middle part and co-$B$ and $b$ as adjacent parts,

$$\sin a = \tan (\text{co-}B) \tan b = \cot B \tan b.$$

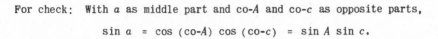

For $A$: With co-$B$ as middle part and co-$A$ and $b$ as opposite parts,

$$\sin (\text{co-}B) = \cos (\text{co-}A) \cos b, \quad \cos B = \sin A \cos b,$$

and $\qquad\qquad \sin A = \cos B \sec b.$

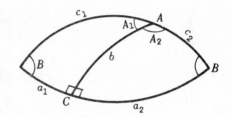

For $c$: With $b$ as middle part and co-$c$ and co-$B$ as opposite parts,

$$\sin b = \cos (\text{co-}c) \cos (\text{co-}B) = \sin c \sin B$$

and $\qquad\qquad \sin c = \sin b \csc B.$

For check: With $a$ as middle part and co-$A$ and co-$c$ as opposite parts,

$$\sin a = \cos (\text{co-}A) \cos (\text{co-}c) = \sin A \sin c.$$

Since every required part is to be found from its sine, two values for each are obtained. The two triangles are shown in the adjacent figure.

Since $b > 90°$, $a_1, A_1 < 90°$ and $c_1 > 90°$

$a_2, A_2 > 90°$ and $c_2 < 90°$.

| | (A) | (c) | (a) |
|---|---|---|---|
| $b = 138°46.4'$ | l sec 0.12372 (n) | l sin 9.81891 | l tan 9.94263 (n) |
| $B = 125°10.6'$ | l cos 9.76050 (n) | l csc 0.08757 | l cot 9.84807 (n) |
| $A_1 = 49°59.7'$ | l sin 9.88422 | | |
| $A_2 = 130° 0.3'$ | | | |
| $c_2 = 53°44.0'$ | l sin 9.90648 | l sin 9.90648 | |
| $c_1 = 126°16.0'$ | | | |
| $a_1 = 38° 8.4'$ | l sin 9.79070 | | l sin 9.79070 |
| $a_2 = 141°51.6'$ | | | |

**11.** Solve the quadrantal spherical triangle $ABC$, given $a = 115°24.6'$, $b = 60°18.4'$, $c = 90°$.

We first solve the polar triangle $A'B'C'$ of the given triangle having $C' = 180° - c = 90°$, $A' = 180° - a = 64°35.4'$, and $B' = 180° - b = 119°41.6'$.

For $a'$: With co-$A'$ as middle part and $a'$ and co-$B'$ as opposite parts,

$$\sin (\text{co-}A') = \cos a' \cos (\text{co-}B'), \quad \cos A' = \cos a' \sin B',$$

and $\qquad\qquad \cos a' = \cos A' \csc B'.$

For $b'$: With co-$B'$ as middle part and $b'$ and co-$A'$ as opposite parts,

$$\sin (\text{co-}B') = \cos b' \cos (\text{co-}A'), \quad \cos B' = \cos b' \sin A',$$

and $\qquad\qquad \cos b' = \cos B' \csc A'.$

For $c'$: With co-$c'$ as middle part and co-$A'$ and co-$B'$ as adjacent parts,

$$\sin (\text{co-}c') = \tan (\text{co-}A') \tan (\text{co-}B')$$

and $\qquad\qquad \cos c' = \cot A' \cot B'.$

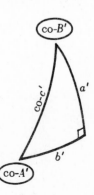

For check:  With co-$c'$ as middle part and $a'$ and $b'$ as opposite parts,

$$\sin (\text{co-}c') = \cos a' \cos b'$$

and
$$\cos c' = \cos a' \cos b'.$$

|         | $(a')$ | $(b')$ | $(c')$ |
|---------|--------|--------|--------|
| $A' = 64°35.4'$ | l cos 9.63255 | l csc 0.04419 | l cot 9.67674 |
| $B' = 119°41.6'$ | l csc 0.06113 | l cos 9.69492 (n) | l cot 9.75605 (n) |
| $a' = 60°24.0'$ | l cos 9.69368 | | |
| $b' = 123°15.5'$ | l cos 9.73911 (n) | l cos 9.73911 (n) | |
| $c' = 105°43.0'$ | l cos 9.43279 (n) | | l cos 9.43279 (n) |

The required parts of the triangle $ABC$ are:

$A = 180° - a' = 119°36.0'$,  $B = 180° - b' = 56°44.5'$,  $C = 180° - c' = 74°17.0'$.

**12.** Solve the isosceles spherical triangle $ABC$, given  $b = c = 54°28.4'$, $A = 112°36.2'$.

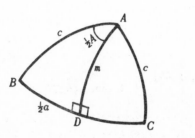
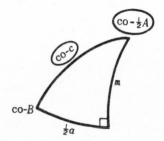

Let the great circle through $A$ and perpendicular to $BC$ meet $BC$ in $D$.  Consider the right spherical triangle $ABD$ with right angle at $D$.

For $B$:  With co-$c$ as middle part and co-$\tfrac{1}{2}A$ and co-$B$ as adjacent parts,
$$\sin (\text{co-}c) = \tan (\text{co-}\tfrac{1}{2}A) \tan (\text{co-}B), \qquad \cos c = \cot \tfrac{1}{2}A \cot B,$$
and
$$\cot B = \cos c \tan \tfrac{1}{2}A.$$

For $\tfrac{1}{2}a$:  With $\tfrac{1}{2}a$ as middle part and co-$c$ and co-$\tfrac{1}{2}A$ as opposite parts,
$$\sin \tfrac{1}{2}a = \cos (\text{co-}\tfrac{1}{2}A) \cos (\text{co-}c) = \sin \tfrac{1}{2}A \sin c.$$

For $m$:  With co-$\tfrac{1}{2}A$ as middle part and co-$c$ and $m$ as adjacent parts,
$$\sin (\text{co-}\tfrac{1}{2}A) = \tan (\text{co-}c) \tan m, \qquad \cos \tfrac{1}{2}A = \cot c \tan m,$$
and
$$\tan m = \cos \tfrac{1}{2}A \tan c.$$

For check:  With $\tfrac{1}{2}a$ as middle part and co-$B$ and $m$ as adjacent parts,
$$\sin \tfrac{1}{2}a = \tan (\text{co-}B) \tan m = \cot B \tan m.$$

|         | $(B)$ | $(m)$ | $(\tfrac{1}{2}a)$ |
|---------|-------|-------|-------------------|
| $\tfrac{1}{2}A = 56°18.1'$ | l tan 0.17596 | l cos 9.74415 | l sin 9.92011 |
| $c = 54°28.4'$ | l cos 9.76424 | l tan 0.14630 | l sin 9.91055 |
| $B = 48°55.9'$ | l cot 9.94020 | | |
| $m = 37°51.0'$ | l tan 9.89045 | l tan 9.89045 | |
| $\tfrac{1}{2}a = 42°37.1'$ | l sin 9.83065 | | l sin 9.83066 |

The required parts of the isosceles triangle are:  $B = C = 48°55.9'$ and $a = 85°14.2'$.

13. The initial course for a great circle track from New York (lat. 40°42.0' N, long. 74°1.0' W) is N 30°10.0' E or 30°10.0'. Locate on the track the point $M$ which is nearest the north pole and find the great circle distance (in nautical miles) of $M$ from the pole and from New York.

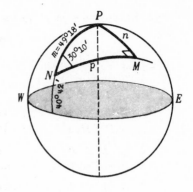

The shortest distance from the track to the north pole $P$ is measured along that meridian which is perpendicular to the track. In the spherical triangle $NPM$, $M = 90°$, $m = NP = 90° - 40°42.0' = 49°18.0'$ and $N = 30°10.0'$; $P$, $n$ and $p$ are required.

For $n$ :   $\sin n = \cos (\text{co-}m) \cos (\text{co-}N)$
$= \sin m \sin N.$

For $P$ :   $\sin (\text{co-}m) = \tan (\text{co-}N) \tan (\text{co-}P)$   and   $\cot P = \cos m \tan N.$

For $p$ :   $\sin (\text{co-}N) = \tan (\text{co-}m) \tan p$   and   $\tan p = \tan m \cos N.$

|  | $(n)$ | $(P)$ | $(p)$ |
|---|---|---|---|
| $m = 49°18.0'$ | l sin 9.87975 | l cos 9.81431 | l tan 0.06543 |
| $N = 30°10.0'$ | l sin 9.70115 | l tan 9.76435 | l cos 9.93680 |
| $n = 22°23.6'$ | l sin 9.58090 | | |
| $P = 69°14.6'$ | | l cot 9.57866 | |
| $p = 45°\ 8.8'$ | | | l tan 0.00223 |

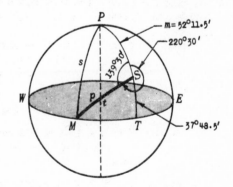

The latitude of $M$ is $90° - n = 67°36.4'$ N; the longitude is $74°1.0' - P = 4°46.4'$ W. The distance of $M$ from $P$ is $n = 22°23.6' = 1343.6' = 1343.6$ miles, and the distance of $M$ from New York is $p = 45°8.8' = 2708.8$ miles.

14. The initial course for a great circle track from San Francisco (lat. 37°48.5' N, long. 122°24.0' W) is S 40°30.0' W or 220°30.0'. Locate the point $M$ where the track crosses the equator and find the great circle distance of $M$ from San Francisco.

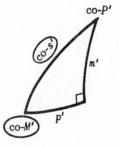

SOLUTION 1. In the spherical triangle $MSP$, $s = MP = 90°$, $S = 360° - 220°30.0' = 139°30.0'$, and $m = SP = 90° - 37°48.5' = 52°11.5'$; $P$ and $p$ are required. Since this triangle is quadrantal, we make use of its polar (right) triangle $M'S'P'$ in which $S' = 90°$, $s' = 180° - S = 40°30.0'$ and $M' = 180° - m = 127°48.5'$ to find $p'$ and $P'$.

For $p'$:   $\sin (\text{co-}M') = \tan p' \tan (\text{co-}s')$
and   $\tan p' = \cos M' \tan s'.$

For $P'$:   $\sin (\text{co-}s') = \tan (\text{co-}M') \tan (\text{co-}P')$
and   $\cot P' = \tan M' \cos s'.$

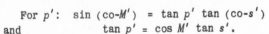

|  | $(p')$ | $(P')$ |
|---|---|---|
| $M' = 127°48.5'$ | l cos 9.78748 (n) | l tan 0.11019 (n) |
| $s' = 40°30.0'$ | l tan 9.93150 | l cos 9.88105 |
| $p' = 152°21.9'$ | l tan 9.71898 (n) | |
| $P' = 134°25.3'$ | | l cot 9.99124 (n) |

Then $P = 180° - p' = 27°38.1'$ and $p = 180° - P' = 45°34.7'$. The longitude of $M$ is $122°24.0' + 27°38.1' = 150°2.1'$ W and the distance of $M$ from San Francisco is $p = 45°34.7' = 2734.7$ miles.

SOLUTION 2. A solution somewhat simpler than that given above is obtained by using the right triangle $SMT$ in which $m = TS = 37°48.5'$, $S = \angle MST = 40°30.0'$, and $T = 90°$. We seek $s = $ arc $MT$, which measures the difference in longitude between $M$ and $S$, and $t = $ arc $MS$.

For $s$:  $\sin m = \tan s \tan (\text{co-}S)$  and  $\tan s = \sin m \tan S$.

For $t$:  $\sin (\text{co-}S) = \tan (\text{co-}t) \tan m$  and  $\cot t = \cot m \cos S$.

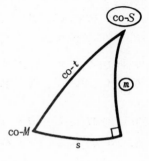

|  | (s) | (t) |
|---|---|---|
| $m = 37°48.5'$ | l sin 9.78748 | l cot 0.11019 |
| $S = 40°30.0'$ | l tan 9.93150 | l cos 9.88105 |
| $s = 27°38.1'$ | l tan 9.71898 |  |
| $t = 45°34.7'$ |  | l cot 9.99124 |

The longitude of $M$ is $122°24.0' + 27°38.1' = 150°2.1'$ W  and the required distance is $t = 45°34.7' = 2734.7$ miles as before.

**15.** Find the initial course and the course on arrival for the great circle track from Chicago (lat. $41°50.0'$ N, long. $87°31.7'$ W) which crosses the equator at $M$ (long. $170°15.0'$ E). Find the distance of $M$ from Chicago.

As in Problem 14, this may be solved by using the polar (right) triangle of the quadrantal triangle $MCP$ or by using the right triangle $MCT$. In this latter triangle, $m = $ arc $TC = 41°50.0'$  and  $c = $ arc $MT = 102°13.3'$.

For $C$:        $\sin m = \tan c \tan (\text{co-}C)$        and
                $\cot C = \cot c \sin m$.

For $M$:        $\sin c = \tan m \tan (\text{co-}M)$        and
                $\cot M = \sin c \cot m$.

For $t$:        $\sin (\text{co-}t) = \cos c \cos m$        and
                $\cos t = \cos c \cos m$.

|  | (C) | (M) | (t) |
|---|---|---|---|
| $c = 102°13.3'$ | l cot 9.33566 (n) | l sin 9.99004 | l cos 9.32571 (n) |
| $m = 41°50.0'$ | l sin 9.82410 | l cot 0.04810 | l cos 9.87221 |
| $C = 98°13.2'$ | l cot 9.15976 (n) |  |  |
| $M = 42°29.2'$ |  | l cot 0.03814 |  |
| $t = 99°4.5'$ |  |  | l cos 9.19792 (n) |

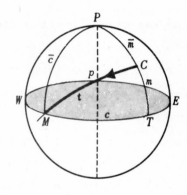

The initial course is $180° + 98°13.2' = 278°13.2'$,  the course on arrival is $270° - 42°29.2' = 227°30.8'$,  and the distance is $t = 99°4.5' = 5944.5$ miles.

## SUPPLEMENTARY PROBLEMS

Solve each of the following right spherical triangles $ABC$, in which $C = 90°$.

16. $a = 125°24.8'$, $b = 32°16.5'$      Ans. $A = 110°47.4'$, $B = 37°46.4'$, $c = 119°20.2'$

17. $a = 30°45.3'$, $B = 135°24.4'$      Ans. $A = 52°53.4'$, $b = 153°14.7'$, $c = 140°7.0'$

18. $c = 70°25.2'$, $A = 52°54.8'$      Ans. $a = 48°43.7'$, $b = 59°28.0'$, $B = 66°5.5'$

19. $a = 35°34.6'$, $c = 45°48.2'$      Ans. $A = 54°14.4'$, $B = 45°55.8'$, $b = 31°0.4'$

20. $a = 46°46.4'$, $A = 57°28.3'$      Ans. $b_1 = 42°43.6'$, $c_1 = 59°47.7'$, $B_1 = 51°43.8'$
      $b_2 = 137°16.4'$, $c_2 = 120°12.3'$, $B_2 = 128°16.2'$

21. $A = 67°38.8'$, $B = 155°12.6'$      Ans. $a = 24°54.2'$, $b = 169°0.0'$, $c = 152°55.2'$

22. $a = 40°44.6'$, $b = 64°48.3'$      Ans. $A = 43°35.5'$, $B = 72°55.8'$, $c = 71°11.1'$

23. $b = 121°42.5'$, $A = 154°8.6'$      Ans. $a = 157°35.7'$, $c = 60°55.6'$, $B = 103°15.1'$

24. $c = 152°24.4'$, $B = 68°38.2'$      Ans. $a = 169°13.2'$, $b = 25°33.2'$, $A = 156°11.1'$

25. $b = 158°22.4'$, $c = 122°36.7'$      Ans. $a = 54°34.0'$, $A = 75°18.3'$, $B = 154°3.2'$

26. $b = 162°53.4'$, $B = 138°14.9'$      Ans. $a_1 = 20°10.4'$, $c_1 = 153°46.8'$, $A_1 = 51°18.8'$
      $a_2 = 159°49.6'$, $c_2 = 26°13.2'$, $A_2 = 128°41.2'$

27. $A = 33°50.5'$, $B = 72°24.2'$      Ans. $a = 29°23.1'$, $b = 57°7.3'$, $c = 61°46.2'$

Solve each of the following quadrantal spherical triangles $ABC$, in which $c = 90°$.

28. $a = 60°34.9'$, $B = 122°18.8'$      Ans. $b = 117°45.0'$, $A = 56°17.3'$, $C = 72°44.5'$

29. $A = 32°53.6'$, $B = 115°24.9'$      Ans. $a = 35°36.3'$, $b = 104°28.2'$, $C = 68°52.7'$

30. $a = 69°15.2'$, $A = 56°45.4'$      Ans. $b_1 = 40°15.3'$, $B_1 = 35°18.2'$, $C_1 = 116°34.5'$
      $b_2 = 139°44.7'$, $B_2 = 144°41.8'$, $C_2 = 63°25.5'$

Solve each of the following isosceles spherical triangles $ABC$.

31. $a = b = 78°23.5'$, $C = 118°54.6'$      Ans. $A = B = 71°10.3'$, $c = 115°2.8'$

32. $B = C = 38°52.5'$, $a = 132°15.0'$      Ans. $b = c = 70°59.2'$, $A = 150°34.0'$

33. A ship leaves New York (lat. $40°42.0'$ N, long. $74°1.0'$ W) with initial course due east and sails on a great circle course. Find its course and position after it has sailed 525 n.m. $Ans.$ course, $97°27.3'$; lat. $40°7.7'$ N, long. $62°32.4'$ W

34. The initial course for a great circle track from Yokohama (lat. $35°37.0'$ N; long. $139°39.0'$ E) is $40°40.0'$. Locate the point on the track which is nearest the north pole. $Ans.$ lat. $58°0.7'$ N, long. $23°4.2'$ W

35. A ship leaves $A$(lat. $36°50.0'$ N, long. $76°20.0'$ W) and, sailing on a great circle arc, crosses the equator at $50°0.0'$ W. Find the initial course and the distance traveled. $Ans.$ course, $140°27.3'$; distance, 2649.9 n.m.

# CHAPTER 21

# Oblique Spherical Triangles – Standard Solutions

AN OBLIQUE SPHERICAL TRIANGLE is a spherical triangle, no one of whose angles is a right angle. When *any* three of its parts are known, the oblique spherical triangle is determined except for possible ambiguities noted later.

There are six cases to be considered:

Case I   : Given the three sides.
Case II  : Given the three angles.
Case III : Given two sides and the included angle.
Case IV  : Given two angles and the included side.
Case V   : Given two sides and an angle opposite one of them.
Case VI  : Given two angles and a side opposite one of them.

THE STANDARD PROCEDURES for solving and checking the several cases involve the use of special formulas listed below.

LAW OF SINES. In any spherical triangle $ABC$, $\dfrac{\sin a}{\sin A} = \dfrac{\sin b}{\sin B} = \dfrac{\sin c}{\sin C}$.

For a derivation, see Problem 1.

LAW OF COSINES FOR SIDES. In any spherical triangle $ABC$,

$$\cos a = \cos b \cos c + \sin b \sin c \cos A$$
$$\cos b = \cos c \cos a + \sin c \sin a \cos B$$
$$\cos c = \cos a \cos b + \sin a \sin b \cos C.$$

For a derivation, see Problem 2.

LAW OF COSINES FOR ANGLES. In any spherical triangle $ABC$,

$$\cos A = -\cos B \cos C + \sin B \sin C \cos a$$
$$\cos B = -\cos C \cos A + \sin C \sin A \cos b$$
$$\cos C = -\cos A \cos B + \sin A \sin B \cos c.$$

For a derivation, see Problem 3.

HALF-ANGLE FORMULAS. In any spherical triangle $ABC$,

$$\tan \tfrac{1}{2}A = \frac{\tan r}{\sin(s-a)}, \qquad \tan \tfrac{1}{2}B = \frac{\tan r}{\sin(s-b)}, \qquad \tan \tfrac{1}{2}C = \frac{\tan r}{\sin(s-c)}$$

where $s = \tfrac{1}{2}(a+b+c)$ and $\tan r = \sqrt{\dfrac{\sin(s-a)\,\sin(s-b)\,\sin(s-c)}{\sin s}}$.

For a derivation, see Problem 4.

**HALF-SIDE FORMULAS.** In any spherical triangle $ABC$,

$$\cot \tfrac{1}{2}a = \frac{\tan R}{\cos(S-A)}, \qquad \cot \tfrac{1}{2}b = \frac{\tan R}{\cos(S-B)}, \qquad \cot \tfrac{1}{2}c = \frac{\tan R}{\cos(S-C)}$$

where $S = \tfrac{1}{2}(A+B+C)$ and $\tan R = \sqrt{\dfrac{\cos(S-A)\,\cos(S-B)\,\cos(S-C)}{-\cos S}}$.

For a derivation, see Problem 5.

**GAUSS' OR DELAMBRE'S ANALOGIES.** In any spherical triangle $ABC$,

$$\frac{\sin \tfrac{1}{2}(A-B)}{\cos \tfrac{1}{2}C} = \frac{\sin \tfrac{1}{2}(a-b)}{\sin \tfrac{1}{2}c} \qquad\qquad \frac{\sin \tfrac{1}{2}(A+B)}{\cos \tfrac{1}{2}C} = \frac{\cos \tfrac{1}{2}(a-b)}{\cos \tfrac{1}{2}c}$$

$$\frac{\cos \tfrac{1}{2}(A-B)}{\sin \tfrac{1}{2}C} = \frac{\sin \tfrac{1}{2}(a+b)}{\sin \tfrac{1}{2}c} \qquad\qquad \frac{\cos \tfrac{1}{2}(A+B)}{\sin \tfrac{1}{2}C} = \frac{\cos \tfrac{1}{2}(a+b)}{\cos \tfrac{1}{2}c}$$

and other forms obtained by cyclic change of the letters.

For a derivation, see Problem 6.

**NAPIER'S ANALOGIES.** In any spherical triangle $ABC$,

$$\frac{\tan \tfrac{1}{2}(A-B)}{\cot \tfrac{1}{2}C} = \frac{\sin \tfrac{1}{2}(a-b)}{\sin \tfrac{1}{2}(a+b)} \qquad \frac{\tan \tfrac{1}{2}(a-b)}{\tan \tfrac{1}{2}c} = \frac{\sin \tfrac{1}{2}(A-B)}{\sin \tfrac{1}{2}(A+B)}$$

$$\frac{\tan \tfrac{1}{2}(A+B)}{\cot \tfrac{1}{2}C} = \frac{\cos \tfrac{1}{2}(a-b)}{\cos \tfrac{1}{2}(a+b)} \qquad \frac{\tan \tfrac{1}{2}(a+b)}{\tan \tfrac{1}{2}c} = \frac{\cos \tfrac{1}{2}(A-B)}{\cos \tfrac{1}{2}(A+B)}$$

and other forms obtained by cyclic change of the letters.

For a derivation, see Problem 7.

## SOLVED PROBLEMS

1. Derive the law of sines.

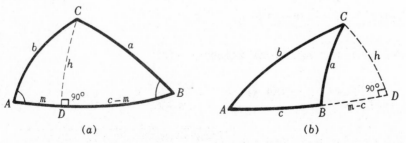

(a)                                (b)

Let $ABC$ be any spherical triangle. (Refer to Figures (a) and (b) above.)  Through $C$ pass a great circle perpendicular to $AB$ meeting it in $D$.  Let $CD = h$.

In right triangle $ACD$, $\sin h = \sin b \sin A$.  In right triangle $BCD$, $\sin h = \sin a \sin B$.  Then

$$\sin a \sin B = \sin b \sin A \qquad \text{and} \qquad \frac{\sin a}{\sin A} = \frac{\sin b}{\sin B}.$$

Similarly, passing a great circle through $B$ perpendicular to $AC$, we find $\dfrac{\sin a}{\sin A} = \dfrac{\sin c}{\sin C}$.

Thus,     $\dfrac{\sin a}{\sin A} = \dfrac{\sin b}{\sin B} = \dfrac{\sin c}{\sin C}$.

**2.** Derive the law of cosines for sides.

In Figures $(a)$ and $(b)$ above, let $AD = m$. In the right triangle $ACD$,
(1) $\sin m = \tan h \cot A$,     (2) $\sin h = \sin b \sin A$,     (3) $\cos b = \cos h \cos m$.

In the right triangle $BCD$,
(4) $\cos a = \cos h \cos(c - m) = \cos h(\cos c \cos m + \sin c \sin m)$, since $\cos(c-m) = \cos(m-c)$.

Substituting in (4) for $\sin m$ from (1) and for $\cos m$ from (3),

$\cos a = \cos h(\cos c \,\dfrac{\cos b}{\cos h} + \sin c \tan h \cot A) = \cos c \cos b + \sin c \sin h \cot A$;

and substituting for $\sin h$ from (2),

$$\cos a = \cos c \cos b + \sin c \sin b \sin A \cot A$$
$$= \cos b \cos c + \sin b \sin c \cos A.$$

The other formulas may be obtained by cyclic change of the letters.

**3.** Derive the law of cosines for angles.

Consider the polar triangle $A'B'C'$ of $ABC$ for which $a' = 180^\circ - A$, $A' = 180^\circ - a$, etc. Applying the formula derived in Problem 2 to this triangle, we have

$$\cos a' = \cos b' \cos c' + \sin b' \sin c' \cos A'$$

and recalling that $\sin(180^\circ - \theta) = \sin\theta$ and $\cos(180^\circ - \theta) = -\cos\theta$, we get

$$-\cos A = (-\cos B)(-\cos C) + \sin B \sin C (-\cos a)$$

or          $$\cos A = -\cos B \cos C + \sin B \sin C \cos a.$$

The other formulas may be obtained by cyclic change of the letters.

**4.** Derive the half-angle formulas.

From the law of cosines for sides (Problem 2),     $\cos a = \cos b \cos c + \sin b \sin c \cos A$,

we have          $\cos A = \dfrac{\cos a - \cos b \cos c}{\sin b \sin c}$.

Then     $1 - \cos A = 1 - \dfrac{\cos a - \cos b \cos c}{\sin b \sin c} = \dfrac{\cos b \cos c + \sin b \sin c - \cos a}{\sin b \sin c}$

$= \dfrac{\cos(b-c) - \cos a}{\sin b \sin c} = \dfrac{-2 \sin \frac{1}{2}(b-c+a) \sin \frac{1}{2}(b-c-a)}{\sin b \sin c}$,

$1 + \cos A = 1 + \dfrac{\cos a - \cos b \cos c}{\sin b \sin c} = \dfrac{\cos a - (\cos b \cos c - \sin b \sin c)}{\sin b \sin c}$

$= \dfrac{\cos a - \cos(b+c)}{\sin b \sin c} = \dfrac{-2 \sin \frac{1}{2}(a+b+c) \sin \frac{1}{2}(a-b-c)}{\sin b \sin c}$, and  $\dfrac{1 - \cos A}{1 + \cos A} = \dfrac{-2 \sin \frac{1}{2}(b-c+a)\ \sin \frac{1}{2}(b-c-a)}{\sin b\ \sin c} \cdot \dfrac{\sin b\ \sin c}{-2 \sin \frac{1}{2}(a+b+c)\ \sin \frac{1}{2}(a-b-c)}$

$$= \dfrac{\sin \frac{1}{2}(a+b-c)\ \sin \frac{1}{2}(c+a-b)}{\sin \frac{1}{2}(a+b+c)\ \sin \frac{1}{2}(b+c-a)}$$

since  $\sin \frac{1}{2}(b-c-a) = -\sin \frac{1}{2}(c+a-b)$  and  $\sin \frac{1}{2}(a-b-c) = -\sin \frac{1}{2}(b+c-a)$.

Let  $s = \frac{1}{2}(a+b+c)$.  Then  $\frac{1}{2}(a+b-c) = s-c$,  $\frac{1}{2}(c+a-b) = s-b$,  $\frac{1}{2}(b+c-a) = s-a$,  and

$$\tan \tfrac{1}{2}A = \sqrt{\dfrac{1-\cos A}{1+\cos A}} = \sqrt{\dfrac{\sin(s-c)\ \sin(s-b)}{\sin s\ \sin(s-a)}} = \dfrac{1}{\sin(s-a)}\sqrt{\dfrac{\sin(s-a)\ \sin(s-b)\ \sin(s-c)}{\sin s}}.$$

Finally, defining  $\sqrt{\dfrac{\sin(s-a)\ \sin(s-b)\ \sin(s-c)}{\sin s}} = \tan r$,  we obtain

$$\tan \tfrac{1}{2}A = \dfrac{\tan r}{\sin(s-a)}.$$

The other formulas may be obtained by cyclic change of letters.  Note that a cyclic change of letters does not affect $\tan r$.

## 5. Derive the half-side formulas.

Consider the polar triangle $A'B'C'$ of the spherical triangle $ABC$.

Let  $s' = \frac{1}{2}(a'+b'+c')$   and   $\tan r' = \sqrt{\dfrac{\sin(s'-a')\ \sin(s'-b')\ \sin(s'-c')}{\sin s'}}.$

Since  $a' = 180^\circ - A$,  $A' = 180^\circ - a$,  etc.,

$s' = \frac{1}{2}\{(180^\circ - A) + (180^\circ - B) + (180^\circ - C)\} = 270^\circ - \frac{1}{2}(A+B+C) = 270^\circ - S$,   where $S = \frac{1}{2}(A+B+C)$.

Then    $\sin s' = \sin(270^\circ - S) = -\cos S$,

$\sin(s'-a') = \sin\{270^\circ - S - (180^\circ - A)\} = \sin\{90^\circ - (S-A)\} = \cos(S-A)$,

$\sin(s'-b') = \cos(S-B)$,

$\sin(s'-c') = \cos(S-C)$,

and     $\tan r' = \sqrt{\dfrac{\cos(S-A)\ \cos(S-B)\ \cos(S-C)}{-\cos S}} = \tan R$.

From Problem 4,  $\tan \frac{1}{2}A' = \dfrac{\tan r'}{\sin(s'-a')}$.   Then, since $A' = 180^\circ - a$,

$$\tan \tfrac{1}{2}(180^\circ - a) = \cot \tfrac{1}{2}a = \dfrac{\tan R}{\cos(S-A)}$$

and similarly for the other formulas.

## 6. Derive Gauss' or Delambre's analogies.

From Problem 4,

$$\sin \tfrac{1}{2}A = \sqrt{\tfrac{1}{2}(1-\cos A)} = \sqrt{\dfrac{-\sin \frac{1}{2}(b-c+a)\ \sin \frac{1}{2}(b-c-a)}{\sin b\ \sin c}} = \sqrt{\dfrac{\sin(s-b)\ \sin(s-c)}{\sin b\ \sin c}}$$

and $\quad \cos \frac{1}{2}A \;=\; \sqrt{\frac{1}{2}(1 + \cos A)} \;=\; \sqrt{\dfrac{-\sin \frac{1}{2}(a+b+c)\,\sin \frac{1}{2}(a-b-c)}{\sin b \,\sin c}} \;=\; \sqrt{\dfrac{\sin s \,\sin(s-a)}{\sin b \,\sin c}} .$

Similarly, $\quad \sin \frac{1}{2}B \;=\; \sqrt{\dfrac{\sin(s-c)\,\sin(s-a)}{\sin c \,\sin a}} \quad$ and $\quad \cos \frac{1}{2}B \;=\; \sqrt{\dfrac{\sin s \,\sin(s-b)}{\sin c \,\sin a}} .$

Then

$$\sin \tfrac{1}{2}A \,\cos \tfrac{1}{2}B \;=\; \sqrt{\dfrac{\sin(s-b)\,\sin(s-c)}{\sin b \,\sin c}}\;\sqrt{\dfrac{\sin s \,\sin(s-b)}{\sin c \,\sin a}} \;=\; \dfrac{\sin(s-b)}{\sin c}\,\sqrt{\dfrac{\sin s \,\sin(s-c)}{\sin a \,\sin b}}$$

$$=\; \dfrac{\sin(s-b)}{\sin c}\,\cos \tfrac{1}{2}C ,$$

$$\cos \tfrac{1}{2}A \,\sin \tfrac{1}{2}B \;=\; \sqrt{\dfrac{\sin s \,\sin(s-a)}{\sin b \,\sin c}}\;\sqrt{\dfrac{\sin(s-c)\,\sin(s-a)}{\sin c \,\sin a}} \;=\; \dfrac{\sin(s-a)}{\sin c}\,\sqrt{\dfrac{\sin s \,\sin(s-c)}{\sin a \,\sin b}}$$

$$=\; \dfrac{\sin(s-a)}{\sin c}\,\cos \tfrac{1}{2}C ,$$

and $\quad \sin \tfrac{1}{2}(A - B) \;=\; \sin \tfrac{1}{2}A \,\cos \tfrac{1}{2}B \;-\; \cos \tfrac{1}{2}A \,\sin \tfrac{1}{2}B \;=\; \dfrac{\sin(s-b) - \sin(s-a)}{\sin c}\,\cos \tfrac{1}{2}C$

$$=\; \dfrac{2\,\cos \tfrac{1}{2}\{(s-b)+(s-a)\}\,\sin \tfrac{1}{2}\{(s-b)-(s-a)\}}{2\,\sin \tfrac{1}{2}c \,\cos \tfrac{1}{2}c}\,\cos \tfrac{1}{2}C$$

$$=\; \dfrac{\cos \tfrac{1}{2}c \,\sin \tfrac{1}{2}(a-b)}{\sin \tfrac{1}{2}c \,\cos \tfrac{1}{2}c}\,\cos \tfrac{1}{2}C \;=\; \dfrac{\sin \tfrac{1}{2}(a-b)}{\sin \tfrac{1}{2}c}\,\cos \tfrac{1}{2}C .$$

Thus $\quad \dfrac{\sin \frac{1}{2}(A - B)}{\cos \frac{1}{2}C} \;=\; \dfrac{\sin \frac{1}{2}(a - b)}{\sin \frac{1}{2}c} .$

In a similar manner, we may obtain formulas for $\sin \frac{1}{2}(A+B)$, $\cos \frac{1}{2}(A-B)$, and $\cos \frac{1}{2}(A+B)$.

**7.** Derive Napier's analogies.

Using the Delambre analogies

$$\sin \tfrac{1}{2}(A-B) \;=\; \dfrac{\sin \frac{1}{2}(a-b)}{\sin \frac{1}{2}c}\,\cos \tfrac{1}{2}C \quad\text{and}\quad \cos \tfrac{1}{2}(A-B) \;=\; \dfrac{\sin \frac{1}{2}(a+b)}{\sin \frac{1}{2}c}\,\sin \tfrac{1}{2}C ,$$

$$\tan \tfrac{1}{2}(A-B) \;=\; \dfrac{\sin \frac{1}{2}(A-B)}{\cos \frac{1}{2}(A-B)} \;=\; \dfrac{\sin \frac{1}{2}(a-b)\,\cos \frac{1}{2}C}{\sin \frac{1}{2}c} \cdot \dfrac{\sin \frac{1}{2}c}{\sin \frac{1}{2}(a+b)\,\sin \frac{1}{2}C} \;=\; \dfrac{\sin \frac{1}{2}(a-b)}{\sin \frac{1}{2}(a+b)}\,\cot \tfrac{1}{2}C$$

and $$\dfrac{\tan \frac{1}{2}(A - B)}{\cot \frac{1}{2}C} \;=\; \dfrac{\sin \frac{1}{2}(a - b)}{\sin \frac{1}{2}(a + b)} .$$

Using the Delambre analogies

$$\sin \tfrac{1}{2}(a-b) \;=\; \dfrac{\sin \frac{1}{2}(A-B)}{\cos \frac{1}{2}C}\,\sin \tfrac{1}{2}c \quad\text{and}\quad \cos \tfrac{1}{2}(a-b) \;=\; \dfrac{\sin \frac{1}{2}(A+B)}{\cos \frac{1}{2}C}\,\cos \tfrac{1}{2}c ,$$

$$\tan \tfrac{1}{2}(a-b) \;=\; \dfrac{\sin \frac{1}{2}(A-B)\,\sin \frac{1}{2}c}{\sin \frac{1}{2}(A+B)\,\cos \frac{1}{2}c} \;=\; \dfrac{\sin \frac{1}{2}(A-B)}{\sin \frac{1}{2}(A+B)}\,\tan \tfrac{1}{2}c \quad\text{and}\quad \dfrac{\tan \frac{1}{2}(a - b)}{\tan \frac{1}{2}c} \;=\; \dfrac{\sin \frac{1}{2}(A - B)}{\sin \frac{1}{2}(A + B)} .$$

The other analogies may be obtained in a similar manner.

## CASE I.

**8.** Solve the oblique spherical triangle $ABC$, given $a = 121°15.4'$, $b = 104°54.7'$, $c = 65°42.5'$.

$$(1)\ \tan r = \sqrt{\frac{\sin(s-a)\ \sin(s-b)\ \sin(s-c)}{\sin s}} \qquad\qquad s = \tfrac{1}{2}(a+b+c) = 145°56.3'$$

$$(2)\ \tan \tfrac{1}{2}A = \frac{\tan r}{\sin(s-a)} \qquad (3)\ \tan \tfrac{1}{2}B = \frac{\tan r}{\sin(s-b)} \qquad (4)\ \tan \tfrac{1}{2}C = \frac{\tan r}{\sin(s-c)}$$

Check: $\dfrac{\sin a}{\sin A} = \dfrac{\sin b}{\sin B} = \dfrac{\sin c}{\sin C}$

|  | (1) | (2) | (3) | (4) |
|---|---|---|---|---|
| $s-a = 24°40.9'$ | l sin 9.62073 | l csc 0.37927 | | |
| $s-b = 41°\ 1.6'$ | l sin 9.81717 | | l csc 0.18283 | |
| $s-c = 80°13.8'$ | l sin 9.99366 | | | l csc 0.00634 |
| $s = 145°56.3'$ | l csc 0.25175 | | | |
| | 2 \| 9.68331 | | | |
| $\tan r$ | log 9.84166 | log 9.84166 | log 9.84166 | log 9.84166 |
| $\tfrac{1}{2}A = 58°59.0'$ | | l tan 0.22093 | | |
| $A = 117°58.0'$ | | | | |
| $\tfrac{1}{2}B = 46°36.9'$ | | | l tan 0.02449 | |
| $B = 93°13.8'$ | | | | |
| $\tfrac{1}{2}C = 35°10.3'$ | | | | l tan 9.84800 |
| $C = 70°20.6'$ | | | | |

Check:

| | | | | | | | | |
|---|---|---|---|---|---|---|---|---|
| $a$ | l sin 9.93189 | | $b$ | l sin 9.98512 | | $c$ | l sin 9.95974 | |
| $A$ | l csc 0.05393 | | $B$ | l csc 0.00069 | | $C$ | l csc 0.02608 | |
| | 9.98582 | | | 9.98581 | | | 9.98582 | |

## CASE II.

**9.** Solve the oblique spherical triangle $ABC$, given $A = 117°22.8'$, $B = 72°38.6'$, $C = 58°21.2'$.

*Solution 1.* (Half-side formulas)                    $S = \tfrac{1}{2}(A+B+C) = 124°11.3'$

$$(1)\ \tan R = \sqrt{\frac{\cos(S-A)\ \cos(S-B)\ \cos(S-C)}{-\cos S}} = \sqrt{\frac{\cos(S-A)\ \cos(S-B)\ \cos(S-C)}{\cos(180°-S)}}$$

$$(2)\ \cot \tfrac{1}{2}a = \frac{\tan R}{\cos(S-A)} \qquad (3)\ \cot \tfrac{1}{2}b = \frac{\tan R}{\cos(S-B)} \qquad (4)\ \cot \tfrac{1}{2}c = \frac{\tan R}{\cos(S-C)}$$

Check: $\dfrac{\sin a}{\sin A} = \dfrac{\sin b}{\sin B} = \dfrac{\sin c}{\sin C}$

|  |  | (1) | (2) | (3) | (4) |
|---|---|---|---|---|---|
| $S-A$ = | $6°48.5'$ | l cos 9.99692 | l sec 0.00308 |  |  |
| $S-B$ = | $51°32.7'$ | l cos 9.79372 |  | l sec 0.20628 |  |
| $S-C$ = | $65°50.1'$ | l cos 9.61211 |  |  | l sec 0.38789 |
| $180°-S$ = | $55°48.7'$ | l sec 0.25033 |  |  |  |
|  |  | 2 ⎸9.65308 |  |  |  |
| $\tan R$ |  | log 9.82654 | log 9.82654 | log 9.82654 | log 9.82654 |
| $\tfrac{1}{2}a$ = | $55°57.7'$ |  | l cot 9.82962 |  |  |
| $a$ = | $111°55.4'$ |  |  |  |  |
| $\tfrac{1}{2}b$ = | $42°50.2'$ |  |  | l cot 0.03282 |  |
| $b$ = | $85°40.4'$ |  |  |  |  |
| $\tfrac{1}{2}c$ = | $31°23.8'$ |  |  |  | l cot 0.21443 |
| $c$ = | $62°47.6'$ |  |  |  |  |

Check:

| | | | | | | | |
|---|---|---|---|---|---|---|---|
| $a$ | l sin 9.96740 | | $b$ | l sin 9.99876 | | $c$ | l sin 9.94908 |
| $A$ | l csc 0.05160 | | $B$ | l csc 0.02024 | | $C$ | l csc 0.06992 |
| | 0.01900 | | | 0.01900 | | | 0.01900 |

*Solution 2.* (Polar triangle)

Consider the polar triangle $A'B'C'$ in which $a' = 62°37.2'$, $b' = 107°21.4'$, $c' = 121°38.8'$.

(1) $\tan r' = \sqrt{\dfrac{\sin(s'-a')\,\sin(s'-b')\,\sin(s'-c')}{\sin s'}}$ $\qquad$ $s' = \tfrac{1}{2}(a'+b'+c') = 145°48.7'$

(2) $\tan \tfrac{1}{2}A' = \dfrac{\tan r'}{\sin(s'-a')}$ $\qquad$ (3) $\tan \tfrac{1}{2}B' = \dfrac{\tan r'}{\sin(s'-b')}$ $\qquad$ (4) $\tan \tfrac{1}{2}C' = \dfrac{\tan r'}{\sin(s'-c')}$

Check: $\dfrac{\sin a'}{\sin A'} = \dfrac{\sin b'}{\sin B'} = \dfrac{\sin c'}{\sin C'}$

|  |  | (1) | (2) | (3) | (4) |
|---|---|---|---|---|---|
| $s'-a'$ = | $83°11.5'$ | l sin 9.99692 | l csc 0.00308 |  |  |
| $s'-b'$ = | $38°27.3'$ | l sin 9.79372 |  | l csc 0.20628 |  |
| $s'-c'$ = | $24°\ 9.9'$ | l sin 9.61211 |  |  | l csc 0.38789 |
| $s'$ = | $145°48.7'$ | l csc 0.25033 |  |  |  |
|  |  | 2 ⎸9.65308 |  |  |  |
| $\tan r'$ |  | log 9.82654 | log 9.82654 | log 9.82654 | log 9.82654 |
| $\tfrac{1}{2}A'$ = | $34°\ 2.3'$ |  | l tan 9.82962 |  |  |
| $A'$ = | $68°\ 4.6'$ |  |  |  |  |
| $\tfrac{1}{2}B'$ = | $47°\ 9.8'$ |  |  | l tan 0.03282 |  |
| $B'$ = | $94°19.6'$ |  |  |  |  |
| $\tfrac{1}{2}C'$ = | $58°36.2'$ |  |  |  | l tan 0.21443 |
| $C'$ = | $117°12.4'$ |  |  |  |  |

Check:

| | | | | | | | |
|---|---|---|---|---|---|---|---|
| $a'$ | l sin 9.94840 | | $b'$ | l sin 9.97976 | | $c'$ | l sin 9.93008 |
| $A'$ | l csc 0.03260 | | $B'$ | l csc 0.00124 | | $C'$ | l csc 0.05092 |
| | 9.98100 | | | 9.98100 | | | 9.98100 |

Then $\quad a = 180° - A' = 111°55.4'$, $\quad b = 85°40.4'$, $\quad c = 62°47.6'$.

**CASE III.**

**10.** Solve the spherical triangle $ABC$, given $a = 106°25.3'$, $c = 42°16.7'$, $B = 114°53.2'$.

*(Napier analogies)*

For $A,C$:   (1)   $\tan \frac{1}{2}(A+C) = \cos \frac{1}{2}(a-c) \sec \frac{1}{2}(a+c) \cot \frac{1}{2}B$
          (2)   $\tan \frac{1}{2}(A-C) = \sin \frac{1}{2}(a-c) \csc \frac{1}{2}(a+c) \cot \frac{1}{2}B$

For $b$:     (3)      $\tan \frac{1}{2}b = \tan \frac{1}{2}(a-c) \sin \frac{1}{2}(A+C) \csc \frac{1}{2}(A-C)$

Check:    $\dfrac{\sin a}{\sin A} = \dfrac{\sin b}{\sin B} = \dfrac{\sin c}{\sin C}$

|  |  | (1) | (2) | (3) |
|---|---|---|---|---|
| $\frac{1}{2}(a-c) =$ | $32°\ 4.3'$ | l cos 9.92808 | l sin 9.72508 | l tan 9.79699 |
| $\frac{1}{2}(a+c) =$ | $74°21.0'$ | l sec 0.56902 | l csc 0.01641 | |
| $\frac{1}{2}B =$ | $57°26.6'$ | l cot 9.80513 | l cot 9.80513 | |
| $\frac{1}{2}(A+C) =$ | $63°29.9'$ | l tan 0.30223 | | l sin 9.95178 |
| $\frac{1}{2}(A-C) =$ | $19°23.7'$ | | l tan 9.54662 | l csc 0.47876 |
| $A =$ | $82°53.6'$ | | | |
| $C =$ | $44°\ 6.2'$ | | | |
| $\frac{1}{2}b =$ | $59°22.0'$ | | | l tan 0.22753 |
| $b =$ | $118°44.0'$ | | | |

Check:

| | | | | | | | |
|---|---|---|---|---|---|---|---|
| $a$ | l sin 9.98191 | $b$ | l sin 9.94293 | $c$ | l sin 9.82784 |
| $A$ | l csc 0.00335 | $B$ | l csc 0.04232 | $C$ | l csc 0.15742 |
| | 9.98526 | | 9.98525 | | 9.98526 |

**11.** Solve the spherical triangle $ABC$, given $b = 119°41.4'$, $c = 81°17.6'$, $A = 66°37.8'$.

*(Napier analogies)*

For $B,C$:   (1)   $\tan \frac{1}{2}(B+C) = \cos \frac{1}{2}(b-c) \sec \frac{1}{2}(b+c) \cot \frac{1}{2}A$
          (2)   $\tan \frac{1}{2}(B-C) = \sin \frac{1}{2}(b-c) \csc \frac{1}{2}(b+c) \cot \frac{1}{2}A$

For $a$:     (3)      $\tan \frac{1}{2}a = \tan \frac{1}{2}(b-c) \sin \frac{1}{2}(B+C) \csc \frac{1}{2}(B-C)$

Check:    $\dfrac{\sin a}{\sin A} = \dfrac{\sin b}{\sin B} = \dfrac{\sin c}{\sin C}$

|  |  | (1) | (2) | (3) |
|---|---|---|---|---|
| $\frac{1}{2}(b-c) =$ | $19°11.9'$ | l cos 9.97515 | l sin 9.51698 | l tan 9.54183 |
| $\frac{1}{2}(b+c) =$ | $100°29.5'$ | l sec 0.73971 (n) | l csc 0.00732 | |
| $\frac{1}{2}A =$ | $33°18.9'$ | l cot 0.18227 | l cot 0.18227 | |
| $\frac{1}{2}(B+C) =$ | $97°13.3'$ | l tan 0.89713 (n) | | l sin 9.99654 |
| $\frac{1}{2}(B-C) =$ | $26°58.1'$ | | l tan 9.70657 | l csc 0.34342 |
| $B =$ | $124°11.4'$ | | | |
| $C =$ | $70°15.2'$ | | | |
| $\frac{1}{2}a =$ | $37°17.8'$ | | | l tan 9.88179 |
| $a =$ | $74°35.6'$ | | | |

Check:

| | | | | | | | |
|---|---|---|---|---|---|---|---|
| $a$ | l sin 9.98411 | $b$ | l sin 9.93888 | $c$ | l sin 9.99496 |
| $A$ | l csc 0.03717 | $B$ | l csc 0.08240 | $C$ | l csc 0.02632 |
| | 0.02128 | | 0.02128 | | 0.02128 |

CASE IV.

**12.** Solve the oblique spherical triangle $ABC$, given $A = 48°44.6'$, $B = 60°42.6'$, $c = 76°22.4'$.

*Solution 1.*                    *(Napier analogies)*

For $a, b$:    (1) $\tan \frac{1}{2}(b+a) = \cos \frac{1}{2}(B-A) \sec \frac{1}{2}(B+A) \tan \frac{1}{2}c$
                  (2) $\tan \frac{1}{2}(b-a) = \sin \frac{1}{2}(B-A) \csc \frac{1}{2}(B+A) \tan \frac{1}{2}c$

For $C$:        (3)        $\cot \frac{1}{2}C = \sin \frac{1}{2}(b+a) \csc \frac{1}{2}(b-a) \tan \frac{1}{2}(B-A)$

Check:    $\dfrac{\sin a}{\sin A} = \dfrac{\sin b}{\sin B} = \dfrac{\sin c}{\sin C}$

|  | | (1) | (2) | (3) |
|---|---|---|---|---|
| $\frac{1}{2}(B-A) = 5°59.0'$ | | l cos 9.99763 | l sin 9.01803 | l tan 9.02040 |
| $\frac{1}{2}(B+A) = 54°43.6'$ | | l sec 0.23847 | l csc 0.08810 | |
| $\frac{1}{2}c = 38°11.2'$ | | l tan 9.89572 | l tan 9.89572 | |
| $\frac{1}{2}(b+a) = 53°33.9'$ | | l tan 0.13182 | | l sin 9.90554 |
| $\frac{1}{2}(b-a) = 5°44.1'$ | | | l tan 9.00185 | l csc 1.00031 |
| $a = 47°49.8'$ | | | | |
| $b = 59°18.0'$ | | | | |
| $\frac{1}{2}C = 49°50.5'$ | | | | l cot 9.92625 |
| $C = 99°41.0'$ | | | | |

Check :

| | | $a$ | l sin 9.86991 | $b$ | l sin 9.93442 | $c$ | l sin 9.98760 |
|---|---|---|---|---|---|---|---|
| | | $A$ | l csc 0.12392 | $B$ | l csc 0.05941 | $C$ | l csc 0.00623 |
| | | | 9.99383 | | 9.99383 | | 9.99383 |

*Solution 2.* (Polar Triangle)

Consider the polar triangle $A'B'C'$ in which $a' = 131°15.4'$, $b' = 119°17.4'$, $C' = 103°37.6'$.

*(Napier analogies)*

For $A', B'$:    (1) $\tan \frac{1}{2}(A'+B') = \cos \frac{1}{2}(a'-b') \sec \frac{1}{2}(a'+b') \cot \frac{1}{2}C'$
                    (2) $\tan \frac{1}{2}(A'-B') = \sin \frac{1}{2}(a'-b') \csc \frac{1}{2}(a'+b') \cot \frac{1}{2}C'$

For $c'$:        (3)        $\tan \frac{1}{2}c' = \tan \frac{1}{2}(a'-b') \sin \frac{1}{2}(A'+B') \csc \frac{1}{2}(A'-B')$

Check:    $\dfrac{\sin a'}{\sin A'} = \dfrac{\sin b'}{\sin B'} = \dfrac{\sin c'}{\sin C'}$

|  | | (1) | (2) | (3) |
|---|---|---|---|---|
| $\frac{1}{2}(a'-b') = 5°59.0'$ | | l cos 9.99763 | l sin 9.01803 | l tan 9.02040 |
| $\frac{1}{2}(a'+b') = 125°16.4'$ | | l sec 0.23847 (n) | l csc 0.08810 | |
| $\frac{1}{2}C' = 51°48.8'$ | | l cot 9.89572 | l cot 9.89572 | |
| $\frac{1}{2}(A'+B') = 126°26.1'$ | | l tan 0.13182 (n) | | l sin 9.90554 |
| $\frac{1}{2}(A'-B') = 5°44.1'$ | | | l tan 9.00185 | l csc 1.00031 |
| $A' = 132°10.2'$ | | | | |
| $B' = 120°42.0'$ | | | | |
| $\frac{1}{2}c' = 40° 9.5'$ | | | | l tan 9.92625 |
| $c' = 80°19.0'$ | | | | |

Then the required parts of triangle $ABC$ are :
    $a = 180° - A' = 47°49.8'$,    $b = 180° - B' = 59°18.0'$,    $C = 180° - c' = 99°41.0'$.

CASE V.

**13.** Solve the oblique spherical triangle $ABC$, given $a = 80°26.2'$, $c = 115°30.6'$, $A = 72°24.4'$.

For $C$ :   $\sin C = \sin c \csc a \sin A$
For $B$ :   $\cot \tfrac{1}{2}B = \sin \tfrac{1}{2}(c+a) \csc \tfrac{1}{2}(c-a) \tan \tfrac{1}{2}(C-A)$
For $b$ :   $\tan \tfrac{1}{2}b = \sin \tfrac{1}{2}(C+A) \csc \tfrac{1}{2}(C-A) \tan \tfrac{1}{2}(c-a)$

Check :   $\sin \tfrac{1}{2}(C-A) \sin \tfrac{1}{2}b = \sin \tfrac{1}{2}(c-a) \cos \tfrac{1}{2}B$

$$
\begin{array}{llll}
c = & 115°30.6' & \text{l sin } 9.95545 & \\
a = & 80°26.2' & \text{l csc } 0.00608 & \\
A = & 72°24.4' & \text{l sin } 9.97920 & \text{Since } a < c, \text{ then } A < C. \\
\hline
C = & 119°15.4' & \text{l sin } 9.94073 & \text{One Solution}
\end{array}
$$

$$
\begin{array}{lll}
\tfrac{1}{2}(C+A) = 95°49.9' & & \text{l sin } 9.99775 \\
\tfrac{1}{2}(C-A) = 23°25.5' & \text{l tan } 9.63674 & \text{l csc } 0.40061 \\
\tfrac{1}{2}(c+a) = 97°58.4' & \text{l sin } 9.99578 & \\
\tfrac{1}{2}(c-a) = 17°32.2' & \text{l csc } 0.52098 & \text{l tan } 9.49969 \\
\quad\quad \tfrac{1}{2}B = 35°\ 4.7' & \text{l cot } \overline{0.15350} & \\
\quad\quad\quad B = 70°\ 9.4' & & \overline{\phantom{xxxx}} \\
\quad\quad \tfrac{1}{2}b = 38°20.2' & & \text{l tan } 9.89805 \\
\quad\quad\quad b = 76°40.4' & &
\end{array}
$$

Check :

$$
\begin{array}{ll}
\begin{array}{ll}
\tfrac{1}{2}(C-A) = 23°25.5' & \text{l sin } 9.59939 \\
\tfrac{1}{2}b = 38°20.2' & \text{l sin } 9.79259 \\
\hline
& 9.39198
\end{array}
&
\begin{array}{ll}
\tfrac{1}{2}(c-a) = 17°32.2' & \text{l sin } 9.47902 \\
\tfrac{1}{2}B = 35°\ 4.7' & \text{l cos } 9.91295 \\
\hline
& 9.39197
\end{array}
\end{array}
$$

**14.** Solve the oblique spherical triangle $ABC$, given $b = 81°42.3'$, $c = 52°19.8'$, $C = 47°25.1'$.

For $B$ :   $\sin B = \sin b \csc c \sin C$
For $A$ :   $\cot \tfrac{1}{2}A = \sin \tfrac{1}{2}(b+c) \csc \tfrac{1}{2}(b-c) \tan \tfrac{1}{2}(B-C)$
For $a$ :   $\tan \tfrac{1}{2}a = \sin \tfrac{1}{2}(B+C) \csc \tfrac{1}{2}(B-C) \tan \tfrac{1}{2}(b-c)$

Check :   $\sin \tfrac{1}{2}(B-C) \sin \tfrac{1}{2}a = \sin \tfrac{1}{2}(b-c) \cos \tfrac{1}{2}A$

$$
\begin{array}{llll}
b = & 81°42.3' & \text{l sin } 9.99544 & \\
c = & 52°19.8' & \text{l csc } 0.10153 & \\
C = & 47°25.1' & \text{l sin } 9.86706 & \text{Since } b > c, \text{ then } B > C. \\
\hline
B_1 = & 67°\ 0.0' & \text{l sin } 9.96403 & \text{Two Solutions} \\
B_2 = & 113°\ 0.0' & &
\end{array}
$$

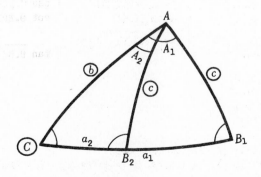

|  | $(A_1)$ | $(a_1)$ | $(A_2)$ | $(a_2)$ |
|---|---|---|---|---|
| $\tfrac{1}{2}(B_1+C) = 57°12.6'$ |  | l sin 9.92462 |  |  |
| $\tfrac{1}{2}(B_1-C) = 9°47.4'$ | l tan 9.23691 | l csc 0.76946 |  |  |
| $\tfrac{1}{2}(b+c) = 67° 1.0'$ | l sin 9.96408 |  | l sin 9.96408 |  |
| $\tfrac{1}{2}(b-c) = 14°41.2'$ | l csc 0.59596 | l tan 9.41846 | l csc 0.59596 | l tan 9.41846 |
| $\tfrac{1}{2}A_1 = 57°55.9'$ | l cot 9.79695 |  |  |  |
| $A_1 = 115°51.8'$ |  |  |  |  |
| $\tfrac{1}{2}a_1 = 52°20.5'$ |  | l tan 0.11254 |  |  |
| $a_1 = 104°41.0'$ |  |  |  |  |
| $\tfrac{1}{2}(B_2+C) = 80°12.6'$ |  |  |  | l sin 9.99363 |
| $\tfrac{1}{2}(B_2-C) = 32°47.4'$ |  |  | l tan 9.80903 | l csc 0.26635 |
| $\tfrac{1}{2}A_2 = 23° 8.8'$ |  |  | l cot 0.36907 |  |
| $A_2 = 46°17.6'$ |  |  |  |  |
| $\tfrac{1}{2}a_2 = 25°29.8'$ |  |  |  | l tan 9.67844 |
| $a_2 = 50°59.6'$ |  |  |  |  |

Check :

| $\tfrac{1}{2}(B_1-C) = 9°47.4'$ | l sin 9.23054 | $\tfrac{1}{2}(b-c) = 14°41.2'$ | l sin 9.40404 |
|---|---|---|---|
| $\tfrac{1}{2}a_1 = 52°20.5'$ | l sin 9.89854 | $\tfrac{1}{2}A_1 = 57°55.9'$ | l cos 9.72504 |
|  | 9.12908 |  | 9.12908 |

| $\tfrac{1}{2}(B_2-C) = 32°47.4'$ | l sin 9.73365 | $\tfrac{1}{2}(b-c) = 14°41.2'$ | l sin 9.40404 |
|---|---|---|---|
| $\tfrac{1}{2}a_2 = 25°29.8'$ | l sin 9.63393 | $\tfrac{1}{2}A_2 = 23° 8.8'$ | l cos 9.96355 |
|  | 9.36758 |  | 9.36759 |

## CASE VI.

**15.** Solve the oblique spherical triangle $ABC$, given $A = 35°52.5'$, $B = 56°10.7'$, $a = 40°38.6'$.

For $b$ :   $\sin b = \sin B \csc A \sin a$

For $c$ :   $\tan \tfrac{1}{2}c = \sin \tfrac{1}{2}(B+A) \csc \tfrac{1}{2}(B-A) \tan \tfrac{1}{2}(b-a)$

For $C$ :   $\cot \tfrac{1}{2}C = \sin \tfrac{1}{2}(b+a) \csc \tfrac{1}{2}(b-a) \tan \tfrac{1}{2}(B-A)$

Check :   $\cos \tfrac{1}{2}(B+A) \cos \tfrac{1}{2}c = \cos \tfrac{1}{2}(b+a) \sin \tfrac{1}{2}C$

| $B = 56°10.7'$ | l sin 9.91948 |
|---|---|
| $A = 35°52.5'$ | l csc 0.23209 |
| $a = 40°38.6'$ | l sin 9.81381 |
| $b_1 = 67°25.5'$ | l sin 9.96538 |
| $b_2 = 112°34.5'$ |  |

Since $A < B$, then $a < b$.

Two Solutions

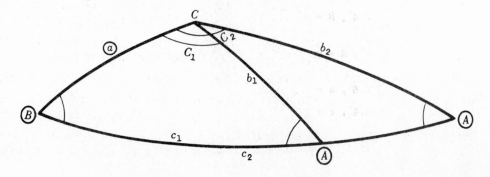

|  | $(c_1)$ | $(C_1)$ | $(c_2)$ | $(C_2)$ |
|---|---|---|---|---|
| $\frac{1}{2}(b_1 + a) = 54°\,2.0'$ |  | l sin 9.90814 |  |  |
| $\frac{1}{2}(b_1 - a) = 13°23.4'$ | l tan 9.37666 | l csc 0.63530 |  |  |
| $\frac{1}{2}(B + A) = 46°\,1.6'$ | l sin 9.85713 |  | l sin 9.85713 |  |
| $\frac{1}{2}(B - A) = 10°\,9.1'$ | l csc 0.75386 | l tan 9.25299 | l csc 0.75386 | l tan 9.25299 |
| $\frac{1}{2}c_1 = 44°11.1'$ | l tan 9.98765 |  |  |  |
| $c_1 = 88°22.2'$ |  |  |  |  |
| $\frac{1}{2}C_1 = 57°57.7'$ |  | l cot 9.79643 |  |  |
| $C_1 = 115°55.4'$ |  |  |  |  |
| $\frac{1}{2}(b_2 + a) = 76°36.6'$ |  |  |  | l sin 9.98803 |
| $\frac{1}{2}(b_2 - a) = 35°58.0'$ |  |  | l tan 9.86073 | l csc 0.23113 |
| $\frac{1}{2}c_2 = 71°21.0'$ |  |  | l tan 0.47172 |  |
| $c_2 = 142°42.0'$ |  |  |  |  |
| $\frac{1}{2}C_2 = 73°28.8'$ |  |  |  | l cot 9.47215 |
| $C_2 = 146°57.6'$ |  |  |  |  |

The details of the check have been omitted.

## SUPPLEMENTARY PROBLEMS

Solve the following oblique spherical triangles $ABC$.

16. $a = 56°22.3'$, $b = 65°54.9'$, $c = 78°27.4'$     *Ans.* $A = 58°8.4'$, $B = 68°37.8'$, $C = 91°57.2'$

17. $a = 108°56.4'$, $b = 58°34.8'$, $c = 122°15.6'$     *Ans.* $A = 93°40.8'$, $B = 64°12.4'$, $C = 116°51.0'$

18. $a = 126°29.6'$, $b = 128°1.6'$, $c = 30°46.6'$     *Ans.* $A = 99°20.9'$, $B = 104°47.7'$, $C = 38°54.4'$

19. $A = 71°2.8'$, $B = 119°25.2'$, $C = 60°45.6'$     *Ans.* $a = 83°35.4'$, $b = 113°45.8'$, $c = 66°28.0'$

20. $A = 116°1.8'$, $B = 103°17.6'$, $C = 94°21.2'$     *Ans.* $a = 115°44.2'$, $b = 102°40.6'$, $c = 88°21.8'$

21. $A = 138°40.7'$, $B = 67°23.8'$, $C = 101°50.4'$     *Ans.* $a = 156°42.2'$, $b = 33°34.4'$, $c = 144°6.6'$

22. $a = 136°2.9'$, $c = 21°46.3'$, $B = 75°31.4'$     *Ans.* $b = 127°10.4'$, $A = 122°30.1'$, $C = 26°47.3'$

23. $b = 86°45.2'$, $c = 108°36.8'$, $A = 67°40.2'$     *Ans.* $a = 70°2.2'$, $B = 79°17.1'$, $C = 111°8.7'$

24. $a = 61°51.7'$, $c = 67°55.4'$, $B = 111°57.9'$     *Ans.* $b = 97°37.5'$, $A = 55°36.0'$, $C = 60°7.3'$

25. $B = 66°42.7'$, $C = 84°57.5'$, $a = 107°8.4'$     *Ans.* $b = 67°8.4'$, $c = 92°7.6'$, $A = 107°43.4'$

26. $A = 47°13.3'$, $B = 120°9.9'$, $c = 123°31.6'$     *Ans.* $a = 37°43.7'$, $b = 133°52.9'$, $C = 90°31.8'$

27. $B = 104°30.7'$, $C = 62°52.1'$, $a = 56°6.4'$     *Ans.* $b = 88°20.8'$, $c = 66°46.0'$, $A = 53°30.4'$

28. $a = 98°53.2'$, $c = 64°35.8'$, $A = 95°23.4'$     *Ans.* $b = 99°29.6'$, $C = 65°32.3'$, $B = 96°21.0'$

29. $b = 37°47.2'$, $c = 103°1.4'$, $B = 24°25.6'$     *Ans.* $a_1 = 73°58.0'$, $A_1 = 40°26.4'$, $C_1 = 138°53.2'$
$a_2 = 134°32.6'$, $A_2 = 151°14.8'$, $C_2 = 41°6.8'$

30. $a = 80°5.3'$, $b = 82°4.0'$, $A = 83°34.2'$     *Ans.* $c_1 = 52°27.2'$, $B_1 = 87°34.5'$, $C_1 = 53°6.6'$
$c_2 = 25°12.0'$, $B_2 = 92°25.5'$, $C_2 = 25°26.2'$

31. $A = 117°54.4'$, $C = 45°8.6'$, $a = 76°37.5'$     *Ans.* $b = 41°4.6'$, $c = 51°17.9'$, $B = 36°38.8'$

32. $A = 96°12.8'$, $C = 45°34.4'$, $c = 27°20.3'$     *Ans.* $a_1 = 140°15.7'$, $B_1 = 121°7.6'$, $b_1 = 146°36.0'$
$a_2 = 39°44.3'$, $B_2 = 44°53.8'$, $b_2 = 26°59.6'$

33. $A = 104°40.0'$, $B = 80°13.6'$, $a = 126°50.4'$     *Ans.* $b_1 = 54°36.8'$, $c_1 = 147°36.8'$, $C_1 = 139°39.0'$
$b_2 = 125°23.2'$, $c_2 = 6°51.2'$, $C_2 = 8°17.6'$

## CHAPTER 22

# Oblique Spherical Triangles – Alternate Solutions

THE ALTERNATE SOLUTIONS discussed here include the use of an additional function, called the haversine function, and the separation of an oblique spherical triangle into two right spherical triangles.

A disadvantage in using haversines is that tables of natural and logarithmic haversines are needed along with the usual tables of trigonometric functions. An advantage is that, since there is a single positive angle less than 180° having a given haversine, the error of determining an angle or side in the wrong quadrant is impossible.

The importance of the right triangle procedure lies in the fact that solutions of right spherical triangles may be tabulated and used to solve oblique spherical triangles. For example, if the solutions of the right triangles, with given parts circled, of Fig. (a) and (b) have been tabulated, the solution of the oblique triangle of Fig. (c) may be found as follows:

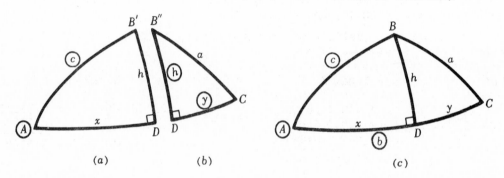

(a)                          (b)                          (c)

1. Obtain the tabulated values of x, h, and B' for the triangle of Fig. (a).
2. In Fig. (b), h and y = b - x are now known; obtain the tabulated values of a, C, and B''.
3. The required parts of the oblique triangle of Fig. (c) are a, C, and B = B' + B''.


THE HAVERSINE FUNCTION. The *haversine* of an angle θ, (hav θ), is defined by

$$\text{hav } \theta = \tfrac{1}{2}(1 - \cos \theta).$$

For certain properties of this function, see Problem 1.

The haversine function is useful in solving Cases I and III directly and in solving the associated polar triangle in Cases II and IV. For this purpose the following formulas are needed.

$$\text{hav } A = \sin(s-b) \, \sin(s-c) \, \csc b \, \csc c,$$

1)　　　$$\text{hav } B = \sin(s-c) \, \sin(s-a) \, \csc c \, \csc a,$$

$$\text{hav } C = \sin(s-a) \, \sin(s-b) \, \csc a \, \csc b,$$

where $s = \tfrac{1}{2}(a + b + c)$, and

$$\text{hav } a = \text{hav}(b-c) + \sin b \sin c \text{ hav } A,$$

2) $$\text{hav } b = \text{hav}(c-a) + \sin c \sin a \text{ hav } B,$$

$$\text{hav } c = \text{hav}(a-b) + \sin a \sin b \text{ hav } C.$$

For derivations, see Problems 2 and 3.

The use of logarithmic haversines is illustrated in

EXAMPLE 1. Use $\text{hav } A = \sin(s-b) \sin(s-c) \csc b \csc c$ to find $A$ when $a = 55°28.0'$, $b = 77°6.0'$, and $c = 49°18.0'$.

| | | |
|---|---|---|
| $a = 55°28.0'$ | $s-b = 13°50.0'$ | l sin 9.37858 |
| $b = 77°\ 6.0'$ | $s-c = 41°38.0'$ | l sin 9.82240 |
| $c = 49°18.0'$ | $b = 77°\ 6.0'$ | l csc 0.01110 |
| $2s =181°52.0'$ | $c = 49°18.0'$ | l csc 0.12025 |
| $s = 90°56.0'$ | $A = 55°14.5'$ | l hav 9.33233 |

Both natural and logarithmic haversines are required when formulas 2) are used.

EXAMPLE 2. Use $\text{hav } a = \text{hav}(b-c) + \sin b \sin c \text{ hav } A$ to find $a$ when $b = 132°46.7'$, $c = 59°50.1'$, and $A = 56°28.4'$.

For convenience, let $x = \sin b \sin c \text{ hav } A$ so that the above formula becomes $\text{hav } a = \text{hav}(b-c) + x$.

| $x = \sin b \sin c \text{ hav } A$ | | $\text{hav } a = \text{hav}(b-c) + x$ | |
|---|---|---|---|
| $b = 132°46.7'$ | l sin 9.86569 | $b\text{-}c = 72°56.6'$ | hav 0.35334 |
| $c = 59°50.1'$ | l sin 9.93681 | $x$ | 0.14205 |
| $A = 56°28.4'$ | l hav 9.34993 | $a = 89°28.3'$ | hav 0.49539 |
| $x = 0.14205$ | log 9.15243 | | |

RIGHT TRIANGLE METHOD. This method of solving an oblique spherical triangle consists in applying Napier's rules to the two right spherical triangles formed by passing a great circle through one vertex of the given triangle perpendicular to the side opposite the vertex. This perpendicular meets the great circle of which the side is an arc in two points. Since each side of an oblique spherical triangle is less than 180°, either just one of these points falls within the triangle or both points fall outside. The two cases are shown in Fig. (d) and (e). In the first case, the point within the triangle is called D and the right triangles are ACD and BCD; in the second case, either of the points may be called D; in Fig. (e), the first intersection reached in moving from A through B is labeled D and the two right triangles are again ACD and BCD.

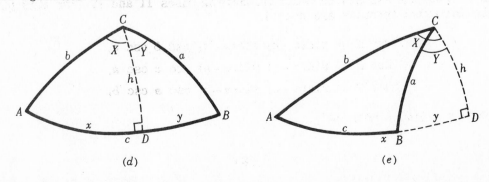

(d)                                        (e)

For convenience the formulas needed for solving the six cases of oblique triangles are listed below. A derivation of the formulas for Case I is given in Problem 6. It is to be noted that the formulas required for Cases III and V consist simply of the formulas necessary to solve completely the two right triangles.

**CASE I.** Given $a, b, c$.

1)     $\tan \frac{1}{2}(b+a) \ \tan \frac{1}{2}(b-a) = \tan \frac{1}{2}(x+y) \ \tan \frac{1}{2}(x-y)$

For Fig. $(d)$, $\frac{1}{2}(x+y) = \frac{1}{2}c$  and  $\frac{1}{2}(x-y)$ is to be determined; for Fig. $(e)$, $\frac{1}{2}(x-y) = \frac{1}{2}c$  and  $\frac{1}{2}(x+y)$ is to be determined.

2)     $x = \frac{1}{2}(x+y) + \frac{1}{2}(x-y), \qquad y = \frac{1}{2}(x+y) - \frac{1}{2}(x-y)$

3)     $\begin{array}{ll} \cos A = \tan x \cot b & \sin X = \sin x \csc b \\ \cos B = \tan y \cot a & \sin Y = \sin y \csc a \end{array}$   for Fig. $(d)$

$\begin{array}{ll} \cos A = \tan x \cot b & \sin X = \sin x \csc b \\ \cos(180° - B) = \tan y \cot a & \sin Y = \sin y \csc a \end{array}$   for Fig. $(e)$

4)     $C = X + Y$ for Fig. $(d)$,    $C = X - Y$ for Fig. $(e)$

5) Check: Use the law of sines  or  construct a perpendicular through $A$ or $B$.

**CASE II.** Given $A, B, C$.

Proceed as in Case I, using the polar triangle $A'B'C'$ in which $a' = 180° - A$, $b' = 180° - B$, $c' = 180° - C$.

**CASE III.** Given $b, c, A$.

1)                    $\tan x = \tan b \cos A$
$\cot X = \cos b \tan A$
$\sin h = \sin b \sin A$

For Fig. $(d)$                          For Fig. $(e)$

2)        $y = c - x$                        $y = x - c$

3)    $\cos a = \cos h \cos y$              $\cos a = \cos h \cos y$
$\cot Y = \sin h \cot y$              $\cot Y = \sin h \cot y$
$\cot B = \cot h \sin y$       $\cot(180° - B) = \cot h \sin y$

4)          $C = X + Y$                      $C = X - Y$

5) Check: Use check formula for each right triangle en route.

**CASE IV.** Given $B, C, a$.

Proceed as in Case III, using the polar triangle $A'B'C'$ in which $b' = 180° - B$, $c' = 180° - C$, $A' = 180° - a$.

**CASE V.** Given $a, b, A$.

1)   $\tan x = \tan b \cos A$          2)   $\cos Y = \tan h \cot a$
$\cot X = \cos b \tan A$               $\cos y = \sec h \cos a$
$\sin h = \sin b \sin A$               $\sin B = \sin h \csc a$

When $\log \sin B < 0$, there is a possibility of two solutions.

3) For Fig. $(d)$: $c = x + y$, $C = X + Y$.    For Fig. $(e)$: $c = x - y$, $C = X - Y$.

CASE VI. Given $A, B, a$.

Proceed as in Case V, using the polar triangle $A'B'C'$ with $a' = 180° - A$, $b' = 180° - B$, $A' = 180° - a$.

## SOLVED PROBLEMS

**1.** Prove: *a*) hav $0° = 0$,   *b*) hav $180° = 1$,   *c*) hav $(-\theta) = $ hav $\theta$,   *d*) $\cos \theta = 1 - 2$ hav $\theta$.

*a*) hav $0° = \frac{1}{2}(1 - \cos 0°) = \frac{1}{2}(1 - 1) = 0$
*b*) hav $180° = \frac{1}{2}(1 - \cos 180°) = \frac{1}{2}[1 - (-1)] = 1$
*c*) hav $(-\theta) = \frac{1}{2}[1 - \cos (-\theta)] = \frac{1}{2}(1 - \cos \theta) = $ hav $\theta$
*d*) Since $2$ hav $\theta = 1 - \cos \theta$,   $\cos \theta = 1 - 2$ hav $\theta$.

**2.** Prove:  hav $A = \sin(s - b) \sin(s - c) \csc b \csc c$.

From Chapter 21, Problem 6,   $\sin \frac{1}{2}A = \sqrt{\dfrac{\sin(s - b) \sin(s - c)}{\sin b \sin c}}$.   Then

hav $A = \frac{1}{2}(1 - \cos A) = \sin^2 \frac{1}{2}A = \dfrac{\sin(s - b) \sin(s - c)}{\sin b \sin c} = \sin(s - b) \sin(s - c) \csc b \csc c$.

**3.** Prove:  hav $a = $ hav$(b - c) + \sin b \sin c$ hav $A$.

Using the law of cosines   $\cos a = \cos b \cos c + \sin b \sin c \cos A$,

$$\begin{aligned}
\text{hav } a = \tfrac{1}{2}(1 - \cos a) &= \tfrac{1}{2}(1 - \cos b \cos c - \sin b \sin c \cos A) \\
&= \tfrac{1}{2}[1 - \cos b \cos c - \sin b \sin c \, (1 - 2 \text{ hav } A)] \\
&= \tfrac{1}{2}[1 - (\cos b \cos c + \sin b \sin c) + 2 \sin b \sin c \text{ hav } A] \\
&= \tfrac{1}{2}[1 - \cos(b - c) + 2 \sin b \sin c \text{ hav } A] \\
&= \tfrac{1}{2}[2 \text{ hav}(b - c) + 2 \sin b \sin c \text{ hav } A] = \text{hav}(b - c) + \sin b \sin c \text{ hav } A.
\end{aligned}$$

**4.** Solve, using haversines, the spherical triangle $ABC$, given $a = 121°15.4'$, $b = 104°54.7'$, $c = 65°42.5'$.   (Case I; Problem 8, Chapter 21.)

For $A$: hav $A = \sin(s - b) \sin(s - c) \csc b \csc c$
For $B$: hav $B = \sin(s - c) \sin(s - a) \csc c \csc a$          Check: $\dfrac{\sin a}{\sin A} = \dfrac{\sin b}{\sin B} = \dfrac{\sin c}{\sin C}$
For $C$: hav $C = \sin(s - a) \sin(s - b) \csc a \csc b$

$$s = \tfrac{1}{2}(a + b + c) = 145°56.3'$$

|  |  | (A) | (B) | (C) |
|---|---|---|---|---|
| $a =$ | $121°15.4'$ |  | l csc 0.06811 | l csc 0.06811 |
| $b =$ | $104°54.7'$ | l csc 0.01488 |  | l csc 0.01488 |
| $c =$ | $65°42.5'$ | l csc 0.04026 | l csc 0.04026 |  |
| $s-a =$ | $24°40.9'$ |  | l sin 9.62073 | l sin 9.62073 |
| $s-b =$ | $41° 1.6'$ | l sin 9.81717 |  | l sin 9.81717 |
| $s-c =$ | $80°13.8'$ | l sin 9.99366 | l sin 9.99366 |  |
| $A =$ | $117°57.9'$ | l hav 9.86597 | ———— |  |
| $B =$ | $93°13.7'$ |  | l hav 9.72276 |  |
| $C =$ | $70°20.6'$ |  |  | l hav 9.52089 |

|  |  |  |  |  |  |  |  |
|---|---|---|---|---|---|---|---|
|  | $a$ | l sin 9.93189 | $b$ | l sin 9.98512 | $c$ | l sin 9.95974 |
| Check: | $A$ | l csc 0.05392 | $B$ | l csc 0.00069 | $C$ | l csc 0.02608 |
|  |  | 9.98581 |  | 9.98581 |  | 9.98582 |

**5.** Using haversines, find side $b$ of the spherical triangle $ABC$, given $a = 106°25.3'$, $c = 42°16.7'$, $B = 114°53.2'$.   (Case III, Problem 10, Chapter 21.)

$$\text{hav } b = \text{hav}(a-c) + \sin a \sin c \text{ hav } B = \text{hav}(a-c) + x$$

| | | |
|---|---|---|
| $a = 106°25.3'$ | l sin 9.98191 | |
| $c = 42°16.7'$ | l sin 9.82784 | |
| $B = 114°53.2'$ | l hav 9.85151 | |
| | ———— | |
| $x = 0.45842$ | log 9.66126 | 0.45842 |
| $a - c = 64° 8.6'$ | | hav 0.28194 |
| | | ———— |
| $b = 118°43.9'$ | | hav 0.74036 |

**6.** Derive the formulas used in solving Case I by the right triangle method.

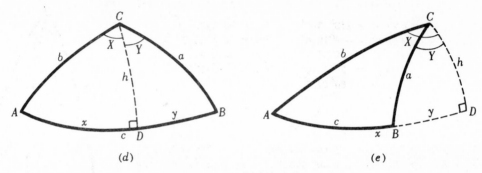

$$(d) \qquad\qquad\qquad (e)$$

Fig. (d) illustrates the case in which the perpendicular through $C$ meets the opposite side within the triangle, and Fig. (e) illustrates the case in which the perpendicular meets the opposite side extended.

In either figure,    $\cos b = \cos h \cos x$    and    $\cos a = \cos h \cos y$.

Then
$$\frac{\cos b}{\cos a} = \frac{\cos x}{\cos y}, \qquad \frac{\cos b}{\cos a} \pm 1 = \frac{\cos x}{\cos y} \pm 1,$$

$$\frac{\cos b + \cos a}{\cos a} = \frac{\cos x + \cos y}{\cos y}, \qquad \frac{\cos b - \cos a}{\cos a} = \frac{\cos x - \cos y}{\cos y},$$

and
$$\frac{\cos b - \cos a}{\cos b + \cos a} = \frac{\cos x - \cos y}{\cos x + \cos y}.$$

Using formulas of Chapter 12, this becomes

$$\frac{2 \sin \tfrac{1}{2}(b+a) \sin \tfrac{1}{2}(b-a)}{2 \cos \tfrac{1}{2}(b+a) \cos \tfrac{1}{2}(b-a)} = \frac{2 \sin \tfrac{1}{2}(x+y) \sin \tfrac{1}{2}(x-y)}{2 \cos \tfrac{1}{2}(x+y) \cos \tfrac{1}{2}(x-y)}$$

or        $(A)$        $\tan \tfrac{1}{2}(b+a) \tan \tfrac{1}{2}(b-a) = \tan \tfrac{1}{2}(x+y) \tan \tfrac{1}{2}(x-y)$.

For the case illustrated in Fig. (d),  $\tfrac{1}{2}(x+y) = \tfrac{1}{2}c$; for the case illustrated in Fig. (e), $\tfrac{1}{2}(x-y) = \tfrac{1}{2}c$.  Having obtained $\tfrac{1}{2}(x-y)$ or $\tfrac{1}{2}(x+y)$ by means of $(A)$,

$$x = \tfrac{1}{2}(x+y) + \tfrac{1}{2}(x-y) \qquad \text{and} \qquad y = \tfrac{1}{2}(x+y) - \tfrac{1}{2}(x-y).$$

Then, in Fig. (d),
$$\cos A = \tan x \cot b, \quad \sin X = \sin x \csc b \quad \text{(triangle } ACD)$$
$$\cos B = \tan y \cot a, \quad \sin Y = \sin y \csc a \quad \text{(triangle } BCD)$$
$$C = X + Y$$

and, in Fig. (e),
$$\cos A = \tan x \cot b, \quad \sin X = \sin x \csc b \quad \text{(triangle } ACD)$$
$$\cos (180° - B) = \tan y \cot a, \quad \sin Y = \sin y \csc a \quad \text{(triangle } BCD)$$
$$C = X - Y.$$

## CASE I.

**7.** Solve the spherical triangle $ABC$, given $a = 121^\circ 15.4'$, $b = 104^\circ 54.7'$, $c = 65^\circ 42.5'$. (Problem 8, Chapter 21.)

$$\tfrac{1}{2}(x+y) = \tfrac{1}{2}c = 32^\circ 51.2'$$

(1)  $\tan \tfrac{1}{2}(x-y) = \tan \tfrac{1}{2}(b+a) \tan \tfrac{1}{2}(b-a) \cot \tfrac{1}{2}(x+y)$

|      Triangle $ACD$      |      Triangle $BCD$      |
|--------------------------|--------------------------|
| (2)  $\cos A = \tan x \cot b$ | (4)  $\cos B = \tan y \cot a$ |
| (3)  $\sin X = \sin x \csc b$ | (5)  $\sin Y = \sin y \csc a$ |

$$C = X + Y$$

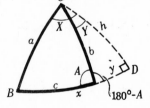

(1)

| | |
|---|---|
| $\tfrac{1}{2}(b+a) = 113^\circ\ 5.0'$ | l tan 0.37039 $(n)$ |
| $\tfrac{1}{2}(b-a) = -8^\circ 10.4'$ | l tan 9.15724 $(n)$ |
| $\tfrac{1}{2}(x+y) = 32^\circ 51.2'$ | l cot 0.18992 |
| $\tfrac{1}{2}(x-y) = 27^\circ 33.5'$ | l tan $\overline{9.71755}$ |
| $x = 60^\circ 24.7'$ | |
| $y = 5^\circ 17.7'$ | |

| (2) | | (3) |
|---|---|---|
| $x = 60^\circ 24.7'$ | l tan 0.24580 | l sin 9.93932 |
| $b = 104^\circ 54.7'$ | l cot $\underline{9.42537\ (n)}$ | l csc 0.01488 |
| $A = 117^\circ 58.2'$ | l cos 9.67117 $(n)$ | |
| $X = 64^\circ\ 8.8'$ | | l sin 9.95420 |

| (4) | | (5) |
|---|---|---|
| $y = 5^\circ 17.7'$ | l tan 8.96698 | l sin 8.96512 |
| $a = 121^\circ 15.4'$ | l cot $\underline{9.78317\ (n)}$ | l csc 0.06811 |
| $B = 93^\circ 13.5'$ | l cos 8.75015 $(n)$ | |
| $Y = 6^\circ 11.8'$ | | l sin 9.03323 |

$$C = X + Y = 70^\circ 20.6'$$

## CASE III.

**8.** Solve the spherical triangle $ABC$, given $a = 106^\circ 25.3'$, $c = 42^\circ 16.7'$, $B = 114^\circ 53.2'$.  (Problem 10, Chapter 21.)

| Triangle $BCD$ | Triangle $ACD$ |
|---|---|
| $\tan x = \tan a \cos B$ | $\cos b = \cos h \cos y$ |
| $\cot X = \cos a \tan B$ | $\cot Y = \sin h \cot y$ |
| $\sin h = \sin a \sin B$ | $\cot(180^\circ - A) = \cot h \sin y$ |
| $y = x - c$ | $C = X - Y$ |

| | | | | |
|---|---|---|---|---|
| $a = 106^\circ 25.3'$ | l tan 0.53058 $(n)$ | l cos 9.45133 $(n)$ | l sin 9.98191 | |
| $B = 114^\circ 53.2'$ | l cos 9.62410 $(n)$ | l tan 0.33357 $(n)$ | l sin 9.95768 | |
| $x = 54^\circ 59.7'$ | l tan $\overline{0.15468}$ | | | |
| $X = 58^\circ 38.5'$ | | l cot 9.78490 | | |
| $*h = 119^\circ 31.5'$ | l cos 9.69268 $(n)$ | l cot 9.75308 $(n)$ | l sin 9.93959 | |
| $y = 12^\circ 43.0'$ | l cos 9.98921 | l sin 9.34268 | l cot 0.64653 | |
| $b = 118^\circ 44.0'$ | l cos $\overline{9.68189}$ $(n)$ | | | |
| $180^\circ - A = 97^\circ\ 6.4'$ | | l cot 9.09576 $(n)$ | | |
| $A = 82^\circ 53.6'$ | | | | |
| $Y = 14^\circ 32.3'$ | | | l cot 0.58612 | |

$$C = X - Y = 44^\circ 6.2'$$

\* $a > 90^\circ$, $x < 90^\circ$; then $h > 90^\circ$ (see Laws of Quadrants, Chapter 20).

CASE V.

9. Solve the spherical triangle $ABC$, given $a = 80°26.2'$, $c = 115°30.6'$,
$A = 72°24.4'$.  (Problem 13, Chapter 21.)

|  Triangle $ABD$  |  Triangle $BCD$  |
|---|---|
| $\tan x = \tan c \cos A$ | $\cos Y = \tan h \cot a$ |
| $\cot X = \cos c \tan A$ | $\cos y = \sec h \cos a$ |
| $\sin h = \sin c \sin A$ | $\sin C = \sin h \csc a$ |
| $b = x - y$ | $B = X - Y$ |

| | | | |
|---|---|---|---|
| $c = 115°30.6'$ | l tan 0.32131 (n) | l cos 9.63414 (n) | l sin 9.95545 |
| $A = 72°24.4'$ | l cos 9.48038 | l tan 0.49882 | l sin 9.97920 |
| $x = 147°39.0'$ | l tan 9.80169 (n) | | |
| $X = 143°38.2'$ | | l cot 0.13296 (n) | |
| *$h = 59°21.0'$ | l tan 0.22726 | l sec 0.29261 | l sin 9.93465 |
| $a = 80°26.2'$ | l cot 9.22655 | l cos 9.22047 | l csc 0.00608 |
| $Y = 73°28.9'$ | l cos 9.45381 | | |
| $y = 70°58.8'$ | | l cos 9.51308 | |
| **$C = 119°15.4'$ | | | l sin 9.94073 |

$b = x - y = 76°40.2'$
$B = X - Y = 70° 9.3'$

\* $A < 90°$, then $h < 90°$.      \*\* $c > a$, then $C > A$; there is one solution.

10. Solve the spherical triangle $ABC$, given $b = 81°42.3'$, $c = 52°19.8'$,
$C = 47°25.1'$.

|  Triangle $ACD$  |  Triangle $ABD$  |
|---|---|
| $\tan x = \tan b \cos C$ | $\cos Y = \tan h \cot c$ |
| $\cot X = \cos b \tan C$ | $\cos y = \sec h \cos c$ |
| $\sin h = \sin b \sin C$ | $\sin B = \sin h \csc c$ |

| | | | |
|---|---|---|---|
| $b = 81°42.3'$ | l tan 0.83626 | l cos 9.15918 | l sin 9.99544 |
| $C = 47°25.1'$ | l cos 9.83036 | l tan 0.03670 | l sin 9.86706 |
| $x = 77°50.4'$ | l tan 0.66662 | | |
| $X = 81° 4.7'$ | | l cot 9.19588 | |
| *$h = 46°46.2'$ | l tan 0.02685 | l sec 0.16436 | l sin 9.86250 |
| $c = 52°19.8'$ | l cot 9.88764 | l cos 9.78612 | l csc 0.10153 |
| $Y = 34°47.2'$ | l cos 9.91449 | | |
| $y = 26°50.7'$ | | l cos 9.95048 | |
| **$B = 67° 0.0'$ | | | l sin 9.96403 |

\* $C < 90°$, then $h < 90°$.      \*\* $h < 90°$, then $B < 90°$.

There are two solutions, $ACB_1$ and $ACB_2$, as shown in the figure.  The required parts are:

Triangle $ACB_1$,   $B_1 = 67°0.0'$,   $a_1 = x + y = 104°41.1'$,   $A_1 = X + Y = 115°51.9'$.

Triangle $ACB_2$,   $B_2 = 180° - B_1 = 113°0.0'$,   $a_2 = x - y = 50°59.7'$,   $A_2 = X - Y = 46°17.5'$.

## SUPPLEMENTARY PROBLEMS

Solve, using haversines.

11. $a = 69°23.6'$, $b = 57°51.3'$, $c = 39°39.7'$.      Ans. $A = 96°7.2'$, $B = 64°4.9'$, $C = 42°41.2'$

12. $a = 59°9.4'$, $b = 101°53.9'$, $c = 98°47.7'$.      Ans. $A = 60°9.7'$, $B = 98°39.7'$, $C = 93°13.2'$

13. $A = 51°44.4'$, $B = 59°31.8'$, $C = 76°20.2'$.      Ans. $a = 28°4.0'$, $b = 31°6.0'$, $c = 35°36.5'$

14. $A = 134°35.4'$, $B = 108°13.6'$, $C = 79°57.0'$. Ans. $a = 143°59.9'$, $b = 128°22.3'$, $c = 54°22.2'$

Using haversines, find the required part.

15. $a = 103°44.7'$, $b = 64°12.3'$, $C = 98°33.8'$ ; find $c = 103°30.6'$.

16. $b = 156°12.2'$, $c = 112°48.6'$, $A = 76°32.4'$; find $a = 63°48.8'$.

17. $a = 67°28.4'$, $b = 34°15.2'$, $C = 24°12.6'$  ; find $c = 37°44.1'$.

18. Solve Supplementary Problems 16-33, Chapter 21, using the right triangle method.

# CHAPTER 23

# Course and Distance

TWO PROBLEMS OF NAVIGATION will be considered in this chapter:

a) The position of the starting point together with the course and the distance made good at a given time being known, to find the position at that time.

b) The positions of the starting point and destination being known, to find the distance between the two positions and the course in sailing from one to the other.

(Throughout this chapter the term 'mile' will mean nautical mile.)

PARALLEL SAILING. Suppose that a ship sails due east or due west (due east in the adjoining figure) from a known position $D$ for a distance of $p$ miles to $B$. Since the trip is along a parallel of latitude, the latitude of $B$ is that of the starting point $D$. Problem a) is thus reduced to that of finding the longitude of $B$.

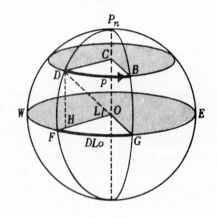

The *difference in longitude (DLo)* of $B$ and and $D$ is measured by the arc $FG$ intercepted on the equator by the meridians through $D$ and $B$. The distance traveled in sailing from $D$ to $B$ (that is, the length in miles of arc $DB$) is called the *departure (p)* of $B$ from $D$. The difference in longitude is marked $E$ or $W$ according as the departure is $E$ or $W$.

To find the longitude of $B$ it is necessary to convert the departure $p$ miles into minutes. (Note that since the departure is now measured along a small circle, the relation 1 mile = 1 minute does not hold.) In the figure, join $D$ and $B$ to the center $C$ of small circle (parallel of latitude), join $D$, $F$, and $G$ to the center $O$ of the earth, and draw $DH$ perpendicular to $OF$. Denote $\angle FOD$, which measures the latitude of $D$, by $L$. Since $\angle FOG = \angle DCB$, the arcs $FG$ and $DB$ are proportional to their radii; hence,

$$\frac{\text{arc } FG}{\text{arc } DB} = \frac{OF}{CD} = \frac{OD}{OH} = \sec L, \qquad \text{arc } FG = (\text{arc } DB) \sec L, \qquad \text{or} \qquad DLo = p \sec L.$$

Thus,

difference in longitude (minutes) = departure (miles) × secant of latitude.

EXAMPLE 1. A ship in latitude 44°30′ N sails 55 miles due east. Find the resulting change in longitude.

Using the above figure, the departure is $p = 55$ miles E and $L = 44°30′$.

Then $DLo = 55 \sec 44°30′ = 77.1$ miles E $= 77.1′$ E or $1°17.1′$ E.

(See also Problems 1 and 2.)

188

**PLANE SAILING.** Suppose that a ship sails for a distance $d$ miles along a great circle from $A$ to $B$, as in Fig. (a) below. Through $B$ draw the parallel of latitude meeting the meridian through $A$ in $D$ and let the meridians through $A$ and $B$ meet the equator in $F$ and $G$. Then $l$ = arc $AD$ is the *change in latitude* and $p$ = arc $DB$ is the departure.

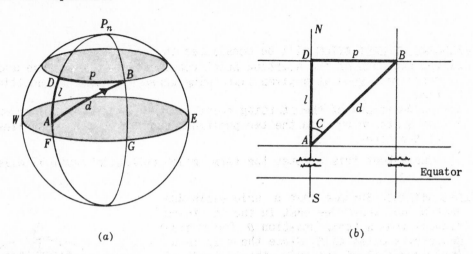

(a)                                              (b)

When comparatively small distances are involved, it is customary to consider the surface of the earth to be a plane in order to take advantage of the simpler formulas of plane trigonometry. We now propose to limit all distance to 200 miles or less and to assume that we are dealing with a plane.

In this plane the equator and the parallels of latitude are represented by parallel (horizontal) lines while the meridians, being perpendicular to the equator, are represented by parallel vertical lines. In Fig. (b) above, *NAS* is the meridian through $A$ and *DB* is a portion of the parallel of latitude through $B$. Then, $d = AB$ is the distance, $p = DB$ is the departure, $l$ is the change in latitude, and $C$ is the course angle.

From the right triangle $ABD$:    $l = d \cos C$,    $p = d \sin C$,    $\tan C = p/l$.

The change in latitude is marked N or S according as $B$ is north or south of $A$. In Fig. (b), the change in latitude is $l$ miles N or $l$ minutes N, the departure is $p$ miles E, and the course is N $C°$ E.

EXAMPLE 2. A ship sails on course 30° (or N 30° E) from $A$ (lat. 45°0' N, long. 70°0' W) for a distance of 120 miles to $B$. Find the departure and latitude of $B$.

From the right triangle $ABD$ of Fig. (b),

$$p = d \sin C = 120 \sin 30° = 60 \text{ miles E}$$

$$l = d \cos C = 120 \cos 30° = 103.9 \text{ miles N}.$$

The change in latitude is 103.9' = 1°44' and the latitude of $B$ is 45°0' + 1°44' = 46°44' N.

(See also Problems 3-5.)

**MIDDLE LATITUDE SAILING.** When the triangle $ABD$ of Fig. (a) is turned into the plane triangle $ABD$ of Fig. (b), arc *DB* must be lengthened. Thus, when the departure $p = DB$ obtained by plane sailing is used in the formula $DLo = p \sec L$ of parallel sailing, the value of $DLo$ is too large. A better approximation is obtained by considering the departure laid out on the parallel of latitude halfway between the parallels of latitude of $A$ and $B$, that is

$$DLo = p \sec \tfrac{1}{2}(\text{lat. } A + \text{lat. } B).$$

This method of converting departure into difference in longitude is called *middle latitude sailing*. It should not be used, however, when the middle latitude exceeds 60°.

EXAMPLE 3. Find the longitude of *B* in Example 2 by the method of middle latitude sailing.

The latitudes of *A* and *B* are 45°0'N and 46°44'N, and the departure is $p = 60$ miles E. The middle latitude is $\frac{1}{2}$(lat. *A* + lat. *B*) = $\frac{1}{2}$(45°0' + 46°44') = 45°52' N  and

$$DLo = 60 \sec 45°52' = 86.2'.$$

Then                                          long. *B* = 70° − 86.2' = 68°34' W.

(See also Problems 6-7.)

**DEAD RECKONING.** The process by which the navigator approximates his present position using as data the last known position of his ship and the courses and distances made good from that position is known as *dead reckoning*. The method of middle-latitude sailing is used in calculating the longitude.

(See Problem 8.)

**GREAT CIRCLE SAILING.** In *great circle sailing* from *A* to *B* (Fig. (c), (d), (e) below), the track of a ship is the shorter arc of the great circle through *A* and *B*. The fundamental problems of great circle navigation are to determine the distance from *A* to *B*, and to determine the direction of the track at any of its points.

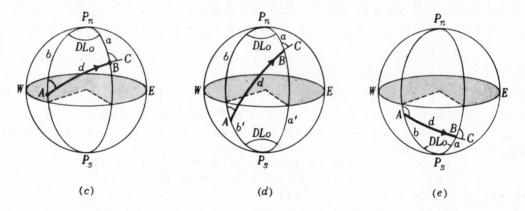

(c)                                    (d)                                    (e)

Problems of great circle sailing require the solution of a spherical triangle (usually oblique) having one of the poles $P_n$ or $P_s$ as a vertex. If *A* and *B* are in the same hemisphere, the pole of that hemisphere is usually taken as this vertex; if *A* and *B* are in different hemispheres, either pole may be used. In Fig.(c), *A* and *B* are in the northern hemisphere; $b = $ arc $AP_n = 90° - $ lat. *A* = colatitude *A*, $a = $ arc $BP_n = 90° - $ lat. *B* = colatitude *B*, and $DLo = \angle AP_n B$ = difference in longitude between *A* and *B*. In Fig.(d), *A* is in the southern hemisphere and *B* is in the northern hemisphere. In triangle $AP_n B$, $b = $ arc $AP_n = 90° + $ lat. *A* and $a = BP_n = 90° - $ lat. *B* while in triangle $AP_s B$, $b' = $ arc $AP_s = 90° - $ lat. *A* and $a' = $ arc $BP_s = 90° + $ lat. *B*. In Fig.(e), *A* and *B* are in the southern hemisphere. In triangle $AP_s B$, $b = $ arc $AP_s = 90° - $ lat. *A* and $a = $ arc $BP_s = 90° - $ lat. *B*.

In each of the figures, $d = $ arc $AB = $ the great circle distance between *A* and *B*, $\angle P_n AB = $ the initial course, and $\angle P_n BC = $ the course on arrival.

(See Problems 9-11.)

## SOLVED PROBLEMS

### PARALLEL SAILING.

1. A ship sails due west for 160 miles in latitude $35°$N. Find the resultant change in its longitude.

Here $p = 160$, $L = 35°$, and $DLo = p \sec L = 160 \sec 35° = 195.3'$ W.

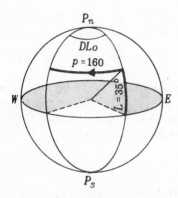

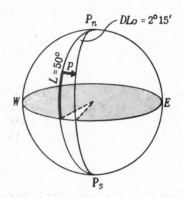

Problem 1                              Problem 2

2. A ship in latitude $50°$N steams due east until it has made good a difference in longitude of $2°15'$. Find the departure.

Here $DLo = 2°15' = 135'$E, $L = 50°$ and, using $DLo = p \sec L$,

$$p = DLo \cos L = 135 \cos 50° = 86.8 \text{ miles E.}$$

### PLANE SAILING.

3. A ship sails on course $245°10'$ (or S $65°10'$ W) from San Francisco (lat. $37°50'$ N) for a distance of 150 miles. Find the departure of the ship and the latitude attained.

Let $A$ be the initial position and $B$ the final position of the ship. In the right triangle $ABD$, $C = 65°10'$; then

$$p = d \sin C = 150 \sin 65°10' = 136.1 \text{ miles W} \quad \text{and}$$
$$l = d \cos C = 150 \cos 65°10' = 63.0' \text{ S.}$$

The latitude of $B$ is $37°50 - 63' = 36°47'$ N.

4. An airplane flies from $A$ to $B$, the consequent difference in latitude being $l = 80$ miles S and the departure being 115 miles E. Find the course and distance.

In the right triangle $ABD$: $\tan C = p/l = 115/80 = 1.4375$, $C = 55°11'$, and the course is S $55°11'$ E or $124°49'$; $d = l \sec C = 80 \sec 55°11' = 140.1$ miles.

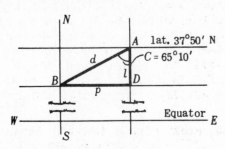

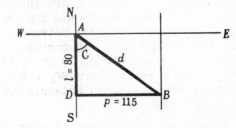

Problem 3                              Problem 4

**5.** A ship steams on a course of 160° from $A$(lat. 53°10′ S) to $B$(lat. 55°40′ S). Find the distance and departure.

Here  $l = 55°40′ - 53°10′ = 2°30′ = 150$ miles S,  $C = $ S 20° E; then

$$d = l \sec C = 150 \sec 20° = 159.6 \text{ miles}  \text{ and}$$
$$p = l \tan C = 150 \tan 20° = 54.6 \text{ miles E}.$$

## MIDDLE LATITUDE SAILING.

**6.** From a position $A$(lat. 54°10′ N, long. 156°0′ W) a ship steams 165 miles on course 220°.  Find the position $B$ reached.

Here  $C = $ S 40° W,  $d = 165$;  then

$$l = d \cos C = 165 \cos 40° = 126.4′ \text{ S}  \text{ and}$$
$$p = d \sin C = 165 \sin 40° = 106.1 \text{ miles W}.$$

The latitude of $B$ is  $54°10′ - 126.4′ = 52°3.6′$ N.  The middle latitude is  $\frac{1}{2}(54°10′ + 52°3.6′) = 53°6.8′$ N.   Then

$$DLo = p \sec \tfrac{1}{2}(\text{lat. } A + \text{lat. } B) = 106.1 \sec 53°6.8′ = 176°8′ \text{ W}$$

and the longitude of $B$ is  $156°0′ + 176.8′ = 158°57′$ W.

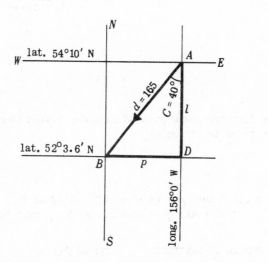

Problem 6                              Problem 7

**7.** A ship leaves $A$(lat. 40°22′ N, long. 47°12′ W) and arrives at $B$(lat. 37°46′ N, long. 44°54′ W). Find the course and distance using middle latitude sailing.

The difference in latitude is $l = 40°22′ - 37°46′ = 2°36′ = 156′$ S.

The difference in longitude is $DLo = 47°12′ - 44°54′ = 2°18′ = 138′$ E.

The middle latitude is  $\frac{1}{2}(40°22′ + 37°46′) = 39°4′$.

From              $DLo = p \sec \tfrac{1}{2}(\text{lat. } A + \text{lat. } B)$,   $p = 138 \cos 39°4′$;

$$\tan C = \frac{p}{l} = \frac{138 \cos 39°4′}{156} = 0.6868  \text{ and}  C = 34°29′ ;$$

and, from $l = d \cos C$,  $d = l \sec C = 156 \sec 34°29′ = 189.3$ miles E.

The course is S 34°29′ E or 145°31′.

## DEAD RECKONING.

**8.** Starting from a position $A$ (lat. $37°49'$ N, long. $123°0'$ W) a ship sails the following courses and distances:

| Course | $230°0'$ | $310°30'$ | $260°0'$ | $340°0'$ | $20°0'$ | $150°30'$ |
|---|---|---|---|---|---|---|
| Distance (miles) | 18.5 | 22.0 | 24.0 | 30.3 | 25.0 | 30.0 |

Find the course and distance made good, and the latitude and longitude of the final position $B$ of the vessel.

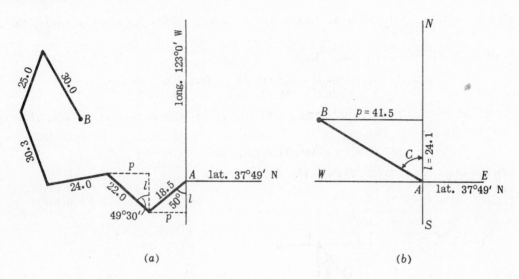

(a)                                    (b)

The first two columns of the table below list the given data. The columns headed **CHANGE IN LATITUDE** and **DEPARTURE** are computed by means of the formulas

$$l = d \cos C \qquad \text{and} \qquad p = d \sin C.$$

For example,

first course: $l = 18.5 \cos 50° = 11.9'$ S and $p = 18.5 \sin 50° = 14.2$ mi W

second course: $l = 22.0 \cos 49°30' = 14.3'$ N and $p = 22.0 \sin 49°30' = 16.7$ mi W.

| COURSE | DISTANCE | CHANGE IN LATITUDE | DEPARTURE |
|---|---|---|---|
| $230°0'$ | 18.5 | 11.9' S | 14.2 W |
| $310°30'$ | 22.0 | 14.3' N | 16.7 W |
| $260°0'$ | 24.0 | 4.2' S | 23.6 W |
| $340°0'$ | 30.3 | 28.5' N | 10.4 W |
| $20°0'$ | 25.0 | 23.5' N | 8.6 E |
| $150°30'$ | 30.0 | 26.1' S | 14.8 E |

Totals $\begin{cases} 66.3' \text{ N} \\ 42.2' \text{ S} \end{cases}$      Totals $\begin{cases} 64.9 \text{ W} \\ 23.4 \text{ E} \end{cases}$

$l = 24.1'$ N         $p = 41.5$ W

In Fig. (b), $\tan C = p/l = 41.5/24.1 = 1.7220$, $C = 59°51'$, and the course made good is **$300°9'$**.

The latitude of the final position of the vessel is $37°49' + 24.1' = 38°13'$ N. The middle-latitude is $\frac{1}{2}(37°49' + 38°13') = 38°1'$; then

$$DLo = 41.5 \sec 38°1' = 52.7' \text{ W}$$

**and** the longitude of the final position is $123°0' + 52.7' = 123°53'$ W.

GREAT CIRCLE SAILING.

9. Find the distance, initial course, and course on **arrival** in traveling from Honolulu (lat. $21°18.3'$ N, long. $157°52.3'$ W) to San Francisco (lat. $37°47.5'$ N, long. $122°25.7'$ W).

In the figure, $A$ is at Honolulu and $B$ is at San Francisco.

Then　$a = 90° - 37°47.5' = 52°12.5'$,　$b = 90° - 21°18.3' = 68°41.7'$, and $C = 157°52.3' - 122°25.7' = 35°26.6'$.

*Standard Solution.*

For $A, B$:　(1)　$\tan \frac{1}{2}(B+A) = \cos \frac{1}{2}(b-a) \sec \frac{1}{2}(b+a) \cot \frac{1}{2}C$
　　　　　　(2)　$\tan \frac{1}{2}(B-A) = \sin \frac{1}{2}(b-a) \csc \frac{1}{2}(b+a) \cot \frac{1}{2}C$

For $c$:　　(3)　$\tan \frac{1}{2}c = \tan \frac{1}{2}(b-a) \sin \frac{1}{2}(B+A) \csc \frac{1}{2}(B-A)$

|  | (1) | (2) | (3) |
|---|---|---|---|
| $\frac{1}{2}(b-a) = \quad 8°14.6'$ | l cos 9.99549 | l sin 9.15648 | l tan 9.16099 |
| $\frac{1}{2}(b+a) = 60°27.1'$ | l sec 0.30701 | l csc 0.06051 | |
| $\frac{1}{2}C = 17°43.3'$ | l cot 0.49545 | l cot 0.49545 | |
| $\frac{1}{2}(B+A) = 80°57.1'$ | l tan 0.79795 | | l sin 9.99456 |
| $\frac{1}{2}(B-A) = 27°16.9'$ | | l tan 9.71244 | l csc 0.33878 |
| $B = 108°14.0'$ | | | |
| $A = \quad 53°40.2'$ | | | |
| $\frac{1}{2}c = 17°20.1'$ | | | l tan 9.49433 |
| $c = 34°40.2'$ | | | |

The required distance is $34°40.2' = 2080.2' = 2080.2$ mi. The initial course is N $53°40.2'$ E or $53°40.2'$ and the course on arrival is N$(180° - 108°14.0')$E $=$ N $71°46.0'$ E or $71°46.0'$.

*Alternate Solution.*

$\text{hav } c = \text{hav}(b-a) + \sin b \sin a \text{ hav } C = \text{hav}(b-a) + x$
$\text{hav } A = \sin(s-b) \sin(s-c) \csc b \csc c$
$\text{hav } B = \sin(s-c) \sin(s-a) \csc c \csc a$

| $b = 68°41.7'$ | l sin 9.96926 | | |
|---|---|---|---|
| $a = 52°12.5'$ | l sin 9.89776 | | |
| $C = 35°26.6'$ | l hav 8.96687 | $b - a = 16°29.2'$ | hav 0.02056 |
| $x = 0.06822$ | log 8.83389 | | 0.06822 |
| $c = 34°40.2'$ | | | hav 0.08878 |

| $a = \quad 52°12.5'$ | $s-a = 25°34.7'$ | | l sin 9.63523 |
|---|---|---|---|
| $b = \quad 68°41.7'$ | $s-b = \quad 9°\ 5.5'$ | l sin 9.19870 | |
| $c = \quad 34°40.2'$ | $s-c = 43°\ 7.0'$ | l sin 9.83473 | l sin 9.83473 |
| $2s = 155°34.4'$ | $a = 52°12.5'$ | | l csc 0.10224 |
| $s = \quad 77°47.2'$ | $b = 68°41.7'$ | l csc 0.03074 | |
| | $c = 34°40.2'$ | l csc 0.24500 | l csc 0.24500 |
| | $A = 53°40.2'$ | l hav 9.30917 | |
| | $B = 108°14.0'$ | | l hav 9.81720 |

The required distance is 2080.2 miles, the initial course is $53°40.2'$, and the course on arrival is $71°46.0'$.

**10.** A ship sails along the great circle track from Dutch Harbor (lat. $53^\circ 53.0' N$, long. $166^\circ 35.0' W$) to Melbourne (lat. $37^\circ 50.0' S$, long. $144^\circ 59.0' E$). (a) Find the distance, the initial course and the course on arrival. (b) Locate the point of intersection of the track and the equator. Find the course at this point, and its distance from Dutch Harbor. (c) Locate the point on the track whose longitude is $180^\circ$. Find the course at this point and its distance from Dutch Harbor.

a) In Fig. (a), A is at Dutch Harbor and B is at Melbourne; then $b = 90^\circ - 53^\circ 53.0' = 36^\circ 7.0'$, $a = 90^\circ + 37^\circ 50.0' = 127^\circ 50.0'$, and $C = 360^\circ - (166^\circ 35.0' + 144^\circ 59.0') = 48^\circ 26.0'$.

For A,B:   (1) $\tan \frac{1}{2}(A+B) = \cos \frac{1}{2}(a-b) \sec \frac{1}{2}(a+b) \cot \frac{1}{2}C$
           (2) $\tan \frac{1}{2}(A-B) = \sin \frac{1}{2}(a-b) \csc \frac{1}{2}(a+b) \cot \frac{1}{2}C$

For c:      (3) $\tan \frac{1}{2}c = \tan \frac{1}{2}(a-b) \sin \frac{1}{2}(A+B) \csc \frac{1}{2}(A-B)$

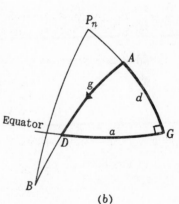

(a)

|  | (1) | (2) | (3) |
|---|---|---|---|
| $\frac{1}{2}(a-b) = 45^\circ 51.5'$ | l cos 9.84288 | l sin 9.85590 | l tan 0.01302 |
| $\frac{1}{2}(a+b) = 81^\circ 58.5'$ | l sec 0.85510 | l csc 0.00427 |  |
| $\frac{1}{2}C = 24^\circ 13.0'$ | l cot 0.34701 | l cot 0.34701 |  |
| $\frac{1}{2}(A+B) = 84^\circ 50.9'$ | l tan 1.04499 |  | l sin 9.99824 |
| $\frac{1}{2}(A-B) = 58^\circ 10.5'$ |  | l tan 0.20718 | l csc 0.07075 |
| $A = 143^\circ 1.4'$ |  |  |  |
| $B = 26^\circ 40.4'$ |  |  |  |
| $\frac{1}{2}c = 50^\circ 22.7'$ |  |  | l tan 0.08201 |
| $c = 100^\circ 45.4'$ |  |  |  |

The required distance is $100^\circ 45.4' = 6045.4' = 6045.4$ miles. The initial course is $S(180^\circ - 143^\circ 1.4')W = S\ 36^\circ 58.6'W$ or $360^\circ - 143^\circ 1.4' = 216^\circ 58.6'$ and the course on arrival is $S\ 26^\circ 40.4'W$ or $180^\circ + 26^\circ 40.4' = 206^\circ 40.4'$.

b) Denote by D the intersection of the track and the equator, and by G the intersection of the meridian through A and the equator. Consider the right spherical triangle AGD of Fig. (b) with $G = 90^\circ$, $d = $ arc $GA = 53^\circ 53.0'$, and $A = \angle DAG = 180^\circ - 143^\circ 1.4' = 36^\circ 58.6'$.

For a:   (4) $\tan a = \sin d \tan A$
For D:   (5) $\cos D = \cos d \sin A$
For g:   (6) $\tan g = \tan d \sec A$

(b)

|  | (4) | (5) | (6) |
|---|---|---|---|
| $d = 53^\circ 53.0'$ | l sin 9.90731 | l cos 9.77043 | l tan 0.13688 |
| $A = 36^\circ 58.6'$ | l tan 9.87675 | l sin 9.77923 | l sec 0.09752 |
| $a = 31^\circ 18.5'$ | l tan 9.78406 |  |  |
| $D = 69^\circ 14.1'$ |  | l cos 9.54966 |  |
| $g = 59^\circ 45.7'$ |  |  | l tan 0.23440 |

The longitude of D is $166^\circ 35.0' + 31^\circ 18.5' = 197^\circ 53.5'W = 162^\circ 6.5'E$.
The course at D is $S(90^\circ - 69^\circ 14.1')W = S\ 20^\circ 45.9'W$ or $200^\circ 45.9'$.
The distance of D from Dutch Harbor is $59^\circ 45.7' = 3585.7' = 3585.7$ mi.

c) Denote by M the point on the track with longitude $180^\circ$ and consider the spherical triangle AMC of Fig. (c) in which $C = 180^\circ - 166^\circ 35.0' = 13^\circ 25.0'$.

For a,c:   (7) $\tan \frac{1}{2}(a+c) = \cos \frac{1}{2}(A-C) \sec \frac{1}{2}(A+C) \tan \frac{1}{2}m$
         (8) $\tan \frac{1}{2}(a-c) = \sin \frac{1}{2}(A-C) \csc \frac{1}{2}(A+C) \tan \frac{1}{2}m$

For M:    (9) $\cot \frac{1}{2}M = \sin \frac{1}{2}(a+c) \csc \frac{1}{2}(a-c) \tan \frac{1}{2}(A-C)$

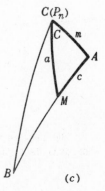

(c)

|  | (7) | (8) | (9) |
|---|---|---|---|
| $\frac{1}{2}(A-C) = 64°48.2'$ | l cos 9.62913 | l sin 9.95658 | l tan 0.32745 |
| $\frac{1}{2}(A+C) = 78°13.2'$ | l sec 0.69004 | l csc 0.00924 | |
| $\frac{1}{2}m = 18° 3.5'$ | l tan 9.51328 | l tan 9.51328 | |
| $\frac{1}{2}(a+c) = 34°12.7'$ | l tan 9.83245 | | l sin 9.74993 |
| $\frac{1}{2}(a-c) = 16°46.3'$ | | l tan 9.47910 | l csc 0.53976 |
| $a = 50°59.0'$ | | | |
| $c = 17°26.4'$ | | | |
| $\frac{1}{2}M = 13°34.5'$ | | | l cot 0.61714 |
| $M = 27° 9.0'$ | | | |

The latitude of $M$ is $(90° - a)$N = $39°1.0'$N. The course at $M$ is S $27°9.0'$ W or $207°9.0'$, and the distance of $M$ from Dutch Harbor is $17°26.4'$ = 1046.4 miles.

**11.** A ship leaves New York (lat. $40°48.6'$N, long. $73°57.5'$W) following a great circle track with initial course $36°0.0'$. (a) Find the latitude and longitude of its position $B$ when it has traveled 500 miles. (b) Locate the northern-most point on the track.

a) In Fig.(a), $A$ is at New York. Then $b = 90° - 40°48.6' = 49°11.4'$, $c = 500$ miles $= 8°20.0'$, and $A = 36°0.0'$.

For $C$:  (1) $\tan \frac{1}{2}(B+C) = \cos \frac{1}{2}(b-c) \sec \frac{1}{2}(b+c) \cot \frac{1}{2}A$
          (2) $\tan \frac{1}{2}(B-C) = \sin \frac{1}{2}(b-c) \csc \frac{1}{2}(b+c) \cot \frac{1}{2}A$

For $a$:  (3) $\tan \frac{1}{2}a = \tan \frac{1}{2}(b-c) \sin \frac{1}{2}(B+C) \csc \frac{1}{2}(B-C)$

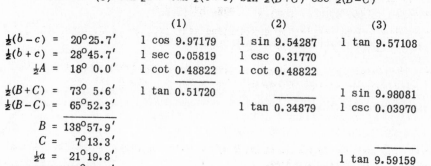

|  | (1) | (2) | (3) |
|---|---|---|---|
| $\frac{1}{2}(b-c) = 20°25.7'$ | l cos 9.97179 | l sin 9.54287 | l tan 9.57108 |
| $\frac{1}{2}(b+c) = 28°45.7'$ | l sec 0.05819 | l csc 0.31770 | |
| $\frac{1}{2}A = 18° 0.0'$ | l cot 0.48822 | l cot 0.48822 | |
| $\frac{1}{2}(B+C) = 73° 5.6'$ | l tan 0.51720 | | l sin 9.98081 |
| $\frac{1}{2}(B-C) = 65°52.3'$ | | l tan 0.34879 | l csc 0.03970 |
| $B = 138°57.9'$ | | | |
| $C = 7°13.3'$ | | | |
| $\frac{1}{2}a = 21°19.8'$ | | | l tan 9.59159 |
| $a = 42°39.6'$ | | | |

The latitude of $B$ is $(90° - 42°39.6')$N = $47°20.4'$ N and the longitude is $(73°57.5' - 7°13.3')$W $= 66°44.2'$ W.

b) The northern-most point on the track is $D$ whose meridian is perpendicular to the track. In the right spherical triangle $ACD$ of Fig.(b), $d = 49°11.4'$ and $A = 36°0.0'$.

For $a$:  (4) $\sin a = \sin d \sin A$
For $C$:  (5) $\tan C = \sec d \cot A$

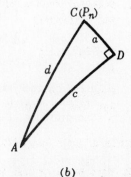

|  | (4) | (5) |
|---|---|---|
| $d = 49°11.4'$ | l sin 9.87902 | l sec 0.18472 |
| $A = 36° 0.0'$ | l sin 9.76922 | l cot 0.13874 |
| $a = 26°24.9'$ | l sin 9.64824 | |
| $C = 64°36.0'$ | | l tan 0.32346 |

The latitude of $D$ is $(90° - 26°24.9')$N = $63°35.1'$ N and the longitude is $(73°57.5' - 64°36.0')$W $= 9°21.5'$ W.

# SUPPLEMENTARY PROBLEMS

## PARALLEL SAILING.

12. A ship sails due east for 200 miles along the parallel of latitude $42^\circ$ N. What is the longitude of its point of arrival if *a*) it starts from longitude $125^\circ$ W, *b*) it starts from longitude $160^\circ$ E?  *Ans.* *a*) $120^\circ 30.9'$ W, *b*) $164^\circ 29.1'$ E

13. A ship in latitude $42^\circ$ N sails due west until it has made good a difference in longitude of $3^\circ 45'$. Find the departure.  *Ans.* 167.2 miles W

14. A ship in latitude $22^\circ$ N sails due west until it has made good a difference in longitude of $3^\circ 45'$. Find the departure.  *Ans.* 208.6 miles W

## PLANE SAILING.

15. A ship sails 125 miles on course $42^\circ 40'$ from $A$(lat. $40^\circ$ N). Find the departure and the latitude attained.  *Ans.* $p$ = 84.7 miles E, lat. = $41^\circ 32'$ N

16. $B$ is 125 miles west and 90 miles north of $A$. Find the distance and course in sailing from $A$ to $B$.  *Ans.* $d$ = 154.0 miles, course = N $54^\circ 15'$ W

## MIDDLE LATITUDE SAILING.

17. From a position $A$(lat. $35^\circ 38'$ N, long. $64^\circ 55'$ W) a ship sails 175 miles on a course S$50^\circ$ E. Find the position $B$ reached.  *Ans.* $B$(lat. $33^\circ 46'$ N, long. $62^\circ 12'$ W)

18. A ship leaves $A$(lat. $45^\circ 15'$ N, long. $140^\circ 38'$ W) and arrives at $B$(lat. $48^\circ 45'$ N, long. $137^\circ 12'$ W). Find the course and distance.  *Ans.* $33^\circ 47'$, 252.7 miles E

19. An airplane flies from San Diego (lat. $32^\circ 42'$ N, long. $117^\circ 10'$ W) to San Francisco (lat. $37^\circ 48'$ N, long. $122^\circ 24'$ W). Find the course and distance.  *Ans.* $320^\circ 2'$, 399.3 miles W

## DEAD RECKONING.

20. Find the course made good and the final position $B$ if a ship, starting at $A$(lat. $47^\circ 24'$ N, long. $75^\circ 45'$ W), sails the following courses:
    *a*) course $189^\circ 0'$, distance 35.0 miles; course $330^\circ 0'$, distance 50.0 miles.
    *b*) course $225^\circ 0'$, distance 105.0 miles; course $50^\circ 0'$, distance 125.0 miles.
    *Ans.* *a*) $285^\circ 55'$;  $47^\circ 33'$ N, $76^\circ 30'$ W
         *b*) $73^\circ 59'$;  $47^\circ 30'$ N, $75^\circ 13'$ W

21. Starting from a position $A$(lat. $40^\circ 50'$ N, long. $125^\circ 0'$ W) a ship sails the following courses and distances:

| Course | $35^\circ 30'$ | $130^\circ 0'$ | $255^\circ 0'$ | $340^\circ 30'$ | $110^\circ 30'$ |
|---|---|---|---|---|---|
| Distance | 30.0 | 45.0 | 100.0 | 40.0 | 30.0 |

Find the course made good and the final position $B$.
*Ans.* $96^\circ 6'$;  $40^\circ 47'$ N, $124^\circ 20'$ W

## GREAT CIRCLE SAILING.

22. Find the shortest distance between:
    *a*) Chicago (lat. $41^\circ 50.0'$ N, long. $87^\circ 37.0'$ W) and Dutch Harbor (lat. $53^\circ 54.0'$ N, long. $166^\circ 30.0'$ W),
    *b*) New York (lat. $40^\circ 43.0'$ N, long. $74^\circ 0.0'$ W) and Rio de Janeiro (lat. $22^\circ 54.0'$ S, long. $43^\circ 11.0'$ W),
    *c*) Dutch Harbor and Rio de Janeiro.
    *Ans.* *a*) 3085.5 miles, *b*) 4186.2 miles, *c*) 7666.4 miles

23. Find the great circle distance, the initial course, and the course on arrival in traveling from Washington (lat. $38°55.0'$ N, long. $77°4.0'$ W) to Moscow (lat. $55°45.0'$ N, long. $37°34.0'$ E). *Ans.* 4219.6 miles, $32°54.6'$, $131°18.8'$

24. Find the great circle distance, initial course, and course on arrival in traveling from Calcutta (lat. $22°35.0'$ N, long. $88°27.0'$ E) to Melbourne (lat. $37°48.0'$ S, long. $144°58.0'$ E). *Ans.* 4822.6 miles, $221°56.7'$, $231°21.5'$

25. Locate the ship in Problem 24 when it crosses the equator and find its distance from Calcutta. *Ans.* long. $107°29.4'$ E, 1752.8 miles

26. An airplane flies from Honolulu (lat. $21°18.0'$ N, long. $157°52.0'$ W) on course $40°43.0'$.
   a) Locate the point on the track nearest the north pole.
   b) Find the position when the longitude is $74°0.0'$ W.
   *Ans.* a) lat. $52°34.4'$ N, long. $85°13.6'$ W
        b) lat. $52°2.3'$ N

# CHAPTER 24

# The Celestial Sphere

TO AN OBSERVER ON THE EARTH'S SURFACE, it appears that he is at the center of a sphere of unlimited radius on which all the other heavenly bodies move from east to west. Such a sphere, of unlimited radius but with its center at the center of the earth, is useful in solving certain problems in astronomy and navigation. This sphere is called the *celestial sphere*.

In order to locate a heavenly body on the celestial sphere (more precisely, in order to locate the point in which a line drawn from the center of the earth through the center of the heavenly body pierces the celestial sphere) certain reference points and great circles are necessary.

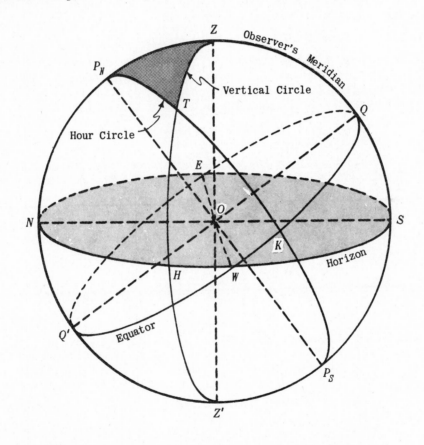

Points and great circles independent of the observer are:

1) the *celestial poles* $P_N$ and $P_S$, being the intersections of the axis of the earth and the celestial sphere.

2) the *celestial equator* $EQWQ'$, being the intersection of the plane of the earth's equator and the celestial sphere.

3) the *celestial meridians*, being half-great circles which pass through $P_N$ and $P_S$.

Points and great circles dependent upon the position of the observer are:

1) the (observer's) *zenith*, being the point Z on the celestial sphere directly above the observer.

2) the (observer's) *nadir*, being the point Z' diametrically opposite Z. (Note that Z and Z' are the intersections with the celestial sphere of the line joining the position of the observer and the center of the earth.)

3) the (observer's) *horizon*, being the great circle NESW whose pole is Z.

4) the (observer's) *celestial meridian*, being the meridian $P_N Z P_S$ through the zenith.

For a heavenly body $T$:

1) the *vertical circle* of $T$ is the half-great circle ZTHZ', H being its intersection with the horizon.

2) the *altitude* of $T$ is its angular distance from the horizon. The altitude (arc HT) is + or − according as $T$ is above or below the horizon.

3) the *zenith distance* of $T$ is 90° − altitude of $T$.

4) the *azimuth* of $T$ is the angle $P_N Z T$ between the observer's meridian and the vertical circle through $T$. It is generally measured along the horizon from the north point N around through the east to H. For a body in the eastern sky, the azimuth is < 180°; for a body in the western sky, it is > 180°.

5) the *hour circle* of $T$ is the half-great circle $P_N T K P_S$, K being its intersection with the equator.

6) the *declination* of $T$ is its angular distance from the equator. The declination (arc KT) is + or − according as $T$ is north or south of the equator.

7) the *polar distance* of $T$ is 90° − declination of $T$.

8) the *hour angle* of $T$ is the angle $Z P_N T$ between the observer's meridian and the hour circle through $T$. It is measured westward from the observer's meridian from 0° to 360°.

Due to the rotation of the earth, an hour circle appears to change by 15° each hour; thus, the hour angle may be measured in time units from $0^h$ to $24^h$.

COORDINATE SYSTEMS.   In the *horizon system*, the axes are the (observer's) horizon and the vertical circle of the heavenly body $T$. The coordinates of $T$ are:
the altitude, HT, measured by a sextant or transit;
the azimuth, $\angle P_N Z T$ or arc NEH, measured by a compass.

In the *equatorial system*, the axes are the celestial equator and the hour circle of $T$. The coordinates of $T$ are:
the declination KT and the hour angle $\angle Z P_N T$.

The declinations of certain heavenly bodies together with their hour angle for an observer on the meridian of Greenwich are given in the American Nautical Almanac.

THE ASTRONOMICAL TRIANGLE for a heavenly body $T$ is the (celestial) spherical triangle $P_N Z T$ formed by the observer's meridian $P_N Z$, the hour circle $P_N T$, and the vertical circle ZT. The parts of this triangle are:

1) side TZ = zenith distance of $T$ = 90° − altitude of $T$,

2) side $T P_N$ = polar distance of $T$ = 90° − declination of $T$,

3) side $ZP_N$ = colatitude of observer = $90°$ – $QZ$
       = $90°$ – latitude of observer (in Northern Hemisphere)
       = $90°$ + latitude of observer (in Southern Hemisphere),

4) angle $P_NZT$ = azimuth of $T$, if $T$ is east of the observer's meridian
       = $360°$ – azimuth of $T$, if $T$ is west of the observer's meridian,

5) angle $ZP_NT$ = hour angle of $T$, if $T$ is west of the observer's meridian
       = $360°$ – hour angle of $T$, if $T$ is east of observer's meridian,

6) angle $ZTP_N$, which is of no special importance.

**SOLAR TIME.** When the center of the sun is on the observer's meridian, $\angle ZP_NT = 0°$, it is *local solar noon* for the observer.

The *local apparent time* of the observer at any instant is $12^h - \angle ZP_NT$ (of the astronomical triangle) when the sun is in the eastern sky, and $12^h + \angle ZP_NT$ when the sun is in the western sky.

**EXAMPLE 1.** Find the local apparent time at New York (lat. $40°42.0'$ N) at the instant (a) in the forenoon and (b) in the afternoon when the altitude of the sun is $34°32.0'$ and its declination is $+12°54.0'$.

In the astronomical triangle, $TZ = 90°$ – altitude = $55°28.0'$, $TP_N = 90°$ – declination = $77°6.0'$, and $ZP_N$ = $90°$ – latitude = $49°18.0'$.

From Example 1, Chapter 22,   $ZP_NT = 55°14.5' = 3^h40^m58^s$.

a) In the forenoon the local apparent time is

$$12^h - 3^h40^m58^s = 8^h19^m2^s = 8:19:2 \text{ AM.}$$

b) In the afternoon the local apparent time is

$$12^h + 3^h40^m58^s = 15^h40^m58^s = 3:40:58 \text{ PM.}$$

**LATITUDE OF AN OBSERVER.** When the altitude, declination and hour angle (or azimuth) of a heavenly body are known, the latitude of the observer may be found by solving the astronomical triangle.

**EXAMPLE 2.** Find the latitude of an observer in the northern hemisphere if, at his local apparent time 10:25:36 AM, the altitude of the sun is $40°10.0'$ and its declination is $+15°38.0'$.

In the astronomical triangle, $TZ = 49°50.0'$, $TP_N = 74°22.0'$,

and $\angle ZP_NT = 12^h - 10^h25^m36^s = 1^h34^m24^s = 23°36.0'$. This is a

Case V triangle for which side $ZP_N$ is required.

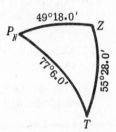

$$\sin P_NZT = \sin TP_N \csc TZ \sin ZP_NT$$

$$\tan \tfrac{1}{2}ZP_N = \sin \tfrac{1}{2}(P_NZT + ZP_NT) \csc \tfrac{1}{2}(P_NZT - ZP_NT) \tan \tfrac{1}{2}(TP_N - TZ)$$

| | | | |
|---|---|---|---|
| •$TP_N = 74°22.0'$ | l sin 9.98363 | $\tfrac{1}{2}(P_NZT + ZP_NT) = 86°39.0'$ | l sin 9.99926 |
| $TZ = 49°50.0'$ | l csc 0.11681 | $\tfrac{1}{2}(P_NZT - ZP_NT) = 63° 3.0'$ | l csc 0.04993 |
| $ZP_NT = 23°36.0'$ | l sin 9.60244 | $\tfrac{1}{2}(TP_N - TZ) = 12°16.0'$ | l tan 9.33731 |
| *$P_NZT = 149°42.0'$ | l sin 9.70288 | $\tfrac{1}{2}ZP_N = 13°41.1'$ | l tan 9.38650 |
| | | $ZP_N = 27°22.2'$ | |

*It is clear from the figure that $\angle P_NZT = 30°18.0'$ when the observer is in the southern hemisphere and is $180° - 30°18.0' = 149°42.0'$ when the observer is in the northern hemisphere.

The observer's latitude is $90° - ZP_N = 62°37.8'$ N.

## SOLVED PROBLEMS

1. Find the azimuth of the sun and the local apparent time at Washington, D.C. (lat. $38°55.0'$ N) at the instant in the afternoon when the sun's altitude is $25°40.0'$ N and its declination is $-19°15.0'$.

In the astronomical triangle, $TZ = 90° -$ altitude of sun $= 64°20.0'$, $TP_N = 90° -$ declination of sun $= 109°15.0'$, and $ZP_N = 90° -$ latitude of observer $= 51°5.0'$.

*Standard Solution.*

$$(1)\ \tan r = \sqrt{\frac{\sin(s-TZ)\ \sin(s-TP_N)\ \sin(s-ZP_N)}{\sin s}} \qquad s = \tfrac{1}{2}(TZ + TP_N + ZP_N)$$

$$(2)\ \tan \tfrac{1}{2}ZP_N T = \frac{\tan r}{\sin(s-TZ)} \qquad\qquad (3)\ \tan \tfrac{1}{2}P_N ZT = \frac{\tan r}{\sin(s-TP_N)}$$

(1)

| | | | |
|---|---|---|---|
| $TZ = 64°20.0'$ | $s-TZ = 48°\ 0.0'$ | l sin | 9.87107 |
| $TP_N = 109°15.0'$ | $s-TP_N = 3°\ 5.0'$ | l sin | 8.73069 |
| $ZP_N = 51°\ 5.0'$ | $s-ZP_N = 61°15.0'$ | l sin | 9.94286 |
| $2s = 224°40.0'$ | $s = 112°20.0'$ | l csc | 0.03386 |
| $s = 112°20.0'$ | | 2 | 8.57848 |
| | $r$ | l tan | 9.28924 |

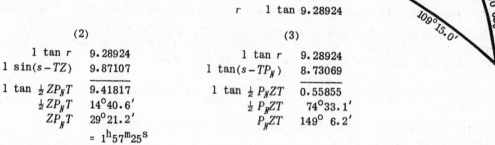

$51°5.0'$  $P_N$  $Z$
$109°15.0'$  $64°20.0'$
$T$

| (2) | | | (3) | |
|---|---|---|---|---|
| l tan $r$ | 9.28924 | l tan $r$ | 9.28924 |
| l sin$(s-TZ)$ | 9.87107 | l tan$(s-TP_N)$ | 8.73069 |
| l tan $\tfrac{1}{2}ZP_N T$ | 9.41817 | l tan $\tfrac{1}{2}P_N ZT$ | 0.55855 |
| $\tfrac{1}{2}ZP_N T$ | $14°40.6'$ | $\tfrac{1}{2}P_N ZT$ | $74°33.1'$ |
| $ZP_N T$ | $29°21.2'$ | $P_N ZT$ | $149°\ 6.2'$ |
| | $= 1^h57^m25^s$ | | |

Since the sun is in the western sky, the azimuth is $360° - P_N ZT = 210°53.8'$ and the local apparent time is 1:57:25 PM.

*Haversine Solution.* $\qquad\qquad s = \tfrac{1}{2}(TZ + TP_N + ZP_N)$

$$(1)\ \text{hav}\ P_N ZT = \sin(s-TZ)\ \sin(s-ZP_N)\ \csc TZ\ \csc ZP_N$$

$$(2)\ \text{hav}\ ZP_N T = \sin(s-TP_N)\ \sin(s-ZP_N)\ \csc TP_N\ \csc ZP_N$$

| | | (1) | | (2) | |
|---|---|---|---|---|---|
| $s-TZ =$ | $48°\ 0.0'$ | l sin | 9.87107 | | |
| $s-TP_N =$ | $3°\ 5.0'$ | | | l sin | 8.73069 |
| $s-ZP_N =$ | $61°15.0'$ | l sin | 9.94286 | l sin | 9.94286 |
| $TZ =$ | $64°20.0'$ | l csc | 0.04512 | | |
| $TP_N =$ | $109°15.0'$ | | | l csc | 0.02499 |
| $ZP_N =$ | $51°\ 5.0'$ | l csc | 0.10899 | l csc | 0.10899 |
| $P_N ZT =$ | $149°\ 6.2'$ | l hav | 9.96804 | | |
| $ZP_N T =$ | $29°21.3'$ | | | l hav | 8.80753 |
| | $= 1^h57^m25^s$ | | | | |

The azimuth is $210°53.8'$ and the local apparent time is 1:57:25 PM, as before.

**2.** Find the local apparent time and the azimuth of sunrise and sunset at Reykjavik, Iceland (lat. $64°9.0'$ N) when the declination of the sun is $+15°45.0'$.

In the astronomical triangle, $TP_N = 74°15.0'$, $ZP_N = 25°51.0'$, and, since at sunrise or sunset the center of the sun is on the horizon, $TZ = 90°$. Its polar triangle $Z'P'T'$ is a right spherical triangle for which $\angle Z' = 105°45.0'$, $\angle T' = 154°9.0'$, $\angle P' = 90°$.

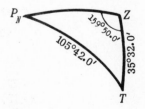

$$(1) \cos p' = \cot T' \cot Z' \qquad (2) \cos z' = \csc T' \cos Z'$$

| | (1) | (2) |
|---|---|---|
| $T' = 154°9.0'$ | l cot 0.31471 (n) | l csc 0.36050 |
| $Z' = 105°45.0'$ | l cot 9.45029 (n) | l cos 9.43367 (n) |
| $p' = 54°24.1'$ | l cos 9.76500 | |
| $z' = 128°30.1'$ | | l cos 9.79417 (n) |

Then $\angle ZP_NT = 125°35.9' = 8^h22^m24^s$ and $\angle P_NZT = 51°29.9'$. At sunrise, the local apparent time $= 12^h - 8^h22^m24^s = 3:37:36$ AM and the azimuth of the sun is $51°29.9'$. At sunset, the local apparent time is 8:22:24 PM and the azimuth is $360° - 51°29.9' = 308°30.1'$.

(Note. A correction must be made in each local apparent time to compensate for the refraction of the rays of the sun by the earth's atmosphere and for the angular radius of the sun.)

**3.** Find the length of the shortest day (declination of the sun $-23°27.7'$) and the azimuth of the rising and setting sun at Reykjavik (lat. $64°9.0'$N).

The astronomical triangle is quadrantal with $TP_N = 113°27.7'$, $ZP_N = 25°51.0'$, and $ZT = 90°$. Solving the polar triangle $Z'P'T'$ for which $T' = 154°9.0'$ and $Z' = 66°32.3'$ as in Problem 2, we find
$$p' = 153°36.7' \quad \text{and} \quad z' = 24°3.5'.$$
Then in the quadrantal triangle, $ZP_NT = 26°23.3' = 1^h45^m33^s$ and $P_NZT = 155°56.5'$.

The local apparent time of sunrise is $12^h - 1^h45^m33^s = 10:14:27$ AM and of sunset is 1:45:33 PM.

The length of the shortest day is $2(1^h45^m33^s) = 3^h31^m6^s$.

The azimuth of the sun is $155°56.5'$ at sunrise and $360° - 155°56.5' = 204°3.5'$ at sunset.

(Note. The length of the longest day is $24^h - 3^h31^m6^s = 20^h28^m54^s$.)

**4.** Find the latitude of an observer in the northern hemisphere when the altitude of the sun is $54°28.0'$, the declination is $-15°42.0'$, and the azimuth is $200°10.0'$.

In the astronomical triangle, $TZ = 35°32.0'$, $TP_N = 105°42.0'$, and $P_NZT = 360° - 200°10.0' = 159°50.0'$ since the sun is in the western sky.

$$\sin TP_NZ = \sin P_NZT \sin TZ \csc TP_N$$

$$\tan \tfrac{1}{2}ZP_N = \sin \tfrac{1}{2}(P_NZT + TP_NZ) \csc \tfrac{1}{2}(P_NZT - TP_NZ) \tan \tfrac{1}{2}(TP_N - TZ)$$

| | | | | |
|---|---|---|---|---|
| $TZ = 35°32.0'$ | l sin 9.76431 | $\tfrac{1}{2}(P_NZT + TP_NZ) = 85°55.4'$ | l sin 9.99890 |
| $TP_N = 105°42.0'$ | l csc 0.01651 | $\tfrac{1}{2}(P_NZT - TP_NZ) = 73°54.6'$ | l csc 0.01736 |
| $P_NZT = 159°50.0'$ | l sin 9.53751 | $\tfrac{1}{2}(P_NT - TZ) = 35°5.0'$ | l tan 9.84657 |
| $TP_NZ = 12°0.8'$ | l sin 9.31833 | $\tfrac{1}{2}ZP_N = 36°5.9'$ | l tan 9.86283 |
| | | $ZP_N = 72°11.8'$ | |

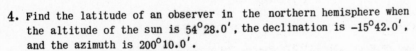

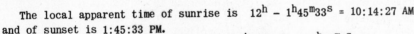

Thus, the latitude is $90° - 72°11.8' = 17°48.2'$ N.

## SUPPLEMENTARY PROBLEMS

5. Find the local apparent time and the azimuth of the sun in the morning at

   a) latitude 39° N when the sun's altitude is 22° and its declination is +20°.
   b) latitude 45°24' N when the sun's altitude is 24°12' and its declination is +13°16'.
   c) latitude 25°14' N when the sun's altitude is 38°26' and its declination is −18°16'.

   *Ans.*  a) 6:50 AM, 81°31'        b) 7:25 AM, 95°36'        c) 10:6 AM, 144°43'

6. Find the local apparent time and the azimuth of the sun in the afternoon at

   a) latitude 40°42' when the altitude of the sun is 28°26' and its declination is −8°16'.
   b) latitude 42°45' when the altitude of the sun is 38°36' and its declination is +18°27'.

   *Ans.*  a) 2:42 PM, 227°3'        b) 3:36 PM, 259°15'

7. Find the local apparent time and the amplitude of sunrise and sunset for that  day  in which the sun's declination is +20°32' at

   a) Acapulco (lat. 16°49' N).       *Ans.*  5:34 AM, 68°30';   6:26 PM, 291°30'
   b) Fairbanks (lat. 64°51' N).      *Ans.*  2:28 AM, 34°23';   9:32 PM, 325°37'
   c) Harrisburg (lat. 40°16' N).     *Ans.*  4:46 AM, 62°38';   7:14 PM, 297°22'

8. Find the duration of daylight on the longest day (dec. +23°28') at

   a) Acapulco,  b) Fairbanks.       *Ans.*   a) $13^h0^m$,  b) $21^h0^m$

9. The declination of a star is +7°24', the hour angle is 48°51', and the latitude of the observer is 64°9' N. Find the azimuth of the star.    *Ans.* 234°36'

10. What is the latitude in the northern hemisphere if

   a) at 8:56 AM the sun's altitude is 36°18' and its declination is +14°35'?
   b) at 3 PM the sun's altitude is 24°42' and its declination is −12°28'?
   c) at 9:15 AM the sun's altitude is 35°23' and its declination is −10°48'?
   d) at 2:10 PM the sun's altitude is 23°26' and its declination is +14°30'?
   e) the sun sets at 10 PM on the longest day of the year?

   *Ans.*  a) 54°58' N,   b) 37°22' N,   c) 26°18' N,   d) 79°18' N,   e) 63°23' N

# Index

205

Catalog

If you are interested in a list of SCHAUM'S
OUTLINE SERIES in Science, Mathematics,
Engineering and other subjects, send your name
and address, requesting your free catalog, to:

SCHAUM'S OUTLINE SERIES, Dept. C
McGRAW-HILL BOOK COMPANY
1221 Avenue of Americas
New York, N.Y. 10020